PROBLEMS IN QUANTUM MECHANICS

I. I. GOL'DMAN
and
V. D. KRIVCHENKOV

Edited by
B. T. GEILIKMAN

Translated from the Russian by
E. MARQUIT and E. LEPA

DOVER PUBLICATIONS, INC.
Mineola, New York

Bibliographical Note

This Dover edition, first published in 1993 and reissued in 2006, is an unabridged republication of the work first published in 1961 by Pergamon Press, London and Addison-Wesley Publishing Company, Inc., Reading, Massachusetts.

Library of Congress Cataloging-in-Publication Data

Gol'dman, I. I. (Iosif Il'ich)
Problems in quantum mechanics / I. I. Gol'dman and V. D. Krivchenkov ; edited by B. T. Geilikman ; translated from the Russian by E. Marquit and E. Lepa.
p. cm.
Originally published: New York : Dover Publications, Inc., 1993.
ISBN 0-486-45322-7 (pbk.)
1. Quantum theory–Problems, exercises, etc. I. Title.

QC174.15.G6513 2006
530.12–dc22

2006047437

Manufactured in the United States of America
Dover Publications, Inc., 31 East 2nd Street, Mineola, N.Y. 11501

CONTENTS

* The numbers in italics are the page numbers of the answers and solutions.

PREFACE TO THE RUSSIAN EDITION

This book contains problems in non-relativistic quantum mechanics which have been solved at seminaria or given as home assignments for 4th year students at the Moscow State University. The selection contains problems of various degrees of difficulty. The problems requiring comparatively lengthy calculations are intended primarily for students of theoretical physics, who as their basic text-book in quantum mechanics have used the book of L. D. Landau and E. M. Lifshitz, *Quantum Mechanics.*

Didactic experience has shown that mastering of the matrix form of quantum mechanics presents the greatest difficulty. Therefore, in preparation of the present book a great deal of attention was paid to the construction of perturbation matrices and their diagonalization. A relatively large amount of space has been devoted to auxiliary problems on angular momentum and spin, since a serious study of quantum mechanics is not possible without an understanding of these basic notions.

The authors would like to express their gratitude to V. V. Tolmachev, to A. R. Frenkin and V. D. Kukin for their aid in preparing the book and also to E. E. Zhabotinskiĭ for his critical remarks.

I. Gol'dman and V. Krivchenkov

TRANSLATORS' PREFACE

While the translation was in progress, the authors proposed a number of important revisions to the original Russian edition. These revisions, all of which have been incorporated into the English edition, involve changes in the formulation of some of the problems and solutions, inclusion of a number of new problems, and deletion of a few problems of lesser importance. The authors also drew attention to a number of typographical errors appearing in the formulae of the original Russian edition.

In the English edition the scalar product of two vectors **A** and **B** is indicated by the symbol **(AB)** and the vector product by **[AB]**.

The translators are indebted to G. Bialkowski, who prepared the translation for the Polish edition, for a number of helpful comments and to Ruth Marquit and Olga Lepa for aid in preparing the book for press.

E. Marquit and E. Lepa

December, 1960

PROBLEMS

§1. One-dimensional motion. Energy spectrum and wave functions

1. Determine the energy levels and normalized wave functions of a particle in a potential box. The potential energy of the particle is $V = \infty$ for $x < 0$ and $x > a$, and $V = 0$ for $0 < x < a$.

2. Show that a particle in a potential box (see the preceding problem) satisfies the relation

$$\bar{x} = \frac{1}{2}a, \quad \overline{(x-\bar{x})^2} = \frac{a^2}{12}\left(1-\frac{6}{n^2\pi^2}\right).$$

Show that for large values of n the above result coincides with the corresponding classical solution.

3. Find the probability distribution for different values of the momentum of a particle in a potential box in the nth energy state.

4. Find the energy levels and the wave functions of a particle in a non-symmetric potential well (Fig. 1). Investigate the case $V_1 = V_2$.

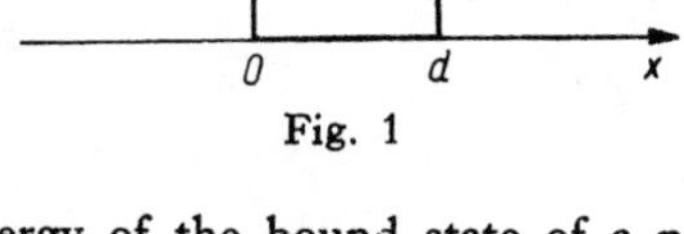

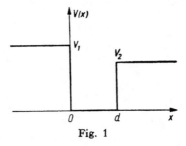

Fig. 1

5. Find the energy of the bound state of a particle in a potential field of the form

$$V = \begin{cases} 0 & x < 0, \quad x > a, \\ -V_0 & 0 < x < a \end{cases}$$

if $V_0 \to \infty$ and $a \to 0$ under the condition that $V_0 a = q$.

6. The Hamiltonian of an oscillator is equal to $\hat{H} = \frac{\hat{p}^2}{2\mu} + \frac{\mu\omega^2\hat{x}^2}{2}$, where $\hat{p}$ and $\hat{x}$ satisfy the commutation relation $\hat{p}\hat{x} - \hat{x}\hat{p} = -i\hbar$. To eliminate the quantities $\hbar$, μ, ω from the subsequent calculations we introduce new variables $\hat{P}$ and $\hat{Q}$ defined by

$$\hat{P} = \frac{1}{\sqrt{\mu\hbar\omega}}\hat{p}, \quad \hat{Q} = \sqrt{\frac{\mu\omega}{\hbar}}\hat{x} \quad (\hat{P}\hat{Q} - \hat{Q}\hat{P} = -i),$$

so that the energy E will be expressed in units of $\hbar\omega$ ($E = \varepsilon\hbar\omega$). The Schrödinger equation for the oscillator in the new variables then takes the form

$$\hat{H}'\psi = \frac{1}{2}(\hat{P}^2 + \hat{Q}^2)\psi = \varepsilon\psi.$$

(a) Making use of the commutation relation $\hat{P}\hat{Q} - \hat{Q}\hat{P} = -i$, show that

$$\frac{1}{2}(\hat{P}^2 + \hat{Q}^2)(\hat{Q} \pm i\hat{P})^n\psi = (\varepsilon \mp n)(\hat{Q} \pm i\hat{P})^n\psi.$$

(b) Find the normalized wave functions and energy levels of the oscillator.

(c) Find the commutation relations for the operator $\hat{a} = \frac{1}{\sqrt{2}}(\hat{Q} + i\hat{P})$ and its Hermitian conjugate $\hat{a}^+ = \frac{1}{\sqrt{2}}(\hat{Q} - i\hat{P})$. Express the wave function of the nth excited state in terms of the wave function of the ground state with the help of the operator $\hat{a}$.

(d) Find the matrix elements of the operators $\hat{P}$ and $\hat{Q}$ in the energy representation.

Hint. $\hat{P}^2 + \hat{Q}^2 - 1 = (\hat{P} + i\hat{Q})(\hat{P} - i\hat{Q})$.

7. On the basis of the results of the previous problem, show by direct multiplication of matrices that for an oscillator in the nth energy state

$$\overline{(\Delta x)^2} = \overline{x^2} = \frac{\hbar}{\mu\omega}\left(n + \frac{1}{2}\right); \quad \overline{(\Delta p)^2} = \overline{p^2} = \mu\hbar\omega\left(n + \frac{1}{2}\right).$$

8. A particle moves in a potential field $V(x) = \frac{\mu\omega^2x^2}{2}$. Determine the probability of finding the particle outside of the classical limits for the ground state.

9. Find the energy levels of a particle moving in a potential field of the shape

$$V(x) = \infty \quad (x < 0); \qquad V(x) = \frac{\mu\omega^2 x^2}{2} \quad (x > 0).$$

10. Write the Schrödinger equation for an oscillator in the p representation and determine the probability distribution for different values of momentum.

11. Find the wave functions and energy levels of a particle in a field of the shape $V(x) = V_0\left(\frac{a}{x} - \frac{x}{a}\right)^2$ $(x > 0)$ (see Fig. 2) and show that the energy spectrum coincides with the spectrum of an oscillator.

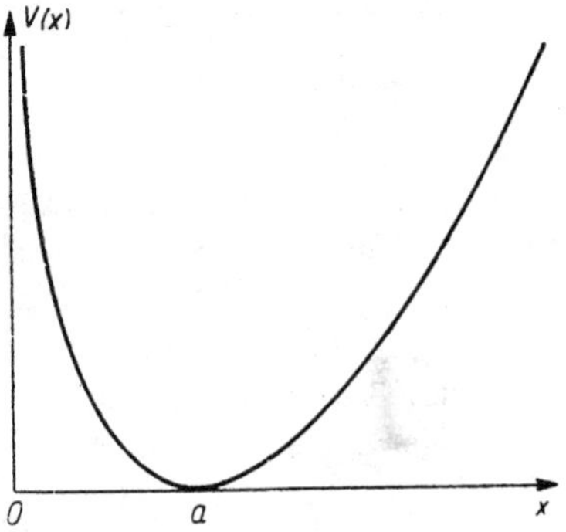

Fig. 2

12. Find the energy levels for particles in a potential field $V = -V_0 \cosh^{-2}\frac{x}{a}$ (see Fig. 3).

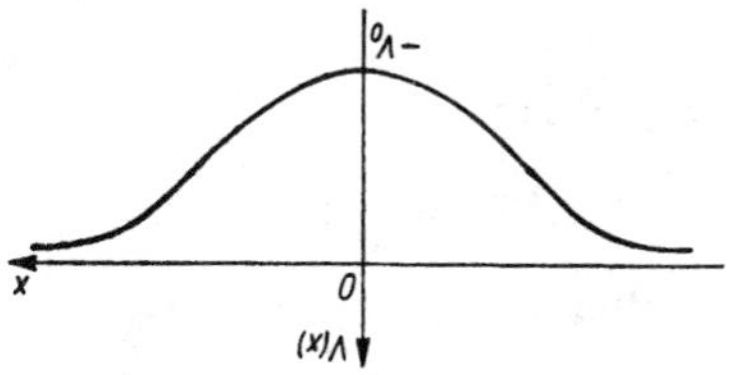

Fig. 3

13. Find the energy levels and wave functions of particles in the field $V = V_0 \cot^2 \frac{\pi}{a} x$ $(0 < x < a)$ (Fig. 4). Normalize the wave function for the ground state.

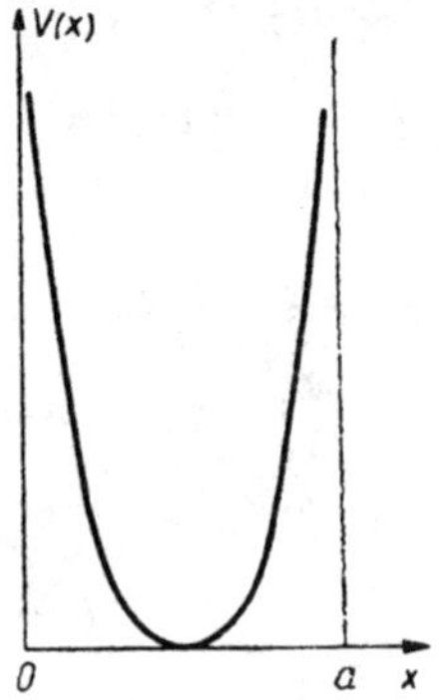

Fig. 4

Consider the limiting cases of small and large values of V_0.

14. Find the wave functions of a charged particle in a uniform field $V(x) = -Fx$.

15. Find the Schrödinger equation in the p representation for a particle moving in a periodic potential field $V(x) = V_0 \cos bx$.

16. Find the Schrödinger equation in the p representation for a particle moving in a periodic potential field $V(x) = V(x + b)$.

17. Determine the zones of allowable energy for a particle moving in the periodic potential field shown in Fig. 5. Investigate the limiting case $V_0 \to \infty$, $b \to 0$ with the condition

$$V_0 b = \text{constant}.$$

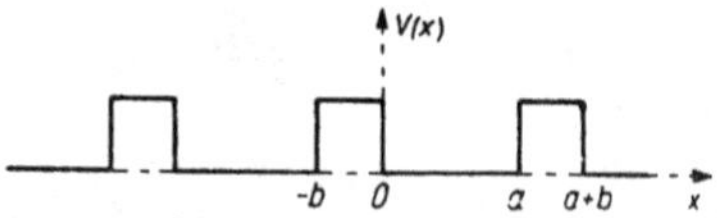

Fig. 5

18. For the potential $V = -V_0 \cosh^{-2} \frac{x}{a}$ find the energy levels and the total number of discrete levels in the quasi-classical approximation.

19. Using the quasi-classical approximation, determine the energy spectrum of a particle in the field:

(a) $V=\dfrac{\mu\omega^2x^2}{2}$ (oscillator),

(b) $V=V_0\cot^2\dfrac{\pi}{a}x \quad (0<x<a)$.

20. Using the quasi-classical approximation, find the mean value of the kinetic energy of a stationary state.

21. Using the result of the preceding problem, find in the quasi-classical approximation the mean kinetic energy of a particle in the field:

(a) $V=\dfrac{1}{2}\mu\omega^2x^2$,

(b) $V=V_0\cot^2\dfrac{\pi}{a}x \quad (0<x<a)$ (see Prob. 19).

22. Determine the energy spectrum of a particle in the field $V(x)=ax^{\nu}$ using the quasi-classical approximation and the virial theorem.

23. Derive an expression for the potential energy $V(x)$ in terms of the energy spectrum E_n in the quasi-classical approximation. Let $V(x)$ be an even function $V(x)=V(-x)$, increasing monotonically for $x>0$.

24. Find the energy of the bound state in the well $V=-q\,\delta(x)$.

§2. Transmission through a potential barrier

1. In studying the emission of electrons from metals it is necessary to take into account the fact that electrons with energy sufficient to escape from the metal can, according to quantum mechanics, undergo reflection at the surface of the metal. Consider a one-dimensional model with the potential $V=-V_0$ for $x<0$ (inside the metal) and $V=0$ for $x>0$ (outside the metal) (Fig. 6) and determine the reflection coefficient of an electron of energy $E>0$ at the surface of the metal.

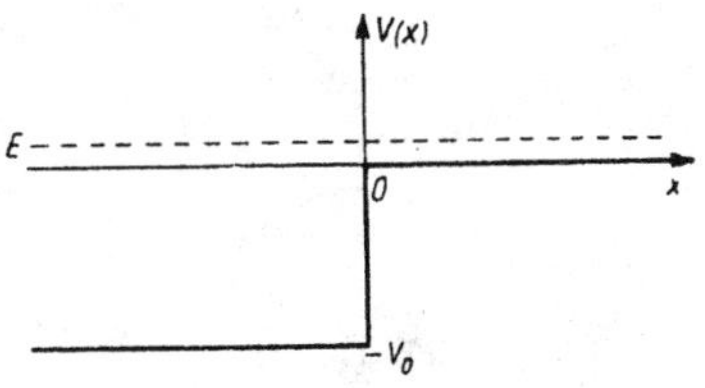

Fig. 6

2. In the preceding problem assume that the potential at the boundary of the metal has a jump. Actually, this change of potential is continuous over a region of the order of the interatomic distance in the metal. Approximating the potential near the surface of the metal by means of the function $V = -\frac{V_0}{e^{x/a}+1}$ (see Fig. 7),

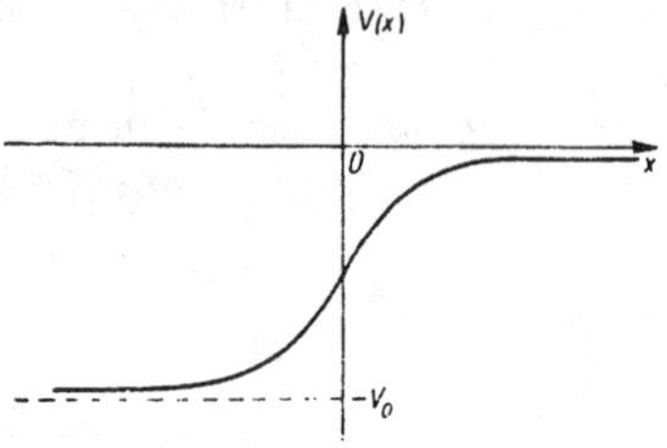

Fig. 7

find the reflection coefficient of an electron of energy $E > 0$.

3. Find the transmission coefficient of a particle through a rectangular barrier (see Fig. 8).

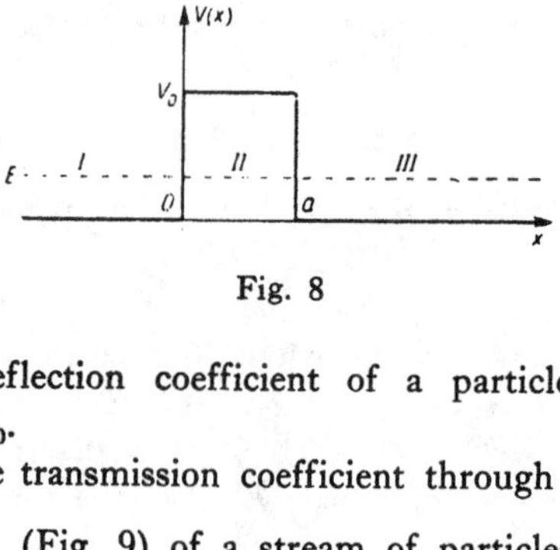

Fig. 8

4. Find the reflection coefficient of a particle by a rectangular barrier for $E > V_0$.

5. Calculate the transmission coefficient through a potential barrier $V(x) = V_0 \cosh^{-2}\frac{x}{a}$ (Fig. 9) of a stream of particles moving with the energy $E < V_0$.

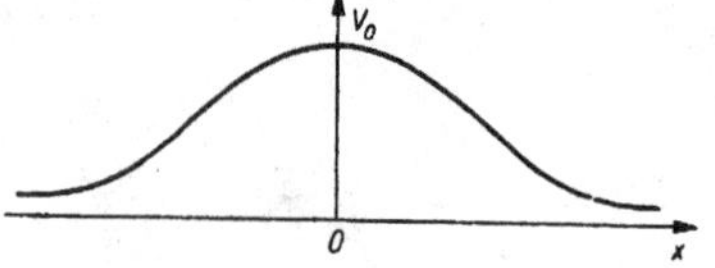

Fig. 9

6. Calculate in the quasi-classical approximation the transmission coefficient of electrons through the surface of a metal in a strong electric field of intensity F (Fig. 10). Determine the limits of applicability of the approximation.

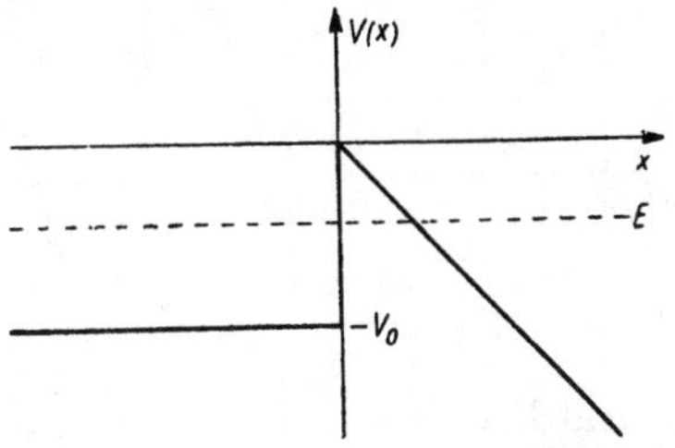

Fig. 10

7. In reality, the change in potential near the surface of a metal is continuous. Thus, e.g., the potential of an electric image $V_{e.i.} = -\frac{e}{4x}$ acts over a large distance from the surface. Taking into account the force due to the electric image, determine the transmission coefficient D through the surface of a metal in an electric field (Fig. 11).

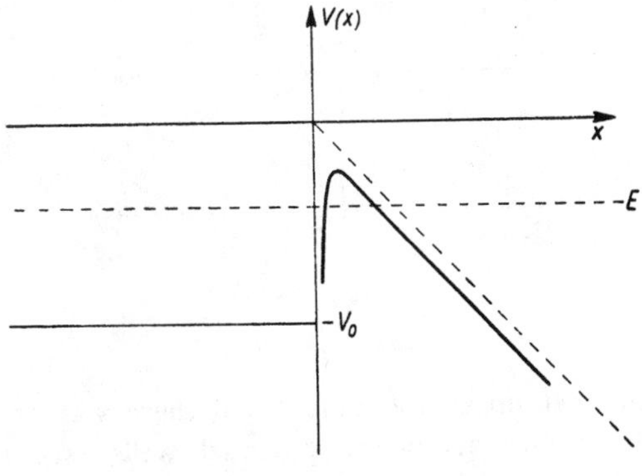

Fig. 11

8. Find approximate expressions for the energy levels and the wave functions of a particle in a symmetric potential field (see Fig. 12) if $E \ll V_0$, the penetrability of the barrier being small $\left(\frac{2\mu V_0}{\hbar^2} b^2 \gg 1\right)$.

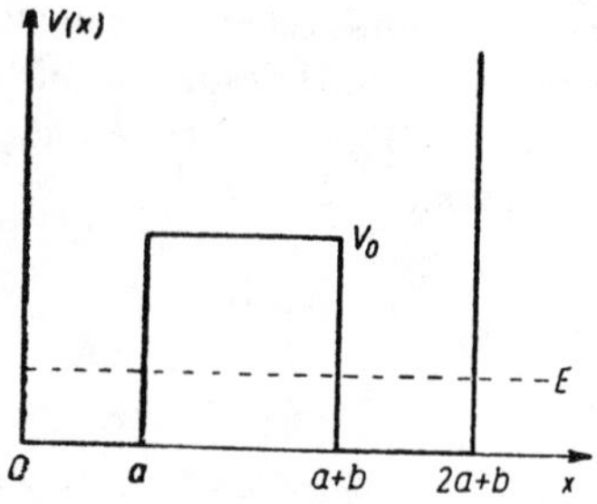

Fig. 12

9. A symmetric field $V(x)$ is represented by two potential wells separated by a barrier (see Fig. 13). Using the quasi-classical approximation, find the energy levels of a particle in the field $V(x)$. Compare the obtained energy spectrum with the energy spectrum of a single well. Calculate the splitting of the energy levels for one of the wells.

Hint. See Appendix I.

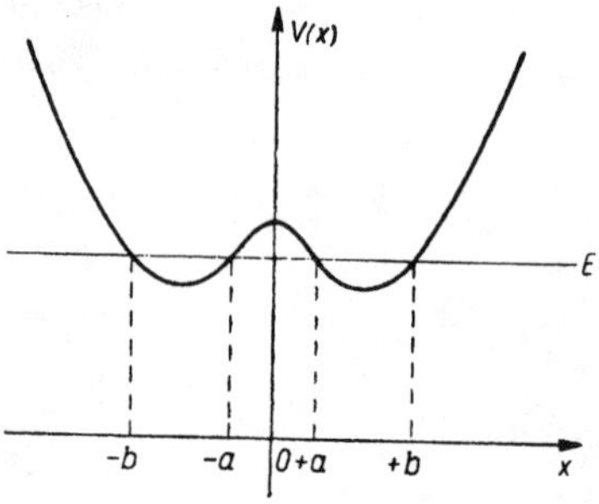

Fig. 13

10. Assume that up to the time $t = 0$ there was an impenetrable partition between two symmetric potential wells (see the preceding problem) and that a particle was in a stationary state in the well on the left. How long after removal of the partition will the particle reach the well on the right?

11. A field $V(x)$ is represented by N identical potential wells separated by equal potential barriers (see Fig. 14). Assuming that the quasi-classical conditions are satisfied, find the energy levels for the field $V(x)$.

Compare the obtained energy spectrum with the energy spectrum for the individual wells.

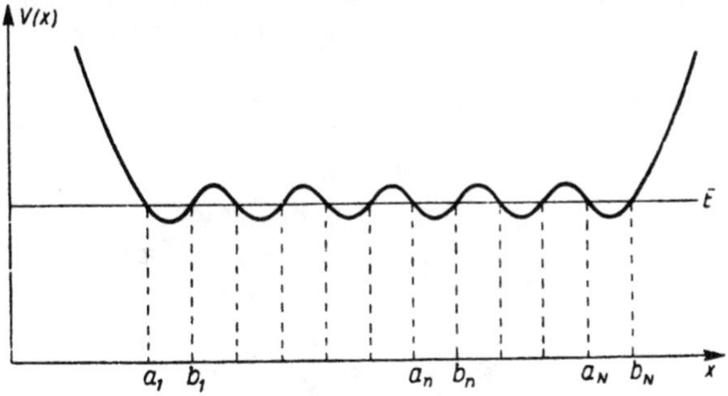

Fig. 14

12. Find for the quasi-classical case the quasi-stationary levels of a particle in the symmetric field shown in Fig. 15. Find also the transmission coefficient $D(E)$ for particles of energy $E < V_0$.

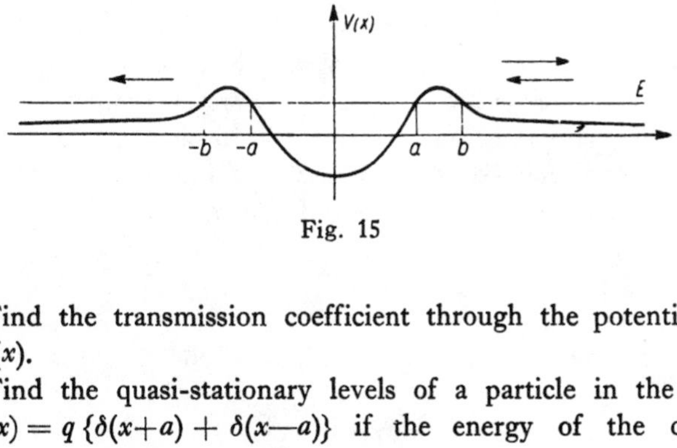

Fig. 15

13. Find the transmission coefficient through the potential barrier $V = q\,\delta(x)$.

14. Find the quasi-stationary levels of a particle in the potential field $V(x) = q\,\{\delta(x+a) + \delta(x-a)\}$ if the energy of the quasi-level satisfies the condition $E \ll \frac{\mu q^2}{\hbar^2}$.

15. Find the transmission coefficient through a potential barrier of the form $V(x) = q\{\delta(x + a) + \delta(x-a)\}$.

16. Consider a one-dimensional model of the scattering of electrons on a fixed particle which can be found in two energy states. Consider the force of interaction to be of the short-range type.

§3. Commutation relations. Uncertainty relation. Spreading of wave packets

1. Show that, if two operators $\hat{A}$ and $\hat{B}$ satisfy the commutation relations $\hat{A}\hat{B} - \hat{B}\hat{A} = i\hat{C}$, where $\hat{A}$ and $\hat{B}$ are Hermitian,

$$\sqrt{\overline{(\Delta\hat{A})^2}\;\overline{(\Delta\hat{B})^2}} \geqslant \frac{|\bar{C}|}{2}$$

is valid.

2. Show that if $\hat{A}$ and $\hat{B}$ are Hermitian, then

$$\overline{(\hat{A})^2}\,\overline{(\hat{B})^2} \geqslant \frac{(\bar{C})^2 + (\bar{D})^2}{4},$$

where $\hat{C} = \frac{1}{i}(\hat{A}\hat{B} - \hat{B}\hat{A})$ and $\hat{D} = \hat{A}\hat{B} + \hat{B}\hat{A}$.

3. Find the uncertainty relation for the operators $\hat{q}$ and $F(\hat{p})$, if $\hat{q}$ and $\hat{p}$ satisfy the commutation relation $\hat{q}\hat{p} - \hat{p}\hat{q} = i\hbar$.

Hint. Expand the function $F(\hat{p})$ in the form of a Taylor series.

4. Estimate the energy of the ground state of an oscillator from the uncertainty relation.

5. Estimate the energy of an electron in the Kth shell of an atom of atomic number Z in the non-relativistic and relativistic cases.

6. Estimate the energy of the ground state of a two-electron atom of nuclear charge Z with the help of the uncertainty relation.

7. The magnetic field produced by a free electron depends on the motion of the particle as well as on its intrinsic magnetic moment.

As is known from electrodynamics, the intensity of the magnetic field of the moving charge is of the order of magnitude

$$H_1 \sim \frac{ev}{cr^2},$$

and the field intensity of a magnetic dipole of moment μ is of the order

$$H_2 \sim \frac{\mu}{r^3}.$$

The magnetic moment μ of a free electron can be determined from a measurement of the intensity of the field produced by it if the following two conditions are satisfied:

$$H_2 \gg H_1 \qquad (1)$$

and

$$\Delta r \ll r. \qquad (2)$$

These conditions mean that the region of localization of an electron Δr should be smaller than the distance from this region to the point of observation of the magnetic field.

Can these two conditions be satisfied simultaneously?

Hint. Use the uncertainty relation and the magnetic moment of the electron $\mu = \dfrac{e\hbar}{2mc}$.

8. What is the physical sense of the quantity p_0 in the expression for the wave function

$$\psi(x) = \varphi(x) \exp\left(\frac{ip_0 x}{\hbar}\right),$$

where the function $\varphi(x)$ is real?

9. Show that in the case of a stationary state with a discrete spectrum the mean value of the momentum is $\bar{\mathbf{p}} = 0$.

10. The wave function of a free particle at the time $t = 0$ is given by the expression $\psi(x, 0) = \varphi(x) \exp\left(\dfrac{ip_0 x}{\hbar}\right)$. The function $\psi(x, 0)$ is real and differs significantly from zero only for values of x lying in the interval $-\delta < x < +\delta$. For which range of values of x will the wave function at time t be be different from zero?

11. The wave function is given for a particle at the time $t = 0$:
(a) free motion

$$\psi(\mathbf{r}, 0) = \frac{1}{(\pi\delta^2)^{3/4}} \exp\left\{\frac{i\mathbf{p}_0\mathbf{r}}{\hbar} - \frac{r^2}{2\delta^2}\right\};$$

(b) motion in a homogeneous field

$$\psi(\mathbf{r}, 0) = \frac{1}{(\pi\delta^2)^{3/4}} \exp\left\{\frac{i\mathbf{p}_0\mathbf{r}}{\hbar} - \frac{r^2}{2\delta^2}\right\};$$

(c) motion of a particle in the potential field $V = \frac{1}{2}\mu\omega^2 x^2$

$$\psi(x, 0) = c \exp\left\{\frac{ip_0 x}{\hbar} - \frac{\alpha^2 (x - x_0)^2}{2}\right\}, \quad \alpha = \left(\frac{\mu\omega}{\hbar}\right)^{1/2}.$$

Discuss the change in the wave function (spreading of the wave packet) for each case.

12. Prove that

$$e^{\hat{L}}\hat{a}e^{-\hat{L}} = \hat{a} + [\hat{L}\hat{a}] + \frac{1}{2!}[\hat{L}[\hat{L}\hat{a}]] + \frac{1}{3!}[\hat{L}[\hat{L}[\hat{L}\hat{a}]]] + \dots$$

13. An oscillator at the time $t = -\infty$ is in the ground state. Find the probability that at time $t = +\infty$ the oscillator will be in the nth excited state if it is subject to a force $f(t)$, where $f(t)$ is an arbitrary function of time ($f \to 0$ as $t \to \pm\infty$).

Treat in full the cases:

(a) $f(t) = f_0 \exp(-t^2/\tau^2)$,

(b) $f(t) = f_0[(t/\tau)^2 + 1]^{-1}$.

14. Show that the problem of determining the motion of an oscillator subject to a force $f(t)$ can be reduced to the simpler problem of determining the motion of a free oscillator by the introduction of a new variable $x_1 = x - \xi(t)$, where $\xi(t)$ satisfies the classical equation $\mu\ddot{\xi} = f(t) - \mu\omega^2\xi$.

15. Use the solution of the classical equation for an oscillator with a variable frequency to express the Green's function for an oscillator whose natural frequency changes with time.

16. Using the Green's function found in the preceding problem, find the time-dependent probability density for a particle moving in the potential field $V(x) = \frac{1}{2}\mu\omega^2x^2$ ($\omega =$ const). The wave function of the particle at the time $t = 0$ is

$$\psi(x, 0) = c \exp\left[-\frac{1}{2}\alpha^2(x - x_0) + \frac{i}{\hbar}p_0 x\right].$$

17. A perturbing force $f(t)$ acts on an oscillator whose natural frequency changes with time. Find the Green's function for the oscillator.

Hint. Use the results of Probs. 14 and 15, §3.

18. A perturbing force $f(t)$ is applied to an oscillator which at time $t = 0$ is in the nth energy state. Find the probability of transition to the mth state. Calculate the mean energy and the energy dispersion at the time t.

19. Since the Schrödinger equation is of the first order in time, $\psi(t)$ is determined uniquely by $\psi(0)$. We write this relation in the form

$$\psi(t) = \hat{S}(t)\psi(0),$$

where $\hat{S}(t)$ is some operator.

(a) Show that the operator $\hat{S}(t)$ satisfies the equation

$$i\hbar\dot{\hat{S}}(t) = \hat{H}\hat{S}(t)$$

and is a unitary operator, i.e., $\hat{S}^+ = \hat{S}^{-1}$.

(b) Show that, if $\hat{H}$ does not depend on time, $\hat{S}(t)$ has the form

$$\hat{S}(t) = \exp\left(-\frac{i}{\hbar}\hat{H}t\right).$$

20. The mean value of some operator $\hat{L}$ at the time t is defined by the following expression:

$$\bar{L}(t) = \int \psi^*(t)\hat{L}\psi(t)\, d\tau.$$

(a) Show that the time-dependence of the operator $\hat{\mathcal{L}} = \hat{S}^{-1}(t)\hat{L}\hat{S}(t)$, where $\hat{S}(t)$ is defined by the relation $\hat{S}(t)\psi(0) = \psi(t)$, satisfies the condition $\int \psi^*(0)\hat{\mathcal{L}}\psi(0)\, d\tau = \bar{L}(t)$.

(b) Establish the relation

$$i\hbar\dot{\hat{\mathcal{L}}} = \hat{\mathcal{L}}\hat{\mathcal{H}} - \hat{\mathcal{H}}\hat{\mathcal{L}},$$

where

$$\hat{\mathcal{H}} = \hat{S}^{-1}\hat{H}\hat{S}.$$

(c) Show that, if the operators $\hat{L}$ and $\hat{M}$ satisfy the commutation rules

$$\hat{L}\hat{M} - \hat{M}\hat{L} = i\hat{N},$$

then for time-dependent operators

$$\hat{\mathcal{L}}\hat{\mathcal{M}} - \hat{\mathcal{M}}\hat{\mathcal{L}} = i\hat{\mathcal{N}}.$$

21. Find the time-dependent operator of the position coordinate $\hat{\mathfrak{x}}$ (in the position-coordinate representation) for

(a) a free particle,
(b) an oscillator.

22. Making use of the results of the previous problem, find the time dependence of the dispersion of the position coordinate in the case of free motion.

23. The wave function of a particle at the time $t = 0$ has the form $\psi(x) = \varphi(x)\exp\left(\frac{i}{\hbar}p_0 x\right)$, where $\varphi(x)$ is a real function normalized to unity. Find the value of the dispersion $\overline{(\Delta x)^2}$ at any time for cases (a) and (b) in Prob. 21, §3. Show that in the case of an oscillator $\overline{(\Delta x)_t^2} = \overline{(\Delta x)_{t=0}^2}$, i.e., there is no spreading if $\varphi(x) = c\exp\left(-\frac{\mu\omega x^2}{\hbar}\right)$ (see Prob. 11c, §3).

§4. Angular momentum. Spin

1. Obtain the expression for the operators $\hat{l}_x$, $\hat{l}_y$, $\hat{l}_z$ in spherical coordinates, where $\hat{l}_x$, $\hat{l}_y$, $\hat{l}_z$ are the operators of an infinitesimal rotation.

2. Prove the commutation relations

(a) $[\hat{l}_i, \hat{x}_k] = ie_{ikl}\hat{x}_l$,

(b) $[\hat{l}_i, \hat{p}_k] = ie_{ikl}\hat{p}_l$,

where e_{ikl} is an antisymmetric unit tensor of rank three whose components change sign with a permutation of any two of its indices, i.e., $e_{ilk} = e_{ilk}$, where $e_{123} = 1$ (1, 2, 3 correspond to x, y, z).

3. Show that

(a) $[\hat{\mathbf{l}}, (\hat{p}_x^2 + \hat{p}_y^2 + \hat{p}_z^2)] = 0$,

(b) $[\hat{\mathbf{l}}, (x^2 + y^2 + z^2)] = 0$.

4. Show that in a state ψ with some definite value of $\hat{l}_z(\hat{l}_z\psi = m\psi)$ the mean values of $\hat{l}_x$ and $\hat{l}_y$ are zero.

Hint. Find the mean value in state ψ of the left-hand and right-hand sides of the commutation relations

$$\hat{l}_y\hat{l}_z - \hat{l}_z\hat{l}_y = i\hat{l}_x,$$
$$\hat{l}_z\hat{l}_x - \hat{l}_x\hat{l}_z = i\hat{l}_y.$$

5. Express the angular momentum about some axis z' in terms of the operators $\hat{l}_x$, $\hat{l}_y$, $\hat{l}_z$.

6. Show that if for some state $\psi\,\hat{l}_z\psi = m\psi$, the mean value of the angular momentum about an axis z' making an angle of θ with the z axis is equal to $m\cos\theta$.

This result can be interpreted physically in the following way: The angular momentum vector in the state ψ_m is uniformly "spread" over a cone whose axis is the z axis, where the generating line is of length $\sqrt{l(l+1)}$, the altitude of the cone being m. The mean value of the projection on the xy plane is zero and the projection on the axis z' after averaging is $m\cos\theta$.

7. What is the form of the eigenfunctions of the operators of the square and projection of the orbital angular momentum in the momentum representation?

8. Find the transformation rule of the spherical functions Y_{11}, Y_{10}, $Y_{1,-1}$ for a rotation of the coordinate system by the Euler angles θ, ψ, φ.

Hint. Represent the spherical functions in the form

$$Y_{11} = \sqrt{\frac{3}{8\pi}}\frac{x+iy}{r}, \quad Y_{10} = \sqrt{\frac{3}{4\pi}}\frac{z}{r}, \quad Y_{1,-1} = \sqrt{\frac{3}{8\pi}}\frac{x-iy}{r}.$$

9. In an experimental arrangement of the type used in the Stern-Gerlach experiment the deflection of atoms with a total angular momentum J depends on the magnitude of the component of the angular momentum along the direction of the instrument's magnetic field. If the beam consists of particles with a given angular momentum about an axis not coinciding with the direction of the magnetic field of the experimental arrangement, it splits into $2J+1$ beams.

Find the relative intensities of these beams if $J = 1$ and if the component of the angular momentum along an axis making an angle of θ with the direction of the magnetic field has the value $M(+1, 0, -1)$.

10. The spin component of an electron along the z axis has been reliably established to be $+\frac{1}{2}$. What is the probability that the projection of the spin on an axis z' making an angle θ with the z axis will have the value $+\frac{1}{2}$ and $-\frac{1}{2}$? Find the mean value of the projection of the spin on the axis z'.

11. The most general form of the spin function of a particle with spin $\frac{1}{2}$ in the z-representation is

$$\begin{pmatrix}\psi_1 \\ \psi_2\end{pmatrix} = \begin{pmatrix}e^{i\alpha}\cos\delta \\ e^{i\beta}\sin\delta\end{pmatrix}.$$

This function describes the state of a particle in which the probability that the component of the spin along the z axis has the value $+\frac{1}{2}$ (or $-\frac{1}{2}$) is $\cos^2\delta$ (or $\sin^2\delta$).

What will be the result of a measurement of the spin component along any arbitrary axis?

12. The spin function in the z-representation has the form

$$\begin{pmatrix}\psi_1 \\ \psi_2\end{pmatrix} = \begin{pmatrix}e^{i\alpha}\cos\delta \\ e^{i\beta}\sin\delta\end{pmatrix}.$$

Does a direction exist in space along which the spin component has for certain the value $+\frac{1}{2}$?

If such a direction does exist, find its spherical coordinates (θ, Φ).

Hint. Find θ and Φ from the condition that the second component of the spin function vanishes.

13. Consider a system of non-interacting identical particles. The particles have the same momentum and their spin is $\frac{1}{2}$. If these particles did not have any spin, we could call such a system a *pure ensemble.* But we do not know if the orientation of the spin of all the particles is the same.

Can one, by means of an experiment of the Stern-Gerlach type, determine whether the beam of particles is a pure or mixed ensemble?

14. Show that the operator transforming the components of the spin function for a rotation by the Euler angles θ, ψ, φ is of the form

$$\hat{T}(\psi, \theta, \varphi) = \exp(i\varphi \hat{s}_z) \exp(i\theta \hat{s}_x) \exp(i\psi \hat{s}_z).$$

15. Show that, for a rotation of the coordinate system by an angle Φ relative to an axis whose direction cosines are α, β, γ, the transformation matrix for components of the spin function can be represented in the form

$$\hat{T} = \exp[i\Phi(\alpha\hat{s}_x + \beta\hat{s}_y + \gamma\hat{s}_z)] = \cos\frac{\Phi}{2} + 2i(\alpha\hat{s}_x + \beta\hat{s}_y + \gamma\hat{s}_z)\sin\frac{\Phi}{2}.$$

Hint. The relation between the Euler angles θ, ψ, φ and α, β, γ and Φ is given by

$$\cos\frac{\Phi}{2} = \cos\frac{\theta}{2}\cdot\cos\frac{1}{2}(\varphi+\psi),$$

$$\alpha\sin\frac{\Phi}{2} = \sin\frac{\theta}{2}\cdot\cos\frac{1}{2}(\varphi-\psi),$$

$$\beta\sin\frac{\Phi}{2} = \sin\frac{\theta}{2}\cdot\sin\frac{1}{2}(\varphi-\psi),$$

$$\gamma\sin\frac{\Phi}{2} = \cos\frac{\theta}{2}\cdot\sin\frac{1}{2}(\varphi+\psi).$$

Note that $\exp[i\Phi(\alpha\hat{s}_x + \beta\hat{s}_y + \gamma\hat{s}_z)]$ is not equal to

$$\exp(i\Phi\alpha\hat{s}_x)\cdot\exp(i\Phi\beta\hat{s}_y)\exp(i\Phi\gamma\hat{s}_z).$$

16. Find the eigenfunctions of the operator $\alpha\hat{s}_x + \beta\hat{s}_y + \gamma\hat{s}_z$, where $\alpha^2 + \beta^2 + \gamma^2 = 1$ and show that the coefficients of the expansion of any spin function $\begin{pmatrix}\psi_1\\ \psi_2\end{pmatrix}$ with respect to these functions give the probability that the spin component along an axis with the direction cosines α, β, γ is $+\frac{1}{2}$ or $-\frac{1}{2}$.

17. Find the matrix transforming the components of the spin function of a particle with spin 1 for an arbitrary rotation of the coordinate system.

18. The angular momentum of a particle is equal to j, the projection of the angular momentum having a maximum on the z axis.

Various values of the component of the angular momentum in a direction making an angle θ with the z axis occur with different probabilities. Find these probabilities.

19. A system having a total angular momentum J is in the state $J_z = M$. What is the probability that a measurement (e.g., in the Stern-Gerlach experiment) of the component of the angular momentum along an axis z' making an angle of ϑ with the z axis will give the value M'?

20. Show that if $\psi_m^{(0)}$ is the eigenfunction of an operator $\hat{J}_z$ corresponding to the eigenvalue m, the function $\psi_m = e^{-i\hat{J}_z\varphi}e^{-i\hat{J}_y\vartheta}\psi_m^{(0)}$ is the eigenfunction of the operator $\hat{J}_\xi = \hat{J}_x \sin\vartheta\cos\varphi + \hat{J}_y \sin\vartheta\sin\varphi + \hat{J}_z\cos\vartheta$, belonging to this eigenvalue, i.e.,

$$\hat{J}_\xi \psi_m = m\psi_m.$$

Hint. Use the relations (see Prob. 12, §3)

$$\exp(-i\hat{J}_y\vartheta)\,\hat{J}_z\exp(i\hat{J}_y\vartheta) = \hat{J}_z\cos\vartheta + \hat{J}_x\sin\vartheta,$$

$$\exp(-i\hat{J}_z\varphi)\,\hat{J}_x\exp(i\hat{J}_z\varphi) = \hat{J}_x\cos\varphi + \hat{J}_y\sin\varphi.$$

21. A particle with spin $\frac{1}{2}$ moves in a central force field. Find the wave functions of this particle which are simultaneously eigenfunctions of the three commutating operators

$$\hat{j}_z = \hat{l}_z + \hat{s}_z, \quad \hat{\mathbf{l}}^2, \quad \hat{\mathbf{j}}^2.$$

22. The state of an electron is specified by the quantum numbers l, j, m. Using the wave functions obtained in the previous problem, find the possible values of the orbital and spin angular momentum components and the corresponding probabilities. Find also the mean values of the components.

23. By the direction of the spin we understand the direction along which the projection of the spin has the value $+\frac{1}{2}$. This will be the direction determined by the polar angles Θ, Φ. Let the state of a particle be described by the wave function ψ (l, $j = l \pm \frac{1}{2}$, m) (see Prob. 21, §4). The direction of the spin of such a particle at different points of space will not, in general, be the same. Determine the relation between the angles Θ, Φ and the space coordinates of the particle.

24. Find the wave functions of a system of two particles of spin $\frac{1}{2}$ which are eigenfunctions of the operators of the square of the total spin and the component of the total spin along the z axis.

25. A system consists of two particles, the angular momentum of the first being $l_1 = 1$ and the second $l_2 = l$. The total angular momentum in this case can take on the values $l+1$, l, and $l-1$. Express the eigenfunctions of the operators $\hat{\mathbf{J}}^2$ and $\hat{J}_z$ in terms of the eigenfunctions of the squares of the angular momenta and the components of the angular momenta along the z axis of the individual particles.

26. Given two particles whose angular momenta are j_1 and j_2, use the properties of the operator $\hat{J}_- = (\hat{j}_{1x} + \hat{j}_{2x}) - i(\hat{j}_{1y} + \hat{j}_{2y})$ to expand the wave function of a system whose angular momentum is $J = j_1 + j_2$ witha projection $M = J - 1$ in terms of the wave functions of both particles.

Construct the wave function of the system whose angular momentum is $J = j_1 + j_2 - 1$ with a projection $M = J$.

27. Show that a wave function whose resultant angular momentum is zero, but which represents a system of two particles, each with an angular momentum of j, can be written in the following form:

$$\Psi_0^0 = \frac{1}{\sqrt{2j+1}} \sum_{m_j=-j}^{j} (-1)^{m_j+\frac{1}{2}} \psi_j^{m_j}(1)\psi_j^{-m_j}(2) \quad (j \text{ is a half-integer}),$$

$$\Psi_0^0 = \frac{1}{\sqrt{2j+1}} \sum_{m_j=-j}^{j} (-1)^{m_j} \psi_j^{m_j}(1)\psi^{-m_j}(2) \quad (j \text{ is an integer}).$$

Hint: Use the properties of the operator $\hat{J}_- = (\hat{j}_{1x} + \hat{j}_{2x}) - i(\hat{j}_{1y} + \hat{j}_{2y})$.

28. Show that the function

$$\Psi_{J=j}^{M=J} = \sqrt{\frac{2}{2j-1}} \sum_{m_j=\frac{1}{2}}^{j-1} (-1)^{m_j+\frac{1}{2}} \frac{1}{\sqrt{3!}} \begin{vmatrix} \psi_j^j(1) & \psi_j^{m_j}(1) & \psi_j^{-m_j}(1) \\ \psi_j^j(2) & \psi_j^{m_j}(2) & \psi_j^{-m_j}(2) \\ \psi_j^j(3) & \psi_j^{m_j}(3) & \psi_j^{-m_j}(3) \end{vmatrix}$$

is an eigenfunction of the total angular momentum of three particles corresponding to a resultant angular momentum $J = j$ and projection $M = J$. The angular momentum of each particle is equal to j (j is a half-integer).

Hint: Use the properties of the operator $\hat{J}_+$.

29. Let us denote by $\hat{\sigma}_1$ and $\hat{\sigma}_2$ the spin operators of two particles, and by $\mathbf{r}$ the radius vector joining these particles. Show that any positive integral power of each of the operators

$$(\hat{\sigma}_1\hat{\sigma}_2) \quad \text{and} \quad \hat{S}_{12} = \frac{3(\hat{\sigma}_1\mathbf{r})(\hat{\sigma}_1\mathbf{r})}{r^2} - (\hat{\sigma}_1\hat{\sigma}_2),$$

and also that any product of these powers can be represented in the form of a linear combination of these operators and a unit matrix.

30. Show that the operator $\hat{S}_{12}$ (see the preceding problem) can be expressed by the total spin operator $\hat{\mathbf{S}}=\frac{1}{2}(\hat{\sigma}_1+\hat{\sigma}_2)$ in the following way:

$$\hat{S}_{12}=\frac{6(\hat{\mathbf{S}}\mathbf{r})^2}{r^2}-2\hat{\mathbf{S}}^2$$

and that, when the total spin of two particles is equal to unity, $\hat{S}_{12}$ can be represented as the tensor of rank three:

$$\hat{S}_{12}=\frac{4\sqrt{\pi}}{\sqrt{5}}\begin{pmatrix} Y_{20} & -\sqrt{3}Y_{2,-1} & \sqrt{6}Y_{2,-2} \\ -\sqrt{3}Y_{21} & -2Y_{20} & \sqrt{3}Y_{2,-1} \\ \sqrt{6}Y_{22} & \sqrt{3}Y_{21} & Y_{20} \end{pmatrix}.$$

31. Show that the normalized part of the wave function for the state ${}^3D_1$ containing the spin and angular variables can be written in the following way:

$$\frac{1}{4\sqrt{2\pi}}\hat{S}_{12}\begin{pmatrix}1\\0\\0\end{pmatrix} \quad (J=1, \quad M=1, \quad L=2, \quad S=1),$$

$$\frac{1}{4\sqrt{2\pi}}\hat{S}_{12}\begin{pmatrix}0\\1\\0\end{pmatrix} \quad (J=1, \quad M=0, \quad L=2, \quad S=1),$$

$$\frac{1}{4\sqrt{2\pi}}\hat{S}_{12}\begin{pmatrix}0\\0\\1\end{pmatrix} \quad (J=1, \quad M=-1, \; L=2, \quad S=1).$$

Hint: See Prob. 25, §4.

32. Two protons are a distance a apart. Calculate the energy of the dipole-dipole interaction.

33. At the time $t=0$ the spin function of two protons (see the preceding problem) has the form $\begin{pmatrix}1\\0\end{pmatrix}_1\begin{pmatrix}0\\1\end{pmatrix}_2$. Through what interval of time will the spin function have the form $\begin{pmatrix}0\\1\end{pmatrix}_1\begin{pmatrix}1\\0\end{pmatrix}_2$?

34. Prove the following relations:

(a) $\{\hat{\mathbf{J}}^2,\ \hat{\mathbf{A}}\} \qquad = i([\hat{\mathbf{A}}\hat{\mathbf{J}}]-[\hat{\mathbf{J}}\hat{\mathbf{A}}])$,

(b) $\{\hat{\mathbf{J}}^2,\ \{\hat{\mathbf{J}}^2,\ \hat{\mathbf{A}}\}\}=2(\hat{\mathbf{J}}^2\hat{\mathbf{A}}+\hat{\mathbf{A}}\hat{\mathbf{J}}^2)-4\hat{\mathbf{J}}(\hat{\mathbf{J}}\hat{\mathbf{A}})$,

(c) $(\hat{\mathbf{A}})^{n'JM'}_{nJM} = \frac{(\hat{\mathbf{J}}\hat{\mathbf{A}})^{n'J}_{nJ}}{J(J+1)}(\mathbf{J})^{JM'}_{JM}.$

Where $\hat{\mathbf{A}}$ is some physical vector quantity satisfying the commutation rules

$$\{\hat{J}_i, \hat{A}_k\} = ie_{ikl}\hat{A}_l.$$

35. Find the mean value of the operator $\hat{\mu} = g_1\hat{\mathbf{J}}_1 + g_2\hat{\mathbf{J}}_2$ in the state characterized by the quantum numbers J, M_J, J_1, J_2 if the total angular momentum is $\hat{\mathbf{J}} = \hat{\mathbf{J}}_1 + \hat{\mathbf{J}}_2$.

Hint: Use the formula given in the preceding problem.

36. Find the magnetic moment (in nuclear magnetons) of the nucleus N^{15}, which needs one proton in the $p_{1/2}$ state to fill the shell. The magnetic moment of a free proton is $\mu_p = 2 \cdot 79$.

37. Calculate the magnetic moment of the nucleus O^{17} containing besides a filled shell, one neutron in the state $d_{5/2}$. The magnetic moment of a free electron is $\mu_n = -1 \cdot 91$.

38. What would be the numerical value of the magnetic moment of the deuteron if the deuteron were in the state:

(a) $^3S_1$, (b) $^1P_1$, (c) $^3P_1$, (d) $^3D_1$?

39. Assuming that the ground state of the deuteron is a superposition of the $^3S_1$ and $^3D_1$ states, find the weight of the D wave if $\mu_p = 2 \cdot 78$, $\mu_n = -1 \cdot 91$, $\mu_d = 0 \cdot 85$.

40. Express the quadrupole moment of the deuteron in terms of the mean square distance under the assumption that the deuteron is in the state: (a) $^1P_1$, (b) $^3P_1$.

41. A nucleus consists of a "core" and an outer proton. In the absence of an interaction with the outer proton the "core" has spherical symmetry and its quadrupole moment is zero.

Neglecting the deformation of the "core" by the outer proton, determine the quadrupole moment of such a nucleus. The angular momentum of the outer proton is j.

42. Under the same assumptions, solve the preceding problem for the case in which there are three outer protons. The radial wave function of the protons are identical, the angular momentum of each of the protons and of the nucleus as a whole is j.

43. Determine the quadrupole energy of a nucleus in an external nonhomogeneous electric field. The spin of the nucleus is (a) $I = 1$, (b) $I = \frac{3}{2}$, (c) $I = 2$. The quadrupole moment of the nucleus is Q_0.

44. Using the expresson for the vector matrix elements* show that the quadrupole moment of a nucleus of atomic number Z is equal to

$$Q = I(2I-1)\sum_{i=1}^{Z}\sum_{n'}\{2(I+1)|(z_i)^{n,\,I}_{n',\,I+1}|^2 - 2I|(z_i)^{n,\,I}_{n',\,I-1}|^2\},$$

the summation being taken over all the protons of the nucleus. I is the spin of the nucleus and n represents the set of all other quantum numbers characterizing the state.

45. We denote by σ_i the spin variable of the ith electron. This variable takes on two values $+1$ and -1. Show that the operators

$$\hat{\sigma}_{lx} = \begin{pmatrix} 0 & 1 \\ 1 & 0 \end{pmatrix}_l; \quad \hat{\sigma}_{ly} = \begin{pmatrix} 0 & -i \\ i & 0 \end{pmatrix}_l; \quad \hat{\sigma}_{lz} = \begin{pmatrix} 1 & 0 \\ 0 & -1 \end{pmatrix}_l,$$

which act on the lth electron satisfy the following relations when acting on a function $f(\sigma_1, \sigma_2, \ldots, \sigma_l, \ldots, \sigma_n)$ of the spin variables of n electrons:

$$\hat{\sigma}_{lx} f = f(\sigma_1, \ldots, \sigma_{l-1}, -\sigma_l, \sigma_{l+1}, \ldots, \sigma_n),$$
$$\hat{\sigma}_{ly} f = -i\sigma_l f(\sigma_1, \ldots, \sigma_{l-1}, -\sigma_l, \sigma_{l+1}, \ldots, \sigma_n),$$
$$\hat{\sigma}_{lz} f = \sigma_l f(\sigma_1, \ldots, \sigma_{l-1}, \sigma_l, \sigma_{l+1}, \ldots, \sigma_n).$$

46. Using the results of the preceding problem, show that the operator representing the square of the total spin angular momentum of the n electrons can be represented in the form

$$\hat{\mathbf{S}}^2 = n - \frac{n^2}{4} + \sum_{k<l} P_{kl},$$

where P_{kl} is an operator which permutes the spin variables σ_k and σ_l, that is

$$P_{kl} f(\sigma_1, \ldots, \sigma_{k-1}, \sigma_k, \sigma_{k+1}, \ldots, \sigma_{l-1}, \sigma_l, \sigma_{l+1}, \ldots, \sigma_n)$$
$$= f(\sigma_1, \ldots, \sigma_{k-1}, \sigma_l, \sigma_{k+1}, \ldots, \sigma_{l-1}, \sigma_k, \sigma_{l+1}, \ldots, \sigma_n).$$

47. The Hamiltonian of a system of two spin $\frac{1}{2}$ particles is symmetric in the spins. Show that the magnitude of the total spin S is a constant of motion.

48. A system consists of two particles. The spin of one is equal to $\frac{1}{2}$ and of the other 0. Show that, for any interaction of these particles, the orbital angular momentum is a constant of motion.

---

* See L. Landau and E. Lifshits, *Quantum Mechanics*, Pergamon Press, 1958, p. 93.

49. An excited nucleus A with spin 1 is in an even state. The reaction

$$A \rightarrow B + \alpha$$

with the emission of an α-particle is energetically possible. The stable nucleus B resulting from this reaction has a spin 0 and is also in an even state. On the basis of the conservation of angular momentum and parity show that such a reaction is forbidden.

50. Show that the orbital angular momentum L of the relative motion of two α-particles is always an even number ($L = 0, 2, 4, \ldots,$).

51. Can an excited ${}_4Be^8$ nucleus with a spin equal to unity decay into two α-particles?

52. Assuming that the only bound state of the neutron-proton system (n, p) is even, that the resultant spin in this state is equal to unity, and that the forces of the (n, n) and (n, p) interactions are the same, show that two neutrons cannot form a bound system.

53. The ${}_2He^6$ nucleus has two neutrons in the level $1p_{3/2}$. Determine on the basis of the Pauli principle the possible values of the total angular momentum of these neutrons.

§5. Centrally symmetric field

1. A particle moves in a centrally symmetric field. Represent the equation for the radial part of the wave function R_{nl} in the form of the Schrödinger equation for one-dimensional motion.

2. Find the radial wave function of a particle in a centrally symmetric field in the quasi-classical approximation.

3. A centrally symmetric field gives rise to a discrete set of energy eigenvalues. Show that the minimum of the energy for a given l (l being the orbital quantum number) increases with l.

4. A system consists of two particles whose masses are μ_1 and μ_2. Express the operator of the resultant orbital angular momentum $\hat{\mathbf{l}}_1 + \hat{\mathbf{l}}_2$ and resultant momentum $\hat{\mathbf{p}}_1 + \hat{\mathbf{p}}_2$ by coordinates of the centre of mass $\mathbf{R} = \dfrac{\mu_1 \mathbf{r}_1 + \mu_2 \mathbf{r}_2}{\mu_1 + \mu_2}$ and the interconnecting distance $\mathbf{r} = \mathbf{r}_2 - \mathbf{r}_1$. Show that if the potential energy of the interaction of the particles is a function of the distance between the particles $U = U(|r_2 - r_1|)$, the Hamiltonian can be given in the form

$$\hat{H} = -\frac{\hbar^2}{2(\mu_1 + \mu_2)} \Delta_R - \frac{\hbar^2(\mu_1 + \mu_2)}{2\mu_1\mu_2} \Delta_r + U(r),$$

where Δ_R and Δ_r are the Laplacians operating on the vectors $\mathbf{R}$ and, $\mathbf{r}$ respectively.

5. Find the wave functions and energy levels of a three-dimensional isotropic oscillator.

6. Solve the preceding problem by separating the variables in Cartesian coordinates. Represent the wave functions for $n_r = 0$, $l = 1$ (see the preceding problem) in the form of a linear combination of the solution in Cartesian coordinates.

7. Assuming that a nucleon in a light nucleus moves in an average potential field of the form $U(r) = -U_0 + \frac{\mu\omega^2}{2} r^2$, find the number of particles of one kind (neutrons or protons) in filled shells. By the term shell is meant states with the same energy.

8. Calculate the theoretical radius of the nuclei $_2He^4$ and $_8O^{16}$ having closed shells using the potential given in the preceding problem. By the theoretical radius is meant the distance from the centre of mass of the nucleus to the point where the "nuclear density" $\varrho(\mathbf{r}) = \sum_{\nu} \psi_\nu^*(\mathbf{r})\psi_\nu(\mathbf{r})$ (summed over all nucleons) falls off most rapidly, that is,

$$\left(\frac{d^2\varrho}{dr^2}\right)_{r=R} = 0.$$

9. The interaction between a proton and neutron can be represented approximately by the potential $U(r) = -A \exp(-r/a)$. Find the wave function of the ground state ($l = 0$). Find the relation between the depth of the well A and the quantity a, characterizing the range of the forces, if the empirical value of the binding energy of the deutron is $E = -2{\cdot}2$ MeV.

10. Using Ritz's variational principle, find the approximate energy of the ground state of the deuteron if the potential $U(r) = -A \exp(-r/a)$ ($A = 32$ MeV, $a = 2{\cdot}2 \times 10^{-13}$ cm). Take the radial wave functions in the form $R = c \exp(-\alpha r/2a)$ dependent on the parameter α. The quantity c is related to α through the normalization condition $\int_0^\infty R^2 r^2\, dr = 1$.

11. Find the energy levels and the wave functions of a particle in the spherical potential box

$$U(r) = 0 \quad (r < a); \qquad U(r) = \infty \quad (r > a).$$

Consider the case $l = 0$.

12. Find the discrete energy spectrum of a particle with angular momentum $l = 0$ in the centrally symmetric potential well

$$U(r) = \begin{cases} -U_0 & (r < a), \\ 0 & (r > a). \end{cases}$$

13. Using perturbation theory, discuss qualitatively the changes in the energy levels in the transition from the potential

$$U(r) = \begin{cases} -U_0 & (r < a), \\ 0 & (r > a) \end{cases}$$

to the potential shown in Fig. 16.

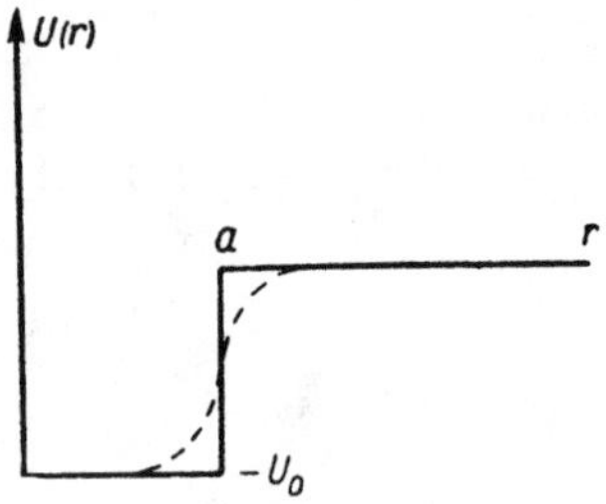

Fig. 16

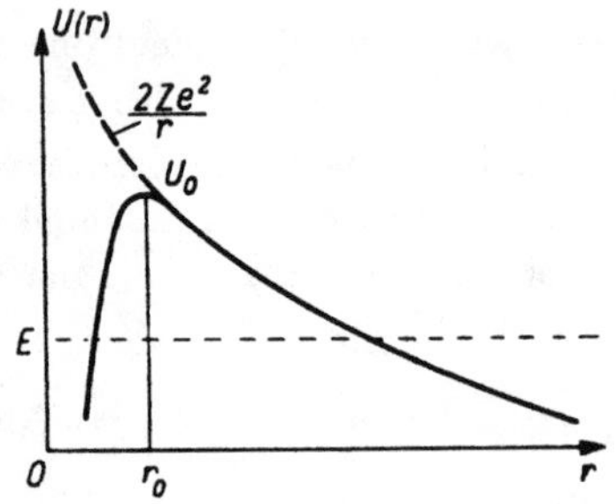

Fig. 17

14. The potential energy of an α-particle in the field of a nucleus consists of two parts, the repulsive Coulomb force and the short-range attractive nuclear force. The form of the potential energy is schematically shown in Fig. 17. The emission of an α-particle is represented as a specific quantum phenomenon resulting from the transparency of the barrier. Discuss the transmission of an α-particle (of angular momentum $l = 0$) through the spherically symmetric potential barrier of simplified shape

$$U(r) = 0 \qquad (r < r_1),$$
$$U(r) = U_0 \qquad (r_1 < r < r_2),$$
$$U(r) = 0 \qquad (r_2 < r).$$

Find the relation between the lifetime of the decay and the energy.

§6. Motion of a particle in a magnetic field

1. Let the wave function of an electron initially have the form

$$\Psi(x, y, z, 0) = \psi(x, y, 0)\, \varphi(z, 0).$$

Then in a uniform magnetic field $\mathcal{H}$ directed along the z axis the wave function at the time t will also have the form of the product

$$\Psi(x, y, z, t) = \psi(x, y, t)\,\varphi(z, t),$$

since in the Schrödinger equation the variable z is separable. Prove that the function $\psi(x, y, T)$, aside from an arbitrary phase factor, takes on its initial value if T is the period of the classical motion of the particle in the magnetic field.

2. Show that in the case of a magnetic field the following commutation rules hold for the operators of the velocity components:

$$\hat{v}_x\hat{v}_y - \hat{v}_y\hat{v}_x = \frac{ie\hbar}{\mu^2 c}\mathcal{H}_z,$$

$$\hat{v}_y\hat{v}_z - \hat{v}_z\hat{v}_y = \frac{ie\hbar}{\mu^2 c}\mathcal{H}_x,$$

$$\hat{v}_z\hat{v}_x - \hat{v}_x\hat{v}_z = \frac{ie\hbar}{\mu^2 c}\mathcal{H}_y.$$

3. Using the results of Probs. 2, §6, and 6, §1, find the energy of a charged particle moving in a uniform magnetic field.

4. Find the energy spectrum of a charged particle moving in uniform electric and magnetic fields whose field intensities are directed perpendicularly to one another.

5. Find the wave functions of a charged particle moving in perpendicular uniform magnetic and electric fields.

6. A charged particle moves in a uniform magnetic field on which is superimposed a centrally symmetric field of the form $U(r) = \frac{1}{2}\mu\omega_0^2 r^2$. Find the energy spectrum of the particle.

7. Find the time dependence of the position-coordinate operators $\hat{x}$, $\hat{y}$ of a charged particle moving in a uniform magnetic field (the vector potential is $A_x = -\frac{\mathcal{H}}{2}y$, $A_y = \frac{\mathcal{H}}{2}x$, $A_z = 0$). Find $\overline{(\hat{x}-\bar{x})^2}$ and $\overline{(\hat{y}-\bar{y})^2}$ as functions of time.

8. Find the energy levels and wave functions of a charged particle moving in a uniform magnetic field. Use a cylindrical coordinate system and take the vector potential in the form $A_\varphi = \frac{\mathcal{H}}{2}\varrho$, $A_\varrho = A_z = 0$.

9. Find the components of the current density for a charged particle moving in a uniform magnetic field for a state with the quantum numbers n, m, k_z (see the preceding problem).

10. Find in the quasi-classical approximation the energy levels of a charged particle in a uniform magnetic field (cylindrical coordinates).

11. Find the region of classical radial motion in a magnetic field (see the preceding problem).

12. Estimate the minimum spread of the orbit in a radial direction for a charged particle in a magnetic field.

13. Using classical mechanics, express the position of the centre of the circle around which a charged particle moves in a uniform magnetic field in terms of the position coordinates x, y and the generalized momenta p_x, p_y. In these expressions consider the position coordinates and momenta to be operators and find the commutation rules for such a "centre of orbit" and the corresponding uncertainty relations. Show that the sum of the squares of the coordinates of the "centre of orbit" takes on the discrete set of values $\frac{2\hbar c}{|e|\mathcal{H}}\left(n+\frac{1}{2}\right)$, where $n = 0, 1, 2, \ldots$

14. Show that in a uniform magnetic field varying in time the wave function of a particle with spin can be separated into the product of the coordinates and the spin functions.

15. A particle with spin $\frac{1}{2}$ moves in a time-varying uniform magnetic field directed along the z axis. The time variation of the field is given by some arbitrary function $(\mathcal{H} = \mathcal{H}(t))$. At the initial time $(t = 0)$ the spin function has the form $\begin{pmatrix} e^{-i\alpha} \cos\delta \\ e^{i\alpha} \sin\delta \end{pmatrix}$. Find the mean value of the projection of the spin on the x and y axes and also the direction along which the spin component has a definite value at the time t.

16. In the region $x > 0$ there is a uniform magnetic field $\mathcal{H}_x = \mathcal{H}_y = 0$, $\mathcal{H}_z = \mathcal{H}$; in the region $x < 0$ the field is zero. A beam of polarized neutrons of momentum $\mathbf{p}$ from the region of zero field is incident on the plane $x = 0$. Find the reflection coefficient at the boundary between the two regions.

17. A particle of spin $\frac{1}{2}$ is moving in a uniform magnetic field. The field intensity is constant in magnitude, but changes direction according to the relations

$$\mathcal{H}_x = \mathcal{H} \sin\vartheta \cos\omega t, \qquad \mathcal{H}_y = \mathcal{H} \sin\vartheta \sin\omega t, \qquad \mathcal{H}_z = \mathcal{H} \cos\vartheta.$$

At the time $t = 0$ the component of the spin along the direction of the magnetic field has the value $+\frac{1}{2}$. What is the probability of the particle making a transition, by the time t, to a state in which the spin component along the direction of the magnetic field is $-\frac{1}{2}$?

18. A particle of spin $\frac{1}{2}$ and magnetic moment μ moves in the non-homogeneous magnetic field

$$\mathcal{H}_z = \mathcal{H}_0 + kz, \quad \mathcal{H}_y = -ky, \quad \mathcal{H}_x = 0 \quad (\operatorname{div} \mathcal{H} = 0).$$

(a) Express the time-dependence of the position-coordinate operators $\hat{x}$, $\hat{y}$, $\hat{z}$.

(b) Find the mean value of the position coordinates and the time-dependence of the dispersion of the position coordinates if the wave function of the particle at the time $t = 0$ has the form

$$\psi = \varphi(x, y, z)\, e^{\frac{ip_0 x}{\hbar}} \begin{pmatrix} \alpha \\ \beta \end{pmatrix}.$$

19. A neutral particle moves in a homogeneous magnetic field whose direction, but not magnitude, changes in time.

Find the equation for the spin function in the ξ-representation, where the axis ξ is directed along the magnetic field. Show that in the case of a sufficiently slow change in the direction of the magnetic field the probabilities for particular values of the angular momentum component along the direction of the field do not change.

20. Two protons are a distance a apart in a magnetic field. The magnetic field intensity $\mathcal{H}$ makes angle θ with the line joining the protons.

Considering the dipole-dipole interaction as a perturbation, determine the energy levels.

21. The operator of the magnetic dipole-dipole interaction of nuclei has the form

$$\hat{H} = \sum_{j>k} \frac{g_j g_k}{r_{jk}^3} \beta^2 \{\hat{\mathbf{I}}_j \hat{\mathbf{I}}_k - 3(\hat{\mathbf{I}}_j \mathbf{r}_{jk})(\hat{\mathbf{I}}_k \mathbf{r}_{jk}) r_{jk}^{-2}\}.$$

Here $\hat{\mathbf{r}}_{jk}$ is the radius vector joining the jth nucleus to the kth nucleus, $\hat{\mathbf{I}}_j$ is the spin operator, and $g_j \beta \hat{\mathbf{I}}_j$ is the magnetic moment operator of the jth nucleus.

Show that the diagonal matrix elements $< m|\hat{H}|m >$, where m is the projection of the total spin of all nuclei on the z axis, are equal to the diagonal matrix elements of the operator $\hat{H}_{\text{eff}}$ where

$$\hat{H}_{\text{eff}} = \sum_{j>k} A_{jk} \{\hat{\mathbf{I}}_j \hat{\mathbf{I}}_k - 3\hat{I}_{zj} I_{zk}\}, \text{ i.e., } < m|\hat{H}|m > = < m|\hat{H}_{\text{eff}}|m >.$$

In the last relation $A_{jk} = \frac{1}{2} g_j g_k \beta^2 r_{jk} (3\cos\theta_{jk} - 1)$, θ_{jk} is the angular component of the line joining the jth and kth nuclei with respect to the z axis.

Hint. Express the operators $\hat{I}_{xj}$ and $\hat{I}_{yj}$ in terms of the operators $\hat{I}_{+j} = \hat{I}_{xj}+i\hat{I}_{yj}$, $\hat{I}_{-j} = \hat{I}_{xj}-i\hat{I}_{yj}$.

22. Three similar nuclei of spin $\frac{1}{2}$ are rigidly fixed in space and are in a uniform magnetic field $\mathcal{H}$. Find the energy levels of such a system whereby the dipole-dipole interactions between nuclei are considered as a small perturbation.

23. A nucleus of spin I and quadrupole moment Q_0 is in a magnetic field of intensity $\mathcal{H}$ and in a nonhomogeneous electric field. Considering the operator of the quadrupole energy as a small perturbation, find the energy levels of the nucleus in the external field in the second-order approximation.

§7. The atom

1. From the inequality

$$\int |\nabla\psi+Z\psi\nabla r|^2\, d\tau \geqslant 0$$

find the minimum energy of a one-electron atom and the wave function corresponding to this energy. Prove that for the ground state of the atom $2\overline{T} \geqslant |\overline{v}|$.

2. An electron in the Coulomb field of a nucleus of charge Z is in the ground state. Show that the average electrostatic potential arising from interaction between the nucleus and the electron is

$$\varphi = \frac{e(Z-1)}{r}+e\left(\frac{Z}{a}+\frac{1}{r}\right)e^{-\frac{2Zr}{a}}, \quad \left(a = \frac{\hbar^2}{\mu e^2}\right).$$

3. Prove that in the ground state of the hydrogen atom:

(a) the most probable value of r is $a = \dfrac{\hbar^2}{\mu e^2}$,

(b) the mean value of $\overline{1/r}$ is $1/a$,

(c) $\overline{\dfrac{1}{r^2}} = \dfrac{2}{a^2}$.

4. The wave function $\psi(\mathbf{r})$ describes the relative motion of two particles, a proton and an electron. Assume that the coordinates of the centre of mass of the hydrogen atom are known exactly and are given by

$$X = 0, \quad Y = 0, \quad Z = 0.$$

Show that in this case the probability density for the proton is

$$w(\mathbf{r}) = \left(\frac{m+M}{m}\right)^3 \left|\psi\left(\frac{m+M}{m}\mathbf{r}\right)\right|^2,$$

where m and M are the electron and proton masses respectively.

5. Find the momentum distribution of the electron in the hydrogen atom for the states 1*s*, 2*s*, and 2*p*.

6. Calculate the mean square deviation of the distance of the electron from the nucleus $\overline{r^2} - \bar{r}^2$ for a hydrogen atom in the state described by the quantum numbers n, l.

7. Express the eigenfunction of the hydrogen atom in parabolic coordinates with $n_1 = 1$, $n_2 = 0$, $m = 0$ by means of the wave functions in spherical coordinates. Also, show that

$$\psi_{n_1=0,\, n_2=0,\, m=n-1}(\xi, \eta, \varphi) = \psi_{n,\, l=n-1,\, m=n-1}(r, \vartheta, \varphi).$$

8. Show directly that the degree of degeneracy of the nth eigenvalue of the energy of the hydrogen atom for a solution of the Schrödinger equation in parabolic coordinates is n^2.

9. Find the correction to the energy levels of the hydrogen atom resulting from the relativistic dependence of the mass on the velocity (consider the term of the order v^2/c^2).

10. From the relativistic equation for the electron (the Dirac equation) it follows that, besides the correction resulting from the velocity dependence of the mass ($\sim v^2/c^2$), there is still another term in the Hamiltonian

$$H_2 = \frac{\hbar^2}{2\mu^2 c^2} \frac{1}{r} \frac{dU}{dr} \hat{\mathbf{l}}\hat{\mathbf{s}},$$

where $\hat{\mathbf{l}}$ is the orbital angular momentum operator, $\hat{\mathbf{s}}$ is the spin operator, $U(\mathbf{r})$ is the potential energy of the electron (we assume that the potential is centrally symmetric). This term can be interpreted physically in the following way: Owing to the motion of the magnetic moment $\boldsymbol{\mu}$ (associated with the spin of the electron) there arises an electric dipole moment $\mathbf{d} = \frac{1}{c}[\mathbf{v}\boldsymbol{\mu}]$ which interacts with the field of the nucleus. Find the correction to the energy levels of the hydrogen atom resulting from the term H_2 (known as spin-orbit coupling).

11. Show that the quadrupole moment of the hydrogen atom is given by

$$Q_0 = -\frac{j-\frac{1}{2}}{j+1}\overline{r^2}\left\{\overline{r^2} = \frac{n^2}{2}(5n^2+1-3l(l+1))\left(\frac{\hbar^2}{\mu e^2}\right)^2\right\}.$$

12. Find the total probability for the excitation and ionization of the tritium atom H^3 by beta decay. Calculate also the probability of excitation to the nth level.

13. Find the energy of the ground state of a two-electron system in the field of a nucleus of charge Z by a variational method. As trial wave functions take the product of the hydrogen functions for an effective charge Z'. The relativistic corrections may be neglected.

14. The wave function of the helium atom can, with a sufficient degree of accuracy, be taken to be

$$\psi = \frac{Z'^3}{\pi a^3} e^{-\frac{Z'(r_1+r_2)}{a}} \quad \left(Z' = \frac{27}{16}\right) \quad \text{(see Prob. 13, §7).}$$

Show that the electrostatic potential of the atom is equal to

$$\varphi(r) = 2e\left(\frac{1}{r} + \frac{Z'}{a}\right) e^{-\frac{2Z'r}{a}}.$$

15. Using the approximate wave function of the ground state of helium (see Prob. 13, §7), calculate the diamagnetic susceptibility of helium.

16. Taking into account exchange effects, find by means of a variational method the energy of the ground state of the lithium atom. Use as eigenfunctions of the electrons the hydrogen functions for an electron in the $1s$ state of the form $\psi_{100} = 2Z_1^{3/2} e^{-Z_1 r}$, for an electron in the $2s$ state

$$\psi_{200} = cZ_2^{3/2} e^{-\frac{Z_2 r}{2}} (1 - \gamma Z_2 r);$$

Z_1 and Z_2 are the variational parameters, c is determined from the condition of the normalization of the wave function ψ_{200}, and γ from the condition of the orthogonality of the functions ψ_{100}, ψ_{200}. If in solving the problem we use the usual perturbation theory, then we set $Z_1 = Z_2 = 3$. By introducing the parameters Z_1 and Z_2, we take into account the screening effects of the electrons.

17. Find the displacement of the energy levels of an atom as a result of the motion of the nucleus. Calculate the magnitude of the displacement for the helium atom for the triplet and singlet states $1\,snp$ using eigenfunctions in the form of the hydrogen-like wave functions of the individual electrons with an effective charge.

18. The potential energy $U(x, y, z)$ is a homogeneous function of the coordinates of the degree ν

$$U(\lambda z, \lambda y, \lambda z) = \lambda^\nu U(x, y, z).$$

Show that the mean value of the kinetic energy in a state belonging to a discrete spectrum is related to the mean value of the potential energy by the relation $2\overline{T} = \nu \overline{U}$ (virial theorem).

19. Estimate the order of the following quantities according to the Thomas-Fermi model:

(a) the mean distance between the electron and nucleus;

(b) the mean energy of the Coulomb interaction between two electrons in the atom;

(c) the mean kinetic energy of one electron;

(d) the energy necessary for the complete ionization of the atom;

(e) the mean velocity of an electron in the atom;

(f) the mean angular momentum of an electron;

(g) the mean radial quantum number of an electron.

20. Express approximately the energy of the atom in terms of the electron density $\varrho(\mathbf{r})$ according to the Thomas-Fermi model.*

21. Show that the Thomas-Fermi equation is obtained as a condition that the total energy be a minimum for the variation of the density $\varrho(\mathbf{r})$.

Hint. Use the result of the preceding problem. In the variation of $E(\varrho)$ consider the condition of normalization $\int \varrho \, d\tau = N$ (for a neutral atom $N = Z$).

22. Find by a variational method the best approximate expression for the electron density in the Thomas-Fermi model taking trial wave functions of the form $\varrho = \frac{Ae^{-x}}{x^3}$, $x = \sqrt{\frac{r}{\lambda}}$, where A is determined from the normalization condition $\int \varrho \, d\tau = N$ (for a neutral atom $N = Z$) and λ is the parameter varied. Find the energy of the atom (ion).

Note. In choosing the trial wave functions it must be borne in mind that the exact solution in the region of small r has a singularity of the type $\varrho \sim \frac{\text{const}}{r^{3/2}}$.

23. Show that the virial theorem is valid for the Thomas-Fermi model.

24. Using the virial theorem, show that in the Thomas-Fermi model the electrostatic interaction energy of the electrons in a neutral atom is $\frac{1}{7}$ of the magnitude of the interaction between the electrons and the nucleus.

25. Calculate the total energy of complete ionization of an atom (ion) in the Thomas-Fermi approximation.

26. Find the displacement of the energy levels of an atom resulting from the finite dimensions of the nucleus. The potential inside the nucleus ($r < a$) can be taken to be constant (this means physically that the electric charge of the nucleus is spread over the surface of a sphere of radius a).

---

* In Probs. 20–25 use the system of units $e = \hbar = \mu = 1$.

27. Calculate the value of $\psi^2(0)$ for a valence electron in the s state in an atom of large Z using the quasi-classical approximation.

28. Find the component of the magnetic field intensity produced at the centre of a hydrogen atom by the orbital motion of the electron. Calculate this quantity for the $2p$ state.

29. How does the expression for the magnetic moment of the hydrogen atom change if the motion of the nucleus is taken into account?

30. Find the interval between the hyperfine-structure terms for the s electron of the hydrogen atom.

31. Find the hyperfine-structure energy of a one-electron atom whose orbital angular momentum is not zero.

32. A diamagnetic atom is in an external magnetic field. Find the intensity of the magnetic field induced at the centre of the atom.

33. Solve the preceding problem for the case of helium.

34. Give the possible values of the total angular momentum for the states ^1S, ^3S, ^3P, ^2D, ^4D.

35. Which states (terms) are possible for the following two-electron configurations: (a) $nsn's$, (b) $nsn'p$, (c) $nsn'd$, (d) $npn'p$?

36. Which terms are possible for the following configurations: (a) $(np)^3$, (b) $(nd)^2$, (c) $ns(n'p)^4$?

37. Find the lowest terms for the following elements: O, Cl, Fe, Co, As, La.*

Hint. To solve the problem it is necessary to use the following empirical rules:

(1) The least energy will occur for the term with the largest value of S for a given electron configuration and the largest value (possible for this S) of L (Hund's rule).

(2) For the normal state of the atom, $J = |L - S|$ if there is not more than half the maximum allowable number of electrons in the unfilled shell, while $J = L + S$ if more than half of this number are present.

38. Find the parity of the lowest terms of the elements K, Zn, B, C, N, O, Cl.

39. A system of N electrons is characterized by N sets of three quantum numbers n, l, m_l. Find the number of states corresponding to a given value of the projection of the total spin M_S.

---

* For the electron configurations of these atoms see D. I. Blokhintsev *Osnovy kvantovoĭ mekhaniki*. (Principles of Quantum Mechanics), Moscow, 1949, pp., 503–505. (See also e.g. Condon and Shortley, *The Theory of Atomic Spectra*, Cambridge, 1953 — *note by translator.*)

40. Find the number of states associated with the configuration nl^x.

41. Show that if $x \leqslant 2l+1$, the term with the greatest value of L for the configuration nl^x will be the singlet with $L = xl - \frac{1}{4}x(x-2)$ if x is even, or the doublet with $L = xl - \frac{1}{4}(x-1)^2$ if x is odd.

42. From the wave functions of the one-electron problem construct the eigenfunctions associated with the quantum numbers S, L, M_S, M_L for the configuration p^3.

Hint. Consider the action of the operators $(\hat{L}_x - i\hat{L}_y)$ and $(\hat{S}_x - i\hat{S}_y)$ on the antisymmetric functions of the zero-order approximation.

43. Find the eigenfunctions for each of the two terms ^2D of the configuration d^3.

44. Two electrons move in a spherically symmetric field. Consider the electrostatic interaction of the electrons to be a perturbation and find the perturbation energy of the first order for terms of the configuration $npn'p$.

Hint. The sum of roots of the secular equation is equal to the sum of the diagonal elements in this equation.

45. In the case of the spin-orbit perturbation given by $\hat{V}_{SL} = A\hat{\mathbf{S}}\hat{\mathbf{L}}$, show that the mean perturbation of all states of a term (the term is characterized by the numbers L and S) is equal to zero.

46. Find the splitting of the energy levels of an atom in the case of a weak magnetic field when $\frac{e\hbar}{2\mu c}\mathcal{H} \ll |\Delta E_{JJ'}|$, where $\Delta E_{JJ'}$ is the distance between levels in the multiplet.

47. Find the limits of variation of the Landé factor g for given values of L and S.

48. Show that for the terms $^4D_{1/2}$, $^5F_1$, $^6G_{3/2}$ there is no splitting of the lines in the presence of a magnetic field.

49. Find the Landé factor for the one-electron atom (hydrogen, the alkali metals) directly from the Pauli eigenfunctions (see Prob. 21, §4).

50. Express the magnetic moment of the atom by means of the Landé factor.

51. Find the splitting of the term of the one-electron atom in the case of the average field $\frac{e\hbar}{2\mu c}\mathcal{H} \sim |\Delta E_{jj}|$.

52. Find the wave functions of an electron in the conditions of the preceding problem.

53. Find the splitting of the levels of a hydrogen atom in a strong magnetic field $\left(\frac{e\hbar}{2\mu c}\mathcal{H} > |E_{nlj} - E_{nlj'}|\right)$. To use perturbation theory it is necessary that the energy of the atom in the magnetic field be small in comparison with the difference in the energy of the various multiplets, i.e.,

$$\frac{e\hbar}{2\mu c}\mathcal{H} < |E_{nlj} - E_{n'lj}|.$$

54. Find the Zeeman components of the hyperfine-structure for the term $^2S_{1/2}$, ($j = \frac{1}{2}$, $l = \theta$) in the case of an average magnetic field $\left(\frac{e\hbar}{2\mu c}\mathcal{H} \sim |\Delta E_{ff'}|\right)$. (The splitting induced by the field is of the same order as the hyperfine-structure intervals.)

55. Show that for an arbitrary value of the intensity of a magnetic field the sum of the energy changes induced by the magnetic field over all states with a given M_J is equal to

$$\frac{e\hbar}{2\mu c}\mathcal{H}M_J\sum\left\{1+\frac{J(J+1)-L(L+1)+S(S+1)}{2J(J+1)}\right\}.$$

Here the sum J runs over the range of values $L+S \leqslant J \leqslant |L-S|$, $J \geqslant M_J$.

56. Show that if a hydrogen atom is placed in a uniform electric field:

(a) the energy of the state with quantum numbers $l = n-1$, $m = n-1$ in the first-order approximation with respect to the field does not change;

(b) the position of the centre of gravity of the split term does not change;

(c) states differing only in the sign of the magnetic quantum number have the same energy.

57. Calculate the splitting of the levels of the hydrogen atom in a weak electric field (the Stark effect is small in comparison with the fine structure).

58. Find the magnetic moment of a hydrogen atom in a weak electric field.

59. Calculate the splitting of the term with $n = 2$ for a hydrogen atom in a medium (as regards intensity) electric field (the Stark effect and the fine structure are of the same order).

60. Consider an atom under the influence of a perturbing potential u. Using perturbation theory, we obtain for the wave function ψ in the first-order approximation an expression of the form:

$$\psi = \psi_0 + \sum_{n\neq 0}\frac{u_{n0}}{E_0 - E_n}\psi_n,$$

and for the energy

$$E = E_0 + u_{00} + \sum_{n=0} \frac{(u_{n0})^2}{E_0 - E_n}.$$

To facilitate the use of a variational method we simplify the form of ψ as far as possible. Since

$$\sum_{n\neq 0}^{\infty} u_{n0}\psi_n = -u_{00}\psi_0 + \sum_{n=0}^{\infty} u_{n0}\psi_n = \psi_0(u - u_{00}),$$

then ψ can be written approximately as

$$\psi \approx \psi_0\left(1 + \frac{u - u_{00}}{E'}\right);$$

where E' in some cases can be considered to be the mean value of $E_0 - E_n$. After establishing the approximate form of the perturbed wave function, we can make use of a variational method to determine the energy.

Using a variational method, find the energy of an atom subject to the perturbing potential u. Seek the minimum energy for the class of trial functions of the form $\psi = \psi_0(1+\lambda u)$, where λ is a variational parameter.

61. Starting from the result of the preceding problem, find the formula for the polarization of the atom. Determine the numerical value of the coefficient of polarization for the hydrogen and helium atoms in the ground state.

62. A hydrogen atoms is in parallel electric and magnetic fields. Find the splitting of the levels in the case of:

(a) weak fields (the energy associated with the Stark and Zeeman effects is smaller than the energy associated with the fine structure);

(b) medium fields, for the term with the principal quantum number $n = 2$.

63. A hydrogen atom in a state with the principal quantum number $n = 2$ is in a magnetic and an electric field directed perpendicularly to one another. Find the splitting of the levels under the assumption that the fields are strong (the energy of the electron in external electric and magnetic fields is greater than the fine-structure energy).

§8. The molecule

1. Obtain the Schrödinger equation for a diatomic molecule assuming by way of approximation that the centre of gravity of the molecule coincides with the centre of gravity of the nuclei. To describe the motion of the electrons use the rotating coordinate system associated with the nuclei. Neglect the spin effects.

2. Solve the preceding problem by taking into account the electronic spin states in the system of rotating coordinates ξ, η, ζ.

3. For small vibrations of the nuclei the wave function of a diatomic molecule can be represented approximately in the form of the product of three functions $\Phi_{el}\,(\xi_i, \eta_i, \zeta_i, \sigma_i, p)$, $f(p)$, $\Theta\,(\theta, \varphi)$. The first function describes the motion of the electrons for fixed nuclei, the second and third describe the vibrational and rotational states of the molecule, respectively. Find the equation describing the vibrational and rotational parts of the wave function for a diatomic molecule.

4. Which terms are possible for the diatomic molecules N_2, Br_2, LiH, HBr, CN formed from the bonding of two atoms in their grionund states?

5. Find the formulae describing the electronic terms for the interaction between a helium and a hydrogen atom if both atoms are in their ground states.

6. Find the vibrational and rotational energy spectrum of a diatomic molecule if the nuclei are moving in a potential field of the form

$$V(r) = -2D\left(\frac{1}{\varrho} - \frac{1}{2\varrho^2}\right), \qquad \text{where } \varrho = \frac{r}{a}.$$

7. Represent the effective potential in the preceding problem $V(r) + \dfrac{\hbar^2}{2\mu r^2} K(K+1)$ near its minimum in the form of an oscillator potential and find the energy levels for small vibrations.

8. Find the moment of interia and the distance between nuclei in the molecule H^1Cl^{35} if the difference in the frequency of two neighbouring lines in the rotational-vibrational infrared band of H^1Cl^{35} is $\Delta\nu = 20{\cdot}9\ \text{cm}^{-1}$.

Calculate the corresponding value of $\Delta\nu$ for the DCl spectrum.

9. Calculate the ratio of the differences in energy between the first two vibrational and the first two rotational levels of the HF molecule. The moment of inertia of the HF molecule is $I = 1{\cdot}35 \times 10^{-40}$ g cm^2 and the frequency of vibration is $\Delta\nu_{\text{vib}} = 3987\ \text{cm}^{-1}$.

10. Find the energy of dissociation of the molecule D_2 if the energy of dissociation and the zero-point vibrational energy of the H_2 molecule are $4{\cdot}46$ eV and $0{\cdot}26$ eV, respectively.

11. In approximating the potential energy curve of a diatomic molecule, one frequently uses the function $V = D(1 - e^{-2\beta\xi})^2$ suggested by Morse, where $\xi = \dfrac{r-a}{a}$. Find the vibrational energy spectrum for $K = 0$.

12. Show that the operator of the square of the total angular momentum of a diatomic molecule can be represented in the form

$$\hat{\mathbf{J}}^2 = -\left\{\frac{1}{\sin\theta}\frac{\partial}{\partial\theta}\left(\sin\theta\frac{\partial}{\partial\theta}\right)+\frac{1}{\sin^2\theta}\left(\frac{\partial}{\partial\varphi}-iM_\zeta\cos\theta\right)^2\right\}+\hat{M}_\zeta^2.$$

13. The axes ξ, η, ζ are the axes of a rectangular coordinate system associated with a rotating rigid body. Find the form of the operators $\hat{J}_\xi$, $\hat{J}_\eta$, $\hat{J}_\zeta$ of the components of the angular momentum along the axes of the body.

14. Show that the operators $\hat{J}_\xi$, $\hat{J}_\eta$, $\hat{J}_\zeta$ obey the following commutation rules:

$$\hat{J}_\xi\hat{J}_\eta-\hat{J}_\eta\hat{J}_\xi = -i\hat{J}_\zeta,$$
$$\hat{J}_\eta\hat{J}_\zeta-\hat{J}_\zeta\hat{J}_\eta = -i\hat{J}_\xi,$$
$$\hat{J}_\zeta\hat{J}_\xi-\hat{J}_\xi\hat{J}_\zeta = -i\hat{J}_\eta,$$

that is, the commutation rules of the operators of the angular momentum components in the rotating system of coordinates differ from the commutation rules in the stationary system only by the sign of the right-hand side of the equations.

15. For the Euler-Poinsot case in classical mechanics we have the following equations:

$$A\frac{dp}{dt}+(C-B)qr = 0,$$

$$B\frac{dq}{dt}+(A-C)rp = 0,$$

$$C\frac{dr}{dt}+(B-A)pq = 0$$

or

$$\frac{dJ_\xi}{dt}+\left(\frac{1}{B}-\frac{1}{C}\right)J_\eta J_\zeta = 0, \quad \text{etc.}$$

Show that in quantum mechanics the above relations take the form

$$\frac{d\hat{J}_\xi}{dt}+\frac{1}{2}\left(\frac{1}{B}-\frac{1}{C}\right)(\hat{J}_\eta\hat{J}_\zeta+\hat{J}_\zeta\hat{J}_\eta) = 0, \quad \text{etc.}$$

16. Molecules having two or more axes of symmetry of the third or higher orders (e.g. CH_4) may be regarded as spherical tops. In such molecules the ellipsoid of intertia degenerates into the sphere $A = B = C$. Find the energy levels of such a spherical top.

17. Molecules with axes of symmetry of order higher than the second (for example, SO_2, NH_3, CH_3Cl) and molecules with lower symmetry or even those not possessing any symmetry at all, but having two identical principal moments of inertia, can be regarded as symmetrical tops of the type $A = B \neq C$.

Find the energy levels of a symmetrical top.

18. Write the Schrödinger equation for a symmetrical top.

19. Find the eigenfunctions of the operator

$$\hat{J}^2 = -\left\{\frac{1}{\sin\theta}\frac{\partial}{\partial\theta}\left(\sin\theta\frac{\partial}{\partial\theta}\right)+\frac{1}{\sin^2\theta}\left(\frac{\partial^2}{\partial\varphi^2}+\frac{\partial^2}{\partial\psi^2}\right)-2\frac{\cos\theta}{\sin^2\theta}\frac{\partial^2}{\partial\varphi\,\partial\psi}\right\}.$$

20. Calculate the matrix elements of the Hamiltonian for an asymmetrical top.

21. Find the energy levels of an asymmetrical top for $J = 1$.

22. Find the wave functions of an asymmetrical top for the case $J = 1$.

23. From the properties of the Pauli matrices show that even if one takes into account the spin-spin interaction the $^2\Sigma$ terms of a diatomic molecule undergo no splitting.

24. Find the multiplet splitting of the $^3\Sigma$ term for case *b*.

25. In the case of the approximate solution of the Schrödinger equation (see Prob. 3, §8) the operator

$$\frac{i\hbar^2}{2M\varrho^2}\left\{\cot\theta(\hat{M}_\xi - i\hat{M}_\eta\hat{M}_\zeta - i\hat{M}_\zeta\hat{M}_\eta)+\frac{2}{\sin\theta}\hat{M}_\eta\frac{\partial}{\partial\varphi}+2\hat{M}_\xi\frac{\partial}{\partial\theta}\right\}=\hat{w},$$

was not taken into account, since the diagonal element of this operator vanishes. If the nondiagonal elements referring to the same electronic (n, Λ) and vibrational (v) states are taken into account, one obtains the effect known as the rotational distortion of the spin. Considering the operator $\hat{w}$ as a perturbation find the change in the levels of the doublet term induced by this perturbation.

26. Find the relation between the total spin of the nuclei of the molecule D_2 in the Σ state and the possible values of the quantum number K.

27. Find the Zeeman components of the term of a diatomic molecule (case *a*). Assume the magnetic field to be small, i.e., the energy of the interaction between the spin and the external magnetic field is small in comparison with the difference in energies between successive rotational levels.

28. Find the Zeeman components of the term of a diatomic molecule for case *b* if the magnetic field is assumed to be such that the energy of the interaction between the spin and the external magnetic field is small in comparison with the spin-axis interaction energy.

29. Solve the preceding problem for the case in which the energy of spin-axis interaction is small in comparison with the energy shifts produced by the external magnetic field.

30. Determine the Zeeman components of the doublet term of a diatomic molecule if the term for case (b) and the magnetic field lead to an interaction energy between the magnetic moment and the field of the same order as the spin-axis interaction energy.

31. A diatomic molecule having a constant dipole moment **p** is placed in an electric field. Determine the splitting of the term for case (a).

32. Solve the preceding problem for the term in case (b).

33. Find the energy of a rigid dipole p in a uniform electric field $\mathcal{E}$, the field being treated as a small perturbation.

Hint. Use the relation

$$\cos\theta P_{lm} = \sqrt{\frac{(l+1)^2-m^2}{(2l+3)(2l+1)}}P_{l+1,\,m}+\sqrt{\frac{l^2-m^2}{(2l+1)(2l-1)}}P_{l-1,\,m}.$$

34. Using perturbation theory find the interaction between two unexcited hydrogen atoms separated by a large distance R.

35. Consider a group of atoms whose charge distribution is spherically symmetric. As shown in the preceding problem, forces known as dispersive forces act between two such atoms separated by a large distance. The dispersive forces have a quantum character and, in contrast to the classical polarization forces, exhibit additive properties. Show that the energy of the interaction between two such atoms does not depend on the presence of other similar atoms, i.e., show that the interaction energy of a group of atoms is the sum of the interaction energies between pairs of atoms.

§9. Scattering

1. Find the cross section for the scattering of a particle by a potential well at small velocities, the de Broglie wavelength being considerably greater than the dimensions of the well.

2. Find the scattering cross section of slow particles in the repulsive field

$$U(r) = U_0 \quad (r < a),$$
$$U(r) = 0 \quad (r > a).$$

3. Express, by means of scattering phase shifts, the first three coefficients of the expansion of the elastic scattering cross section $\frac{d\sigma}{d\Omega}$ in terms of Legendre polynomials.

4. Calculate the differential cross section for scattering in a repulsive field $U = \frac{A}{r^2}$ in the Born approximation. Repeat the calculation for the case of classical mechanics. Find the limits of applicability of the formulae obtained.

5. Find the discrete levels for a particle in the attractive field $U(r) = -U_0 \exp(-r/a)$ for $l = 0$. Find the scattering phase shift δ_0 for this potential and discuss the relation between δ_0 and the discrete spectrum.

6. Show that for a Coulomb field there is a one-to-one correspondence between the poles of the scattering amplitudes and the levels of the discrete spectrum.

Hint. Use the relation

$$e^{2i\delta_l} = \frac{\Gamma\left(l+1+\frac{i}{k}\right)}{\Gamma\left(l+1-\frac{i}{k}\right)}.$$

7. Determine how the scattering phase shift changes for a small change in the scattering potential. Find the expression for the scattering phase shift in the case in which the potential can be considered as a perturbation.

8. Calculate the scattering phase shifts of slow particles in the field $V = a/r^3$. The particles are slow enough for the condition $\mu ak/\hbar^2 \ll 1$ to be satisfied.

9. Find the total cross section for the elastic scattering of fast particles by a perfectly rigid sphere of radius a ($\lambda\!\!\!^{-} \ll a$, where $\lambda\!\!\!^{-}$ is the de Broglie wavelength).

10. Find in the Born approximation the differential and total cross sections for scattering in the fields:

(a) $U(r) = g^2 \frac{e^{-ar}}{r}$,

(b) $U(r) = U_0\, e^{-a^2r^2}$,

(c) $U(r) = U_0\, e^{-ar}$.

11. Using the Born approximation, find the differential and total cross sections for the elastic scattering of fast electrons
(a) by a hydrogen atom, (b) by a helium atom.

12. Calculate the effective scattering cross section of slow particles ($ka \ll 1$) in an attractive field described by a potential of the form

$$V(r) = -V_0 \cosh^{-2}(r/a).$$

13. Calculate the effective cross section for the scattering of slow particles in the repulsive field $V(r) = V_0 \cosh^{-2} \frac{r}{a}$.

14. Consider the collision of identical particles with an interaction energy $U(r)$. Find the effective cross section for the scattering of slow identical particles in the case of short-range forces.

15. Calculate the cross section for the elastic scattering of an electron by an electron and an alpha particle by an alpha particle.

16. Show that the scattering length has the following properties (the scattering amplitude taken with the opposite sign for $E \to 0$ is called the scattering length):

(a) In a repulsive field the scattering length is positive.

(b) In an attractive field, if there are no discrete levels, the scattering length is negative.

(c) The scattering length tends to ∞, if, with an increase in the depth of the potential well, a new level appears.

Hint. If we go from the Schrödinger equation to the equivalent integral equation ($E = 0$),

$$\psi = 1 - \frac{\mu}{2\pi\hbar^2} \int \frac{U(\mathbf{r}')\psi(\mathbf{r}')}{|\mathbf{r}-\mathbf{r}'|}\, d\tau',$$

then for the asymptotic form of ψ we obtain

$$\psi = 1 - \frac{a}{r},$$

where the scattering length

$$a = \frac{\mu}{2\pi\hbar^2} \int \psi U\, d\tau$$

is expressed through the potential energy and $\psi_{E=0}$.

17. For the majority of nuclei the scattering length a of neutrons (see the previous problem) is positive (e.g. Be, C, N, O, F, Mg, Ca, Zn, Cu, Ba, Pb, etc.); for a few nuclei the scattering length is negative (Li, Mn).

Find the probability that the scattering length has a negative sign in the case of neutrons being scattered by a nucleus of atomic weight A.

Use a model for the potential energy of the neutron in the nucleus in the form of a square well of radius $R = 1\cdot 4 \times 10^{-13}\ A^{1/3}$ and depth $V_0 = 40$ MeV.

18. The scattering of neutrons on protons depends on the resultant spin of the neutron and proton. For small energies the cross section in the triplet case ($S = 1$) is $\sigma^{\text{tripl.}} = 4\pi |a_1|^2 \approx 2 \times 10^{-24}$ cm² and the cross section in the singlet case ($S = 0$) is $\sigma^{\text{singl.}} = 4\pi |a_0|^2 \approx 78 \times \times 10^{-24}$ cm².

We introduce the operator

$$\hat{a} = \frac{a_0 + 3a_1}{4} + \frac{a_1 - a_0}{4}(\hat{\sigma}_n \hat{\sigma}_p).$$

As readily seen, the eigenvalues in the triplet and singlet states are a_1 and a_0 respectively. To find the scattering cross section for arbitrary polarization of the neutrons it is necessary to take the average of the operator $\hat{a}^2$:

$$\sigma = 4\pi\overline{\hat{a}^2}.$$

Let the spin states of the incident neutrons be described by the function $\begin{pmatrix} e^{-i\alpha}\cos\beta \\ e^{i\alpha}\sin\beta \end{pmatrix}$ (the direction of the neutron spin is given by the polar angles $\Theta = 2\beta$, $\Phi = 2\alpha + \frac{\pi}{2}$; see Prob. 12, §4) and the spin state of the protons by the function $\begin{pmatrix} 1 \\ 0 \end{pmatrix}$ (proton spins directed along the z axis). Find for this case the cross section for the scattering of neutrons on protons.

19. Find the probability that slow neutrons scattered on protons change their spin orientation if the neutron spins before the collision were directed along the z axis and the proton spins in the opposite direction.

20. Slow neutrons are scattered on a hydrogen molecule. If the energy of the neutrons is much less than that of thermal neutrons ($\lambda\!\!\!^{-} > 10^{-8}$ cm, i.e., the wavelength of the incident neutrons is large in comparison with the distance between protons in the molecule), the scattering amplitude is equal to the sum of the amplitudes for both protons.

Thus

$$\hat{a} = \frac{a_0 + 3a_1}{2} + \frac{a_1 - a_0}{4}\{\hat{\sigma}_n(\hat{\sigma}_{p_1} + \hat{\sigma}_{p_2})\}.$$

The protons in the hydrogen molecule are found in states with parallel spins (orthohydrogen) and in states with antiparallel spins (parahydrogen).

Find the cross section for the scattering of neutrons on parahydrogen and orthohydrogen.

21. A slow neutron is scattered by a nucleus of spin I. Find the cross section for unpolarized neutrons.

22. Consider the elastic scattering of a particle with spin $\frac{1}{2}$ on a scalar (spin zero) particle. Find the differential cross section for scattering with reversal of spin. Investigate the scattering of the S and P waves.

23. Prove that in the general case of inelastic scattering the total cross section $\sigma = \sigma_e + \sigma_s$ and the elastic scattering amplitude for $\vartheta = 0$ are related by the formula

$$\sigma = \frac{4\pi}{k} \operatorname{Im} f(0).$$

Hint. Use the expansion of these quantities in terms of the orbital angular momenta:

$$f(\vartheta) = \frac{1}{2ik} \sum_{l=0}^{\infty} (2l+1)(\eta_l - 1) P_l(\cos\vartheta),$$

$$\sigma_e = \frac{\pi}{k^2} \sum_{l=0}^{\infty} (2l+1) |1 - \eta_l|^2,$$

$$\sigma_s = \frac{\pi}{k^2} \sum_{l=0}^{\infty} (2l+1)(1 - |\eta_l|^2).$$

24. Consider the limiting case of a "black" nucleus. The radius R of the nucleus will be large in comparison with the de Broglie wavelength of the incident neutrons. We assume that all the incident neutrons are absorbed. Find the total cross sections for scattering and absorption.

25. Consider the nuclear reaction

$$A + a \rightarrow C \rightarrow B + b.$$

What will be the angular distribution of the reaction products in the centre-of-mass system, or, what amounts to the same thing, in the system of the intermediate nucleus C for the following cases:

(a) the spin of the intermediate nucleus is zero,

(b) the orbital angular momentum of the relative motion of the reaction products is zero,

(c) the relative orbital angular momentum of the colliding particles is zero (the spin of the intermediate nucleus is not equal to zero).

26. Find the eigenfunctions of the operators of the square of the total isotopic spin I^2 and of the component I_z in a nucleon-meson system.*

27. In the centre-of-mass system the scattering of mesons on nucleons reduces to the scattering of particles by a fixed scattering centre. At a large distance from the centre the incident wave with a given spin projection S_z and isotopic spin projection τ_z can be written in the form

$$e^{ikz}\begin{pmatrix}1\\0\end{pmatrix}\delta(\pi-\pi_i)\,\delta(n-\tau_z).$$

Here π_i takes on the values

$\pi_i = 1$ for a π^+ meson, $\delta(\pi - 1) = \varphi_+$,
$\pi_i = 0$ for a π^0 meson, $\delta(\pi) = \varphi_0$,
$\pi_i = -1$ for a π^- meson, $\delta(\pi + 1) = \varphi_-$;

τ_z takes the values:

$$\tau_z = \tfrac{1}{2} \text{ for a proton, } \delta\left(n-\frac{1}{2}\right) = \psi_p,$$

$$\tau_z = -\tfrac{1}{2} \text{ for a neutron, } \delta\left(n+\frac{1}{2}\right) = \psi_n.$$

Expand the incident wave into eigenfunctions of the operators $\mathbf{J}^2$, J_z, $\mathbf{I}^2$, I_z.

Hint. In the meson-nucleon system the constants of motion are the parity $(-1)^l$, the total angular momentum J, the isotopic spin I.

Since the total spin in such a system is equal to $\frac{1}{2}$, the orbital angular momentum is also conserved. Therefore, in expanding the incident wave in eigenfunctions of the operators of conserved quantities, one may take the sum over l, I instead of over J, I.

28. Find the amplitudes for the scattering of π-mesons on nucleons in terms of the phase shifts for the following reactions:

$$\pi^+ + p \to \pi^+ + p,$$

$$\pi^- + p \to \pi^- + p,$$

$$\pi^- + p \to \pi^0 + n.$$

29. Show that the scattering amplitudes for all possible reactions of a meson and nucleon can be expressed by the amplitudes of the reactions given in Prob. 28 on the basis of the assumption of invariance of isotopic spin.

Express them in terms of the scattering amplitudes for states with isotopic spins $\frac{3}{2}$ and $\frac{1}{2}$.

---

* See Appendix II.

30. Express in terms of phase shifts the total cross sections of the reactions:

$$\pi^+ + p \to \pi^+ + p,$$

$$\pi^- + p \to \pi^- + p,$$

$$\pi^- + p \to \pi^0 + n$$

31. In the region of small energy when the wavelength of a meson is considerably larger than the range of the interaction forces between the meson and nucleon, the basic contribution to the scattering is made only by the S and P waves.

Find the differential scattering cross section expressed in terms of phase shifts for the reactions:

$$\pi^+ + p \to \pi^+ + p,$$

$$\pi^- + p \to \pi^- + p,$$

$$\pi^- + p \to \pi^0 + n.$$

32. A beam of pions is scattered by an unpolarized proton target, i.e., in the target the number of protons with $S_z = \frac{1}{2}$ is equal to the number with $S_z = -\frac{1}{2}$. The initially unpolarized protons become polarized as a result of the scattering. Considering only the S and P waves, find the magnitude of the polarization of the protons.

ANSWERS AND SOLUTIONS

§1. One-dimensional motion. Energy spectrum and wave functions

1.

$$E_n = \frac{\pi^2\hbar^2}{2\mu a^2} n^2, \quad \psi_n(x) = \sqrt{\frac{2}{a}} \sin \frac{\pi n}{a} x.$$

3.

$$|a(p)|^2 = \frac{4n^2\pi a}{\hbar} \frac{1}{\left(\frac{p^2a^2}{\hbar^2} - n^2\pi^2\right)^2} \begin{cases} \cos^2 \frac{pa}{2\hbar} & \text{for odd } n, \\ \sin^2 \frac{pa}{2\hbar} & \text{for even } n. \end{cases}$$

4.

$$\frac{\hbar^2}{2\mu} \frac{d^2\psi}{dx^2} + (E - V_1)\psi = 0 \quad (x < 0),$$

$$\frac{\hbar^2}{2\mu} \frac{d^2\psi}{dx^2} + E\psi = 0 \quad (0 < x < a),$$

$$\frac{\hbar^2}{2\mu} \frac{d^2\psi}{dx^2} + (E - V_2)\psi = 0 \quad (x > a).$$

Introducing the notation

$$\varkappa_1 = \frac{\sqrt{2\mu(V_1 - E)}}{\hbar}, \quad \varkappa = \frac{\sqrt{-2\mu E}}{\hbar}, \quad \varkappa_2 = \frac{\sqrt{2\mu(V_2 - E)}}{\hbar},$$

we find that the general solution in each region has the following form:

$$\psi = A_1 e^{-\varkappa_1 x} + B_1 e^{\varkappa_1 x} \quad (x < 0),$$

$$\psi = A e^{-\varkappa x} + B e^{\varkappa x} \quad (0 < x < a),$$

$$\psi = A_2 e^{-\varkappa_2 x} + B_2 e^{\varkappa_2 x} \quad (x > a).$$

Let us consider the discrete spectrum $E < V_2$. Then $\varkappa_1$ and $\varkappa_2$ are real quantities. Setting $\varkappa = ik$ in the region $0 < x < a$, where k is real, we write the solution in the form

$$\psi = \sin(kx+\delta) \quad (0 < x < a).$$

Since the wave function is finite, $A_1 = 0,\ B_2 = 0$.

The conditions of the continuity of ψ and $\frac{d\psi}{dx}$ can be written conveniently as conditions of the continuity of the logarithmic derivative $\frac{1}{\psi}\frac{d\psi}{dx}$:

$$\varkappa_1 = k \cot \delta,$$

$$-\varkappa_2 = k \cot(ka+\delta).$$

We rewrite the last two conditions, expressing $\varkappa_1$ and $\varkappa_2$ by k

$$\sqrt{\frac{2\mu V_1}{\hbar^2 k^2} - 1} = \cot \delta,$$

$$-\sqrt{\frac{2\mu V_2}{\hbar^2 k^2} - 1} = \cot(ka+\delta).$$

Since the cotangent is a periodic function with the period π, the quantities δ and $ka + \delta$ can be represented in the following form:

$$\delta = \arcsin \frac{\hbar k}{\sqrt{2\mu V_1}} + n_1 \pi,$$

$$ka + \delta = -\arcsin \frac{\hbar k}{\sqrt{2\mu V_2}} + n_2 \pi.$$

Here the values of the arcsine lie in the region 0 to $\pi/2$. Eliminating δ, we obtain a transcendental equation whose solution enables us to obtain the energy levels in the diserete spectrum

$$ka = n\pi - \arcsin \frac{\hbar k}{\sqrt{2\mu V_1}} - \arcsin \frac{\hbar k}{\sqrt{2\mu V_2}}, \qquad k = \frac{\sqrt{2\mu E}}{\hbar} > 0.$$

To find the values of k satisfying this equation, it is convenient to use a graphical method. These values are obtained from the points of intersection of the line $y = ak$ with the curves

$$y = n\pi - \arcsin \frac{\hbar k}{\sqrt{2\mu V_1}} - \arcsin \frac{\hbar k}{\sqrt{2\mu V_2}} \qquad \text{(see Fig. 18).}$$

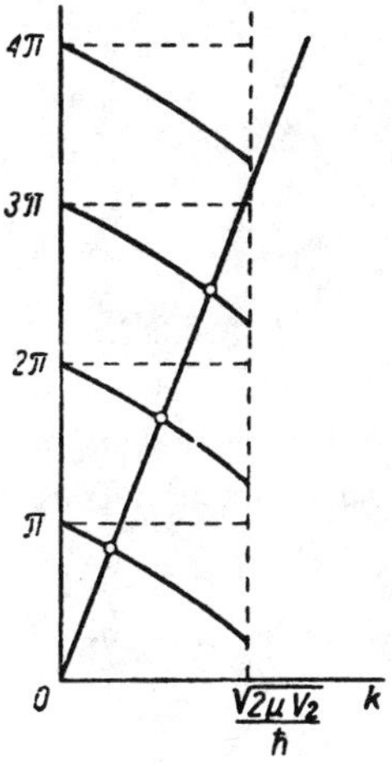

Fig. 18

Let us consider a symmetric potential well $V_1 = V_2 = V$. It is readily seen, in this case, that for any V and a there always exists at least one level. If $\frac{\sqrt{2\mu V}}{\hbar}a \ll 1$, it is not difficult to find the only discrete energy level. Expanding $\pi - 2\arcsin\sqrt{\frac{E}{V}}$ into a series, we obtain $E = V - \frac{\mu a^2}{2\hbar^2}V^2$. The number of levels for any value of V and a will be equal to N, where N is found from the relation

$$N > \frac{\sqrt{2\mu V}a}{\pi\hbar} > N-1.$$

5. We denote by E $(E < 0)$ the energy of the bound state. Using the result of the preceding problem, we have

$$q\frac{\sqrt{2\mu}}{\hbar}\frac{1}{\sqrt{V_0}}\sqrt{1-\frac{|E|}{V_0}} = n\pi - 2\arcsin\sqrt{1-\frac{|E|}{V_0}}$$

The left-hand part of the equation tends to zero as $V_0 \to \infty$. We therefore have only one bound state $(n = 1)$, and we may set $\arcsin\sqrt{1-\frac{|E|}{V_0}}$

$=\frac{\pi}{2}-\alpha$, where α is a small quantity. From the calculation we obtain

$$q=\frac{\sqrt{2\mu}}{\hbar}\sqrt{1-\frac{|E|}{V_0}}=2|E|.$$

Finally, for $V_0\rightarrow\infty$ we have

$$E=-\frac{\mu q^2}{2\hbar^2}.$$

6. (b) The eigenvalues of the operator $\hat{H}'$ are everywhere positive. Therefore, if ψ_0 is the wave function corresponding to the minimum energy, then

$$(\hat{Q}+i\hat{P})\psi_0=0 \quad \text{or} \quad \left(\frac{\partial}{\partial Q}+Q\right)\psi_0=0.$$

We thus find that

$$\psi_0=c_0\exp\left(-\frac{1}{2}Q^2\right), \quad \varepsilon_0=\frac{1}{2}.$$

Hence the energy of the oscillator is equal to

$$\varepsilon_n=n+\frac{1}{2} \quad (n\geqslant 0),$$

and the corresponding wave function will have the form

$$\psi_n=A\left(\frac{\partial}{\partial Q}-Q\right)^n\exp\left(-\frac{1}{2}Q^2\right).$$

The normalizing constant A_n is found from the condition $\int_{-\infty}^{+\infty}\psi_n^2(Q)\,dQ=1$.

For the ground state the normalizing constant A_0 is equal to $\pi^{-1/4}$: therefore

$$\psi_0=\frac{1}{\sqrt[4]{\pi}}\exp\left(-\frac{1}{2}Q^2\right).$$

The wave function of the nth state can be expressed in terms of the wave function of the $(n-1)$th state by the following relation:

$$\psi_n(Q)=c_n(\hat{Q}-i\hat{P})\psi_{n-1}(Q),$$

where c_n is determined from the condition

$$c_n^2\int[(Q-iP)\psi_{n-1}(Q)]^2\,dQ=1.$$

We replace P by $-i\frac{\partial}{\partial Q}$, integrate by parts, and obtain

$$c_n^2\int \psi_{n-1}(P^2+Q^2+1)\psi_{n-1}\,dQ = c_n^2\cdot 2n = 1,$$

from which we have $c_n = \frac{1}{\sqrt{2n}}$ and

$$\psi_n = c_n\left(Q-\frac{\partial}{\partial Q}\right)\psi_{n-1} = c_n\cdot c_{n-1}\cdot\ldots c_1\left(Q-\frac{\partial}{\partial Q}\right)^n\psi_0 = A_n\left(Q-\frac{\partial}{\partial Q}\right)^n\psi_0.$$

Finally, we have

$$\psi_n = \frac{1}{\sqrt{2^n n!}}\frac{1}{\sqrt[4]{\pi}}\left(Q-\frac{\partial}{\partial Q}\right)^n\exp\left(-\frac{1}{2}Q^2\right).$$

The polynomial of nth degree

$$H_n(Q) = e^{\frac{1}{2}Q^2}\left(Q-\frac{\partial}{\partial Q}\right)^n e^{-\frac{1}{2}Q^2}$$

is called the *Hermite-Tchebycheff polynomial.*

(c) $$\hat{a}\hat{a}^+ - \hat{a}^+\hat{a} = 1;\qquad \psi_n = \frac{1}{\sqrt{n!}}(\hat{a}^+)^n\psi_0;$$

(d) $$(\hat{P}+i\hat{Q})^n_{n-1}\cdot[(\hat{P}+i\hat{Q})^n_{n-1}]^* = 2n.$$

We have chosen the wave functions ψ_n to be real, and therefore the matrix elements $\hat{Q}$ and $i\hat{P} = \frac{\partial}{\partial Q}$ will also be real:

$$(\hat{Q})^n_{n-1} = \sqrt{\frac{n}{2}},$$

$$(\hat{P})^n_{n-1} = i\sqrt{\frac{n}{2}};$$

or, returning to the previous variables, we have

$$(\hat{x})^n_{n-1} = (\hat{x})^{n-1}_n = \sqrt{\frac{\hbar n}{2\mu\omega}},$$

$$(\hat{p})^n_{n-1} = -(\hat{p})^{n-1}_n = i\sqrt{\frac{n\mu\hbar\omega}{2}}.$$

8. The probability is

$$w = \frac{\int_1^\infty \exp(-y^2)\,dy}{\int_0^\infty \exp(-y^2)\,dy} \approx 0\cdot 16.$$

9. The wave function should tend to zero as $x \to 0$. For $x > 0$ it satisfies the differential equation of the harmonic oscillator. It is not difficult to see that the wave functions of the oscillator for odd $n = 2k + 1$ tend to zero as $x \to 0$, and in the region $x \geqslant 0$ give the solution to our problem. Consequently

$$E_k = \hbar\omega\left(2k + \frac{3}{2}\right) \qquad (k = 0,\ 1,\ 2,\ \ldots).$$

10.
$$\left(\frac{p^2}{2\mu} - \frac{1}{2}\mu\omega^2\hbar^2\frac{\partial^2}{\partial p^2}\right)a_n(p) = E_n a_n(p),$$

$$|a_n(p)|^2 = \frac{1}{2^n n!\sqrt{\pi\mu\omega\hbar}}\, e^{-\frac{p^2}{\mu\omega\hbar}} H_n^2\left(\frac{p}{\sqrt{\mu\omega\hbar}}\right).$$

11. It is seen from the investigation of the behaviour of the solution of the Schrödinger equation

$$-\frac{\hbar^2}{2\mu}\frac{d^2\psi}{dx^2} - \left[E - V_0\left(\frac{a}{x} - \frac{x}{a}\right)^2\right]\psi = 0$$

that as $x \to \infty$ the function ψ has the asymptotic form $\psi \sim \exp(-\xi/2)$, where ξ is a new independent variable defined as

$$\xi = \frac{\sqrt{2\mu V_0}}{\hbar a}x^2.$$

For $x \to 0$, the function ψ is proportional to $\xi^{\nu/2}$, where

$$\nu = \frac{1}{2}\left(\sqrt{\frac{8\mu V_0 a^2}{\hbar^2} + 1} + 1\right).$$

We make the substitution

$$\psi = e^{-\xi/2}\xi^{\nu/2}u(\xi)$$

and obtain for $u(\xi)$ the following equation:

$$\xi u'' + \left(\nu + \frac{1}{2} - \xi\right)u' - \left[\frac{\nu}{2} + \frac{1}{4} - \frac{\mu a(E + 2V_0)}{2\hbar\sqrt{2\mu V_0}}\right]u = 0. \qquad (1)$$

The solutions of Eq. (1) are degenerate hypergeometric functions. The general solution has the form

$$u(\xi)=c_1F\left(a,\ \nu+\frac{1}{2},\ \xi\right)+c_2F\left(a-\nu+\frac{1}{2},\ \frac{3}{2}-\nu,\ \xi\right)\xi^{\frac{1}{2}-\nu},$$

where a denotes the bracketed expression in Eq. (1).

From the condition that $\psi(0)$ be bound it follows that

$$c_2=0.$$

Moreover, it is necessary to require that as $x\to\infty$ the wave function decreases, that is, that the function $u(\xi)$ reduces to a polynomial. This can be achieved by setting $a=-n$ $(n=0,\ 1,\ 2,\ldots)$, from which we obtain the energy levels

$$E_n=\frac{2\hbar}{a}\sqrt{\frac{2V_0}{\mu}}\left\{n+\frac{1}{2}+\frac{1}{4}\left(\sqrt{\frac{8\mu V_0a^2}{\hbar^2}+1}-\sqrt{\frac{8\mu V_0a^2}{\hbar^2}}\right)\right\}.$$

Thus the energy spectrum (for a suitable choice of the number from which we calculate the energy) will be the same as for an oscillator with an angular frequency of $\omega=\sqrt{\frac{8V_0}{\mu a^2}}$. It is interesting to note that the zero-point energy of the particle for the potential $V_0\left(\frac{a}{x}-\frac{x}{a}\right)^2$ is everywhere higher than the zero-point energy of the corresponding oscillator. The wave functions have the form

$$\psi_n=c_nx^{\nu}\left\{\exp\left[-\sqrt{\frac{\mu V_0}{2\hbar^2a^2}}x^2\right]\right\}F\left(-n,\ \nu+\frac{1}{2},\ \sqrt{\frac{2\mu V_0}{\hbar^2a^2}}x^2\right),$$

where $\nu=\frac{1}{2}\left(\sqrt{\frac{8\mu V_0a^2}{\hbar^2}+1}+1\right)$, and the constants c_n can be found from the normalization conditions.

12. In the Schrödinger equation

$$-\frac{\hbar^2}{2\mu}\frac{d^2\psi}{dx^2}-\left(E+\frac{V_0}{\cosh^2\frac{x}{a}}\right)\psi=0$$

we make the substitution

$$\psi=\left(\cosh\frac{x}{a}\right)^{-2\lambda}u,\qquad \lambda=\frac{1}{4}(\sqrt{8\mu V_0a^2\hbar^{-2}+1}-1).$$

The equation for u takes the following form:

$$\frac{d^2u}{dx^2} - \frac{4\lambda}{a}\tanh\frac{x}{a}\frac{du}{dx} + \frac{4}{a^2}(\lambda^2 - \varkappa^2)u = 0,$$

where

$$\varkappa = \sqrt{-\frac{\mu E a^2}{2\hbar^2}}$$

(we are considering the discrete spectrum $E < 0$).

If we introduce the new independent variable

$$z = -\sinh^2\frac{x}{a},$$

this equation then leads to the hypergeometric equation

$$z(1-z)\frac{d^2u}{dz^2} + \left[\frac{1}{2} - (1-2\lambda)z\right]\frac{du}{dz} - (\lambda^2 - \varkappa^2)u = 0. \qquad (1)$$

The parameters α, β, γ in the hypergeometric equation of the general form

$$z(1-z)\frac{d^2u}{dz^2} + [\gamma - (\alpha+\beta+1)z]\frac{du}{dz} - \alpha\beta u = 0,$$

in our case, have the following values:

$$\gamma = \frac{1}{2}, \quad \alpha = \varkappa - \lambda, \quad \beta = -\varkappa - \lambda.$$

The two solutions of Eq. (1), which give the even and odd wave functions of ψ, have the respective forms

$$u_1 = F\left(-\lambda + \varkappa, -\lambda - \varkappa, \frac{1}{2}; z\right), \qquad (2)$$

$$u_2 = \sqrt{z}\,F\left(-\lambda + \varkappa + \frac{1}{2}, -\lambda - \varkappa + \frac{1}{2}, \frac{3}{2}; z\right). \qquad (3)$$

These solutions lead to finite values of the wave functions at $x = 0$ ($z = 0$).

In order that the wave function

$$\psi = \left(\cosh\frac{x}{a}\right)^{-2\lambda} u$$

vanish as $x \to \pm\infty$ ($z \to -\infty$) the hypergeometric function in Eqs. (2) and (3) should reduce to a polynomial. This condition means, for example, that for u_1 either $\lambda - \varkappa$ or $\lambda + \varkappa$ are non-negative integers.

The latter case, however, must be discarded, since as $x \to \pm\infty$ the wave function increases exponentially. Thus we obtain $\lambda - \varkappa = k$ ($k = 0, 1, 2, \ldots$) and the energy levels are

$$E_k = -\frac{\hbar^2}{2\mu a^2}\left[\frac{1}{2}\sqrt{\frac{8\mu V_0 a^2}{\hbar^2}+1} - 2k - \frac{1}{2}\right]^2.$$

In a similar way, for Eq. (3) we find that the condition that the wave function be finite as $x \to \pm\infty$ will be satisfied if

$$\lambda - \varkappa - \frac{1}{2} = l \quad (l = 0, 1, 2, \ldots),$$

from which we have

$$E_l = -\frac{\hbar^2}{2\mu a^2}\left[\frac{1}{2}\sqrt{\frac{8\mu V_0 a^2}{\hbar^2}+1} - (2l+1) - \frac{1}{2}\right]^2.$$

Combining these results, we obtain

$$E_n = -\frac{\hbar^2}{2\mu a^2}\left[\frac{1}{2}\sqrt{\frac{8\mu V_0 a^2}{\hbar^2}+1} - \left(n+\frac{1}{2}\right)\right]^2$$

$$(n = 0, 1, 2, \ldots).$$

The number of discrete levels is equal to the largest integer N satisfying the inequality

$$N < \frac{1}{2}\sqrt{8\mu V_0 a^2 \hbar^{-2} + 1} - \frac{1}{2}.$$

We note that the energy spectrum is the same as that obtained for the Morse potential (see Prob. 11, § 8) for the corresponding choice of parameters.

13. In the wave equation

$$-\frac{\hbar^2}{2\mu}\frac{d^2\psi}{dx^2} - \left(E - V_0 \cot^2\frac{\pi}{a}x\right)\psi = 0$$

we make the substitution

$$\psi = \left(\sin\frac{\pi}{a}x\right)^{-2\lambda} u.$$

We set

$$\lambda = \frac{1}{4}\left(\sqrt{\frac{8\mu V_0 a^2}{\pi^2\hbar^2}+1} - 1\right),$$

$$\nu = \sqrt{\frac{\mu a^2}{2\hbar^2\pi^2}(E+V_0)},$$

and arrive at the following equation for u:

$$\frac{d^2u}{dx^2} - 4\frac{\pi}{a}\lambda\cot\frac{\pi x}{a}\frac{du}{dx} + \frac{4\pi^2}{a^2}(\nu^2-\lambda^2)u = 0.$$

We introduce into this equation the independent variable

$$z = \cos^2\frac{\pi x}{a},$$

and obtain the hypergeometric equation

$$z(1-z)\frac{d^2u}{dz^2} + \left[\frac{1}{2} - (1-2\lambda)z\right]\frac{du}{dz} + (\nu^2-\lambda^2)u = 0. \tag{1}$$

Comparison with the general form of the hypergeometric equation

$$z(1-z)\frac{d^2u}{dz^2} + [\gamma - (\alpha+\beta+1)z]\frac{du}{dz} - \alpha\beta u = 0,$$

gives us the following values of the parameters:

$$\gamma = \frac{1}{2}, \quad \alpha = -\nu-\lambda, \quad \beta = \nu-\lambda.$$

Eq. (1) has two solutions. One of these solutions is different from zero and finite at $z = 0$ (this value corresponds to $x = a/2$)

$$u_1 = F\left(-\nu-\lambda, \nu-\lambda, \frac{1}{2}; z\right).$$

The other solution

$$u_2 = \sqrt{z}\,F\left(-\nu-\lambda+\frac{1}{2}, \nu-\lambda+\frac{1}{2}, \frac{3}{2}; z\right)$$

tends to zero as $z \to 0$ ($x \to a/2$). In order to determine the behaviour of the solutions at $z = 1$ (this value corresponds to the two values $x = 0$, $x = a$) we make use of the relation

$$F(\alpha, \beta, \gamma; z) = (1-z)^{-\alpha}F\left(\alpha, \gamma-\beta, \gamma; \frac{z}{z-1}\right).$$

For u_1 and u_2 we obtain

$$u_1 = (1-z)^{\nu+\lambda}F\left(-\nu-\lambda, -\nu+\lambda+\frac{1}{2}, \frac{1}{2}; \frac{z}{z-1}\right), \tag{2}$$

$$u_2 = \sqrt{z}(1-z)^{\nu+\lambda-\frac{1}{2}}F\left(-\nu-\lambda+\frac{1}{2}, -\nu+\lambda+1, \frac{3}{2}; \frac{z}{z-1}\right). \tag{3}$$

To satisfy the condition that the wave function ψ vanish as $x \to 0$ and $x \to a$, it is necessary that the series in powers of $z/(z-1)$ break off after a finite number of terms.

The hypergeometric series in Eq. (2) for u_1 terminates if either $\nu + \lambda$ or $\nu - \lambda - \frac{1}{2}$ is a positive integer or zero.

The condition that $\psi = 0$ for $x = 0$, $x = a$ is satisfied, however, only in the second case

$$\nu - \lambda - \frac{1}{2} = k \quad (k = 0, 1, \ldots).$$

The energy levels are then given by

$$E_k = [(2k+1)^2 + 4(2k+1)\lambda - 2\lambda]\frac{\pi^2\hbar^2}{2\mu a^2}. \tag{4}$$

In a similar way, it follows from Eq. (3) that the energy levels are determined by the condition

$$\nu - \lambda = l \quad (l = 1, 2, 3, \ldots). \tag{5}$$

The expressions (4) and (5) for the energy levels may be combined to give

$$E_n = (n^2 + 4n\lambda - 2\lambda)\frac{\pi^2\hbar^2}{2\mu a^2} \quad (n = 1, 2, 3 \ldots).$$

Here the wave functions

$$\psi_n = c_n\left(\sin\frac{\pi x}{a}\right)^{-2\lambda} F\left(-\frac{n}{2} - 2\lambda, \frac{n}{2}, \frac{1}{2}; \cos^2\frac{\pi x}{a}\right),$$

correspond to odd values of n, and the wave functions

$$\psi_n = c_n\left(\sin\frac{\pi x}{a}\right)^{-2\lambda} \cos\frac{\pi x}{a}\, F\left(-\frac{n}{2} - 2\lambda + \frac{1}{2}, \frac{n}{2} + \frac{1}{2}, \frac{3}{2}; \cos^2\frac{\pi x}{a}\right)$$

to even values of n. The normalized wave function of the ground state is

$$\psi = \sqrt{\frac{\pi\Gamma(2\lambda+1)}{a\Gamma\left(\frac{1}{2}\right)\Gamma\left(2\lambda+\frac{3}{2}\right)}}\left(\sin\frac{\pi x}{a}\right)^{2\lambda+1}, \quad \Gamma\left(\frac{1}{2}\right) = \sqrt{\pi}.$$

Let us consider the limiting case $V_0 \to 0$. The problem then reduces to that of a particle in a potential box (see Prob. 1, §1). The value λ

tends to zero, and for the energy levels we obtain, as should be expected, the values

$$E_n = \frac{\pi^2\hbar^2}{2\mu a^2} n^2.$$

In the opposite case $\lambda \gg 1$, we obtain for the lower levels ($n \ll \lambda$),

$$E_n = \hbar\omega\left(n+\frac{1}{2}\right) \quad (n = 1, 2 \ldots),$$

where

$$\omega = \frac{\pi}{a}\sqrt{\frac{2V_0}{\mu}}.$$

This same result can be obtained by expanding the potential energy about the point $x = a/2$ and neglecting terms higher than the second order.

14. In this case we have only a continuous energy spectrum and the eigenfunctions are not degenerate.

In the Schrödinger equation

$$-\frac{\hbar^2}{2\mu}\frac{d^2\psi}{dx^2} - (E+Fx)\psi = 0$$

we go over from position coordinates to momentum coordinates and obtain

$$\frac{p^2}{2\mu}a(p) - Ea(p) = i\hbar F\frac{d}{dp}a(p).$$

The solution of this equation belonging to the eigenvalue E is

$$a_E(p) = c\exp\left[-\frac{i}{\hbar F}\left(\frac{p^3}{6\mu} - Ep\right)\right]$$

and represents a wave function in the momentum representation. We normalize the functions $a(p)$ to $\delta(E-E')$

$$\int a_E^*(p)\, a_{E'}(p)\, dp = \delta(E-E'),$$

i.e.,

$$cc^* \int \exp\left[-\frac{ip}{\hbar F}(E-E')\right] dp = cc^*\, 2\pi\hbar F\delta(E-E'),$$

from which we obtain

$$c = \frac{1}{\sqrt{2\pi\hbar F}}.$$

The wave function in the position-coordinate representation is

$$\psi(x) = \frac{a}{2\pi\sqrt{F}} \int_{-\infty}^{+\infty} \exp\left(i\frac{1}{3}u^3 - iuq\right) du = \frac{a}{\pi\sqrt{F}} \int_0^\infty \cos\left(\frac{u^3}{3} - uq\right) du,$$

$$q = \left(x + \frac{E}{F}\right) a, \qquad a = (2\mu F\hbar^{-2})^{1/3}.$$

This integral can be expressed in terms of the Airy function $\Phi(q)$:

$$\Phi(q) = \frac{1}{\sqrt{\pi}} \int_0^\infty \cos\left(\frac{u^3}{3} + uq\right) du, \qquad \psi(x) = \frac{a}{\sqrt{\pi F}} \Phi(-q).$$

15. The Hamiltonian for the given potential has the form

$$\hat{H} = \frac{1}{2\mu} p^2 + \frac{1}{2} V_0 \exp\left(b\hbar \frac{\partial}{\partial p}\right) + \frac{1}{2} V_0 \exp\left(-b\hbar \frac{\partial}{\partial p}\right).$$

Since

$$\exp\left(b\hbar \frac{\partial}{\partial p}\right) a(p) = a(p + b\hbar),$$

then the Schrödinger equation can be represented in the form of a finite-difference equation

$$\frac{1}{2\mu} p^2 a(p) + \frac{1}{2} V_0 a(p + b\hbar) + \frac{1}{2} V_0 a(p - b\hbar) = Ea(p).$$

16.

$$\frac{\hbar^2}{2\mu} k^2 a(k) + \sum_{-\infty}^{+\infty} V_n a\left(k + \frac{2\pi n}{b}\right) = Ea(k)$$

$$\left(k = \frac{p}{\hbar}, \qquad V(x) = \sum_{-\infty}^{+\infty} V_n \exp(-2\pi nixb^{-1}), \qquad V_n = V^*_{-n}\right).$$

17. The wave function in the region of the well $0 < x < a$ has the form

$$\psi = c_1 \exp(i\varkappa_1 x) + c_2 \exp(-i\varkappa_1 x), \qquad \varkappa_1 = \frac{\sqrt{2\mu E}}{\hbar},$$

and in the region of the barrier $-b < x < 0$,

$$\psi = c_3 \exp(i\varkappa_2 x) + c_4 \exp(-i\varkappa_2 x), \qquad \varkappa_2 = \frac{\sqrt{2\mu(E - V_0)}}{\hbar}.$$

Since $\psi(x)$ is the product of a constant and $\psi(x+l)$ (the constant has a modulus of unity, $l = a + b$), then in the region of the next barrier $a < x < a + b$

$$\{\psi = e^{ikl}\{c_3 \exp[i\varkappa_2(x-l)] + c_4 \exp[-i\varkappa_2(x-l)]\}.$$

From the requirement that the wave function and its first derivative be continuous at the points $x = 0$, $x = a$ we obtain the four equations

$$c_1 + c_2 = c_3 + c_4,$$
$$c_1 e^{i\varkappa_1 a} + c_2 e^{-i\varkappa_1 a} = e^{ikl}(c_3 e^{-i\varkappa_2 b} + c_4 e^{i\varkappa_2 b}),$$
$$\varkappa_1(c_1 - c_2) = \varkappa_2(c_3 - c_4),$$
$$\varkappa_1(c_1 e^{i\varkappa_1 a} - c_2 e^{-i\varkappa_1 a}) = \varkappa_2(c_3 e^{-i\varkappa_2 b} - c_4 e^{i\varkappa_2 b}) e^{ikl}.$$

This system of equations has a nontrivial solution only if

$$\cos kl = \cos \varkappa_1 a \cdot \cos \varkappa_2 b - \frac{\varkappa_1^2 + \varkappa_2^2}{2\varkappa_1 \varkappa_2} \sin \varkappa_1 a \cdot \sin \varkappa_2 b. \tag{1}$$

We shall investigate two cases:

(a) $E < V_0$, $\varkappa_2$ is an imaginary quantity.

We introduce the notation $\varkappa_2 = i\varkappa$ and rewrite Eq. (1) in the form

$$\cos kl = \cos \varkappa_1 a \cdot \cosh \varkappa b + \frac{\varkappa^2 - \varkappa_1^2}{2\varkappa_1 \varkappa} \sin \varkappa_1 a \cdot \sinh \varkappa b. \tag{2}$$

Thus the allowable energy zones are determined from the relation

$$-1 \leqslant \cos \varkappa_1 a \cdot \cosh b + \frac{\varkappa^2 - \varkappa_1^2}{2\varkappa_1 \varkappa} \sin \varkappa_1 a \cdot \sinh \varkappa b \leqslant 1.$$

To investigate the general properties and position of the allowable zones we shall consider the limiting case

$$\frac{\sqrt{2\mu V_0}}{\hbar} b \ll 1, \quad a \gg b. \quad E \ll V_0.$$

We introduce the notation $\dfrac{\mu V_0 ab}{\hbar^2} = \gamma$. Then Eq. (2) can be approximated in the form

$$\cos ka = \gamma \frac{\sin \varkappa_1 a}{\varkappa_1 a} + \cos \varkappa_1 a. \tag{3}$$

Figure 19 shows the function $\left(\gamma \dfrac{\sin \varkappa_1 a}{\varkappa_1 a} + \cos \varkappa_1 a\right)$; the values of $\varkappa_1 a$

are laid off along the axis of abcissae. The allowable energy zones are indicated by thick lines on the axis $\varkappa_1 a$.

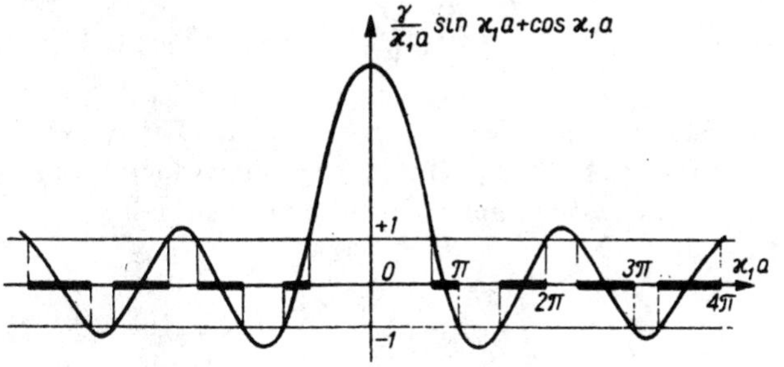

Fig. 19

Each point $\varkappa_1 a = n\pi$ marks the beginning of another forbidden zone. It is seen from the figure that the forbidden zones decrease in width with an increase in the zone number. One may readily estimate the width of the forbidden zones. The left-hand side of (3) takes the values $(-1)^n$ when

$$\cos(\varkappa_1 a - \varphi) = (-1)^n \cos\varphi, \qquad \tan\varphi = \frac{\gamma}{\varkappa_1 a},$$

which is possible for $\varkappa_1 a = n\pi$ and for $\varkappa_1 a = n\pi + 2\varphi$. It thus follows that the width of the forbidden energy zones is 2φ. For large values of n

$$2\varphi \approx \frac{\gamma}{n\pi}.$$

(b) $E > V_0$. In this case the energy zones are determined from the relation

$$-1 \leqslant \cos\varkappa_1 a \cos\varkappa_2 b - \frac{\varkappa_1^2 + \varkappa_2^2}{2\varkappa_1\varkappa_2} \sin\varkappa_1 a \sin\varkappa_2 b \leqslant +1.$$

18. The energy levels E_n are obtained from the Bohr quantization rule

$$\int_{x_1}^{x_2} p\,dx = \pi\left(n + \frac{1}{2}\right)\hbar, \tag{1}$$

where

$$p = \sqrt{2\mu\left(E_n + V_0 \cosh^{-2}\frac{x}{a}\right)};$$

x_1 and x_2 are the turning points determined from the condition $p = 0$ (here $E_n < 0$ and the spectrum is discrete). To evaluate the integral

$$I(E) = \int_{x_1}^{x_2} \sqrt{2\mu\left(E + V_0 \cosh^{-2}\frac{x}{a}\right)}\, dx$$

we differentiate on both sides with respect to E. The derivatives of the integral with respect to the upper and lower limits vanish, since at points x_1 and x_2 the expression under the radical vanishes.
Hence

$$\frac{dI}{dE} = \mu \int_{x_1}^{x_2} \frac{dx}{\sqrt{2\mu\left(E + V_0 \cosh^{-2}\frac{x}{a}\right)}}.$$

A change of variable $\sinh\frac{x}{a} = z$ in the above integral gives

$$\frac{dI}{dE} = \mu a \int_{z_1}^{z_2} \frac{dz}{\sqrt{2\mu\left[E(1+z^2)+V_0\right]}} = \frac{\mu a}{\sqrt{-2\mu E}}\pi.$$

We then obtain

$$I(E) = -\sqrt{-2\mu a^2 E}\,\pi + C.$$

The constant C is found from the condition that for $E = -V_0$ the region of integration shrinks to a point:

$$I(-V_0) = 0,$$

from which it follows that

$$I(E) = \sqrt{2\mu a^2}\,(\sqrt{V_0} - \sqrt{-E})\pi.$$

Thus in the quasi-classical approximation the energy levels are given by:

$$E_n = -\frac{\hbar^2}{2\mu a^2}\left[\sqrt{\frac{2\mu a^2 V_0}{\hbar^2}} - \left(n + \frac{1}{2}\right)\right]^2 \quad (n = 0, 1, 2, \ldots). \qquad (2)$$

For the number of levels we have $N = \frac{\sqrt{2\mu V_0}}{\hbar} a$. We note that the calculation of the energy levels by means of the quantization rule (1) is valid if the number of these levels is large, i.e., if

$$\frac{2\mu a^2 V_0}{\hbar^2} \gg 1.$$

If this condition is satisfied the expression for the energy levels (2) is the same as the exact formula for E_n obtained in Prob. 12, §1.

19.

(a) $$E_n = \left(n + \frac{1}{2}\right)\hbar\omega \quad (n = 0, 1, 2, \ldots),$$

(b) $$E_n = \frac{\pi^2\hbar^2}{2\mu a^2}\left[\sqrt{\frac{2\mu a^2 V_0}{\pi^2\hbar^2}} + \left(n + \frac{1}{2}\right)\right]^2 - V_0 \quad (n = 0, 1, 2, \ldots).$$

20. The mean value of the kinetic energy in the stationary state ψ_n (it is assumed that the wave function is real) is

$$\bar{T} = -\frac{\hbar^2}{2\mu}\int \psi_n \frac{d^2\psi_n}{dx^2}\,dx = \frac{\hbar^2}{2\mu}\int\left(\frac{d\psi_n}{dx}\right)^2 dx.$$

In the quasi-classical approximation the wave function in the allowable classical region $(a < x < b)$ has the form

$$\psi_n = \frac{A_n}{\sqrt{p}}\cos\left(\frac{1}{\hbar}\int_a^x p\,dx - \frac{\pi}{4}\right), \quad p = \sqrt{2\mu(E_n - V)},$$

from which we have

$$\frac{d\psi_n}{dx} = -\frac{\sqrt{p}}{\hbar}A_n \sin\left(\frac{1}{\hbar}\int_a^x p\,dx - \frac{\pi}{4}\right) - \frac{1}{2}\frac{A_n}{p^{3/2}}\frac{dp}{dx}\cos\left(\frac{1}{\hbar}\int_a^x p\,dx - \frac{\pi}{4}\right).$$

If we insert this expression in the integral for $\bar{T}$ we can integrate over the same limits as in the classical case, since outside this region ψ_n decreases exponentially. Replacing the squares of the rapidly oscillating trigonometric functions by their mean value $\frac{1}{2}$ and neglecting the integral containing the oscillating factor

$$\sin\left(\frac{1}{\hbar}\int_a^x p\,dx - \frac{\pi}{4}\right)\cos\left(\frac{1}{\hbar}\int_a^x p\,dx - \frac{\pi}{4}\right) = \frac{1}{2}\sin\left(\frac{2}{\hbar}\int_a^x p\,dx - \frac{\pi}{2}\right),$$

we obtain

$$\bar{T} = \frac{A_n^2}{4\mu}\int_a^b\left[p + \frac{\hbar^2}{4p^3}\left(\frac{dp}{dx}\right)^2\right]dx.$$

The condition of applicability of the quasi-classical approximation $\left|\frac{d\lambda\!\!\!^{-}}{dx}\right| = \frac{\hbar}{p^2}\left|\frac{dp}{dx}\right| \ll 1$ means that the second term of the integrand is small in comparison with the first, and therefore use of the quantization rule gives

$$\overline{T} = \frac{A_n^2}{4\mu}\int_a^b p\,dx = \frac{A_n^2}{4\mu}\pi\hbar\left(n+\frac{1}{2}\right).$$

The constant A_n is determined from the normalization condition

$$\int \psi_n^2\,dx \approx A_n^2\int_a^b \frac{1}{p}\cos^2\left(\frac{1}{\hbar}\int_a^x p\,dx - \frac{\pi}{4}\right)dx \approx \frac{A_n^2}{2}\int_a^b \frac{dx}{p} = 1.$$

On the other hand, if we differentiate the quantization rule

$$\int_a^b p\,dx = \int_a^b \sqrt{2\mu(E_n - V)}\,dx = \pi\hbar\left(n+\frac{1}{2}\right)$$

with respect to n we obtain

$$\mu\frac{dE_n}{dn}\int_a^b \frac{dx}{\sqrt{2\mu(E_n - V)}} = \mu\frac{dE_n}{dn}\int_a^b \frac{dx}{p} = \pi\hbar,$$

from which it follows that

$$A_n^2 = \frac{2\mu}{\pi\hbar}\frac{dE_n}{dn}.$$

Using the above relation, we find that the expression for the mean kinetic energy takes the form

$$\overline{T} = \frac{1}{2}\frac{dE_n}{dn}\left(n+\frac{1}{2}\right).$$

21.

(a)
$$\overline{T} = \frac{1}{2}\hbar\omega\left(n+\frac{1}{2}\right);$$

(b)
$$\overline{T} = \frac{\pi^2\hbar^2}{2\mu a^2}\left[\sqrt{\frac{2\mu a^2 V_0}{\pi^2\hbar^2}} + \left(n+\frac{1}{2}\right)\right]\left(n+\frac{1}{2}\right).$$

22. From the virial theorem it follows that

$$2\overline{T} = \nu\overline{V},$$

which gives

$$E = \frac{2+\nu}{\nu}\overline{T}.$$

Inserting into this relation the mean value of the kinetic energy

$$\overline{T}=\frac{1}{2}\frac{dE_n}{dn}\left(n+\frac{1}{2}\right)$$

(see the preceding problem), we obtain the equation

$$E=\frac{2+\nu}{2\nu}\frac{dE}{dn}\left(n+\frac{1}{2}\right),$$

whose solution has the form

$$E=\text{const}\left(n+\frac{1}{2}\right)^{\frac{2\nu}{2+\nu}}.$$

To find the constant we make use of the quantization condition. From the quantization condition

$$I(E)=2\int_0^{x_0}\sqrt{2\mu(E-ax^\nu)}\,dx=\pi\hbar\left(n+\frac{1}{2}\right).$$

We thus obtain

$$\frac{dI}{dE}=\sqrt{\frac{2\mu}{E}}\int_0^{x_0}\frac{dx}{\sqrt{1-\frac{a}{E}x^\nu}}.$$

The integral is calculated by means of the substitution $u=\frac{a}{E}x^\nu$

$$\frac{dI}{dE}=\frac{1}{\nu}\sqrt{\frac{2\mu}{E}}\left(\frac{E}{a}\right)^{1/\nu}\frac{\Gamma\left(\frac{1}{\nu}\right)\Gamma\left(\frac{1}{2}\right)}{\Gamma\left(\frac{1}{2}+\frac{1}{\nu}\right)},$$

from which we have

$$I(E)=c_\nu E^{1/\nu-1/2},\quad c_\nu=\frac{\sqrt{2\mu}}{\nu a^{1/\nu}}\frac{\Gamma\left(\frac{1}{\nu}\right)\Gamma\left(\frac{1}{2}\right)}{\Gamma\left(\frac{1}{2}+\frac{1}{\nu}\right)}.$$

Hence

$$\text{const}=\left[\frac{\sqrt{\pi}\,\hbar\nu a^{1/\nu}\Gamma\left(\frac{3}{2}+\frac{1}{\nu}\right)}{\sqrt{2\mu}\,\Gamma\left(\frac{1}{\nu}\right)}\right]^{\frac{2\nu}{2+\nu}}.$$

In particular, for $\nu=2$, $a=\frac{\mu\omega^2}{2}$, we have $E=\hbar\omega\left(n+\frac{1}{2}\right)$; for $\nu\to\infty$ we obtain the spectrum of a square well, but it is necessary here to make the change $n+\frac{1}{2}\to n+1$ in the quantization condition.

23. We start from the Bohr quantization rule

$$\int_{x_1}^{x_2}\sqrt{2\mu[E-V(x)]}\,dx=\pi\hbar\left(n+\frac{1}{2}\right),$$

which gives the spectrum, or more precisely, $n(E)$ if the potential energy $V(x)$ is given. Since $V(x)$, according to the conditions of the problem, is an even function,

$$2\int_{0}^{a}\sqrt{2\mu[E-V(x)]}\,dx=\pi\hbar\left(n+\frac{1}{2}\right), \tag{1}$$

where $x_2=-x_1=a$, $E=V(a)$.

The problem therefore reduces to the solution of the integral equation (1) which has the form*

$$x(V)=\frac{\hbar}{\sqrt{2\mu}}\int_{E_0}^{V}\frac{dE}{\frac{dE}{dn}\sqrt{V-E}},$$

where $x(V)$ is the inverse function of $V(x)$ and $\frac{dE}{dn}$ is regarded as a function of E; E_0 is the zero-point energy.

24. From the Schrödinger equation if follows that ψ'' has a singularity of the form $\delta(x)$, and consequently, ψ' undergoes a jump at $x=0$, and ψ is continuous. Integrating the equation in the neighbourhood of the point $x=0$, we find

$$\frac{\hbar^2}{2\mu}\{\psi'(+0)-\psi'(-0)\}=q\psi(0),$$

from which it follows that the normalized wave function of the ground state is

$$\psi=\sqrt{\varkappa}\,e^{-\varkappa|x|},\quad \varkappa=\frac{\mu q}{\hbar^2}.$$

* This problem is closely connected with the following problem of classical mechanics: Given a period of oscillation as a function of the energy of a particle, find the potential energy (see Landau and Pyatigorskiĭ, *Mekhanika* [Mechanics], Gostekhizdat, 1940, where the solution of this problem is given).

The binding energy is

$$E = -\frac{\hbar^2 \varkappa^2}{2\mu} = -\frac{\mu q^2}{2\hbar^2}.$$

This result may be generalized: If there is a potential well such that 1) there is only one bound state with an energy $|E| \ll |V|$, and 2) the dimension a of the region in which the potential is significantly different from zero is much smaller than the distances over which the function ψ changes significantly, then $V(x)$ may be replaced by the equivalent potential $V(x) \to -q\,\delta(x)$, where $q = -\int_{-\infty}^{+\infty} V(x)\,dx$.

Both conditions lead to the same requirement:

$$|V| \ll \frac{\hbar^2}{\mu a^2}.$$

Here, there is only one bound state, whose energy is

$$E = -\frac{\mu}{2\hbar^2}\left(\int_{-\infty}^{+\infty} V(x)\,dx\right)^2.$$

§ 2. Transmission through a potential barrier

1. In the region of the metal ($x<0$) the general form of the wave function belonging to the eigenvalue E is

$$\psi_{\mathrm{I}} = b e^{i\varkappa x} + c e^{-i\varkappa x}, \qquad \varkappa = \frac{1}{\hbar}\sqrt{2\mu(E+V_0)}.$$

In the region $x > 0$ the eigenfunction has the form of a wave travelling away from the metal

$$\psi_{\mathrm{II}} = a e^{ikx}, \quad \text{where} \quad k = \frac{1}{\hbar}\sqrt{2\mu E}.$$

At the boundary of the metal the functions ψ_{I} and ψ_{II} and their derivatives should satisfy the continuity conditions

$$\psi_{\mathrm{II}}(0) = \psi_{\mathrm{I}}(0), \qquad a = b + c,$$
$$\psi'_{\mathrm{II}}(0) = \psi'_{\mathrm{I}}(0), \qquad ak = (b - c)\varkappa.$$

The ratio of the density of the stream of reflected waves to that of the stream of incident waves gives the reflection coefficient

$$R_0 = \left(\frac{\varkappa - k}{\varkappa + k}\right)^2 = \left(\frac{\sqrt{E+V_0} - \sqrt{E}}{\sqrt{E+V_0} + \sqrt{E}}\right)^2 = \frac{V_0^2}{(\sqrt{E+V_0} + \sqrt{E})^4}.$$

If the energy of the electron is $E = 0$ the reflection coefficient is $R_0 = 1$, and, with an increase in energy, R_0 rapidly decreases. For $E \gg V_0$

$$R_0 \approx \frac{V_0^2}{16E^2}.$$

In the other limiting case $E \ll V_0$

$$R_0 \approx 1 - 4\sqrt{\frac{E}{V_0}}.$$

For ordinary metals $V_0 \sim 10$ eV. The reflection coefficient for electrons of energy $E = 0 \cdot 1$ eV is

$$R_0 = 0 \cdot 67.$$

2. In the Schrödinger equation

$$-\frac{\hbar^2}{2\mu}\psi'' - \left[E + V_0\left(e^{\frac{x}{a}} + 1\right)^{-1}\right]\psi = 0 \tag{1}$$

we make the substitution

$$\xi = -e^{-\frac{x}{a}}, \quad \psi = \xi^{-ika} u(\xi),$$

where

$$k = \frac{\sqrt{2\mu E}}{\hbar}.$$

For the function $u(\xi)$ we obtain a hypergeometric equation

$$\xi(1-\xi)u'' + (1 - 2ika)(1-\xi)u' - \varkappa_0^2 a^2 u = 0 \qquad \left(\varkappa_0 = \frac{\sqrt{2\mu V_0}}{\hbar}\right).$$

The solution of Eq. (1) which for $x \to \infty$ ($\xi \to 0$) is finite and asymptotically represents a travelling wave ce^{ikx} is of the form

$$\psi = ce^{ikx} F\left\{i(\varkappa - k)a, -i(\varkappa + k)a, 1 - 2ika, -e^{-\frac{x}{a}}\right\}$$
$$\left(\varkappa = \frac{1}{\hbar}\sqrt{2\mu(E + V_0)}\right).$$

In order to find the reflection coefficient, we must determine the form of the wave function inside the metal ($x \to -\infty$):

$$\psi \approx c \frac{\Gamma(1-2ika)\Gamma(-2ika)}{\Gamma(-i(\varkappa+k)a)\Gamma(1-i(k+\varkappa)a)} e^{i\varkappa x} + $$
$$+ c \frac{\Gamma(1-2ika)\Gamma(2ika)}{\Gamma(i(\varkappa-k)a)\Gamma(1+i(\varkappa-k)a)} e^{-i\varkappa x},$$

from which we find the reflection coefficient

$$R_a = \left| \frac{\Gamma(2ika)\Gamma[-i(\varkappa+k)a]\,\Gamma[1-i(\varkappa+k)a]}{\Gamma(-2ika)\Gamma[i(\varkappa-k)a]\,\Gamma[1+i(\varkappa-k)a]} \right|^2 = \frac{\sinh^2 \pi a\,(\varkappa-k)}{\sinh^2 \pi a\,(\varkappa+k)}.$$

In calculating R_a we must employ the following relations:

$$\Gamma(z+1) = z\Gamma(z),$$

$$\Gamma(z)\Gamma(1-z) = \frac{\pi}{\sin \pi z},$$

$$\Gamma^*(ix) = \Gamma(-ix)$$

(x is a real integer). As $a \to 0$ the formula for the reflection coefficient goes over to the expression for R_0 in the case of a rectangular potential barrier (see the preceding problem).

It is readily seen that $R_a < R_0$, i.e., the reflection coefficient in the case of a smooth variation in the potential is smaller than in the case of a change by steps. When $a = 1$ Å, $V_0 = 10$ eV, $E = 0{\cdot}1$ eV, we find $R_a = 0{\cdot}23$.

3. Let us consider a stream of particles of energy $E < V_0$ moving from left to right. In region III the wave function represents the transmitted wave

$$\psi_{III} = Ce^{ikx}, \quad k = \frac{1}{\hbar}\sqrt{2\mu E}.$$

In region I we have both the incident and reflected waves

$$\psi_I = e^{ikx} + Ae^{-ikx}.$$

In region II the general solution of the Schrödinger equation

$$-\frac{\hbar^2}{2\mu}\psi''_{II} - (E - V_0)\psi_{II} = 0$$

has the form

$$\psi_{II} = B_1 e^{\varkappa x} + B_2 e^{-\varkappa x}, \quad \varkappa = \frac{1}{\hbar}\sqrt{2\mu(V_0 - E)}.$$

The coefficients A, B_1, B_2, C are determined from the continuity conditions of the wave functions and its first derivative.

At the point $x = 0$ these conditions lead to the relations

$$1+A = B_1+B_2,$$
$$ik(1-A) = \varkappa(B_1-B_2).$$

Similarly, at the point $x = a$ we have

$$B_1 e^{\varkappa a}+B_2 e^{-\varkappa a} = Ce^{ika},$$
$$\varkappa(B_1 e^{\varkappa a}-B_2 e^{-\varkappa a}) = ikCe^{ika}.$$

From these equations we find

$$A = C\frac{k^2+\varkappa^2}{4ik\varkappa}e^{ika}(e^{\varkappa a}-e^{-\varkappa a}) = C\frac{2\mu V_0}{\hbar^2}\frac{e^{\varkappa a}-e^{-\varkappa a}}{4ik\varkappa}e^{ika},$$

$$B_1 = C\cdot\frac{1}{2}\left(1+\frac{ik}{\varkappa}\right)e^{ika-\varkappa a},$$

$$B_2 = C\cdot\frac{1}{2}\left(1-\frac{ik}{\varkappa}\right)e^{ika+\varkappa a},$$

$$C = -\frac{4ik}{\varkappa}\frac{e^{-ika}}{e^{\varkappa a}\left(1-\frac{ik}{\varkappa}\right)^2-e^{-\varkappa a}\left(1+\frac{ik}{\varkappa}\right)^2}.$$

Since the expression taken for the incident wave ψ_I had the form e^{ikx}, the transmission coefficient is

$$D = CC^*.$$

The calculation gives

$$D = \frac{4k^2\varkappa^2}{(k^2+\varkappa^2)\sinh^2\varkappa a+4k^2\varkappa^2}.$$

We note that the transmission coefficient tends to zero as we go over to classical mechanics, i.e., as $\hbar \to 0$.

If $\varkappa a \gg 1$ $\left(\text{i.e. } (V_0-E) \gg \frac{\hbar^2}{2\mu a^2}\right)$ then the expression for the transmission coefficient takes a simpler form:

$$D \approx 16\frac{E}{V_0}\left(1-\frac{E}{V_0}\right)\exp\left[-\frac{2}{\hbar}\sqrt{2\mu(V_0-E)}\,a\right].$$

We shall consider two specific cases:

(a) An electron of energy $E = 1$ eV passes through the potential barrier $V_0 = 2$ eV and $a = 1$ Å. We obtain for D the value 0·777.

(b) A proton of the same energy is incident on the same potential barrier. In this case it turns out that the transmission coefficient becomes vanishingly small: $D = 3{\cdot}6\times 10^{-19}$.

4.

$$R = \frac{(k^2-\varkappa^2)\sin^2\varkappa a}{4k^2\varkappa^2+(k^2-\varkappa^2)^2\sin^2\varkappa a} \quad \left(k=\frac{1}{\hbar}\sqrt{2\mu E}, \quad \varkappa=\frac{1}{\hbar}\sqrt{2\mu(E-V_0)}\right).$$

5. For $E < V_0$ we can find the wave function from the expression for the wave functions in Prob. 12, § 1 by changing the signs of E and V_0.

The general form of the wave function corresponding to the energy E is

$$\psi = c_1\left(\cosh\frac{x}{a}\right)^{-2\lambda} F\left(-\lambda+\frac{ika}{2},\ -\lambda-\frac{ika}{2},\ \frac{1}{2};\ -\sinh^2\frac{x}{a}\right)+$$

$$+c_2\left(\cosh\frac{x}{a}\right)^{-2\lambda}\sinh\frac{x}{a}\,F\left(-\lambda+\frac{ika}{2}+\frac{1}{2},\ -\lambda-\frac{ika}{2}+\frac{1}{2},\ \frac{3}{2};\ -\sinh^2\frac{x}{a}\right), \tag{1}$$

where

$$\lambda=\frac{1}{4}\left(\sqrt{1-\frac{8\mu V_0 a^2}{\hbar^2}}-1\right), \quad k=\frac{1}{\hbar}\sqrt{2\mu E}.$$

The coefficients c_1 and c_2 are determined from the condition that when $x\to+\infty$ the wave function has the asymptotic form $\psi\sim e^{ikx}$.

In order to find the asymptotic form of Eq. (1) we use the relation

$$F(\alpha,\beta,\gamma;z)=\frac{\Gamma(\gamma)\Gamma(\beta-\alpha)}{\Gamma(\beta)\Gamma(\gamma-\alpha)}(-z)^{-\alpha}F\left(\alpha,\ \alpha+1-\gamma,\ \alpha+1-\beta;\ \frac{1}{z}\right)+$$

$$+\frac{\Gamma(\gamma)\Gamma(\alpha-\beta)}{\Gamma(\alpha)\Gamma(\gamma-\beta)}(-z)^{-\beta}F\left(\beta,\ \beta+1-\gamma,\ \beta+1-\alpha;\ \frac{1}{z}\right),$$

from which we obtain

$$\psi_{x\to-\infty}\sim(-1)^{2\lambda}\left\{(c_1A_1-c_2A_2)\left(-\frac{1}{2}\right)^{-ika}e^{ikx}+(c_1B_1-c_2B_2)\left(-\frac{1}{2}\right)^{ika}e^{-ikx}\right\}, \tag{2}$$

$$\psi_{x\to+\infty}\sim\left\{(c_1A_1+c_2A_2)\left(\frac{1}{2}\right)^{-ika}e^{-ikx}+(c_1B_1+c_2B_2)\left(\frac{1}{2}\right)^{ika}e^{ikx}\right\}, \tag{3}$$

where, for convenience, we have introduced the notation

$$A_1=\frac{\Gamma\left(\frac{1}{2}\right)\Gamma(-ika)}{\Gamma\left(-\lambda-\frac{ika}{2}\right)\Gamma\left(\lambda+\frac{1}{2}-\frac{ika}{2}\right)},\quad A_2=\frac{\Gamma\left(\frac{3}{2}\right)\Gamma(-ika)}{\Gamma\left(-\lambda+\frac{1}{2}-\frac{ika}{2}\right)\Gamma\left(\lambda+1-\frac{ika}{2}\right)},$$

$$B_1=\frac{\Gamma\left(\frac{1}{2}\right)\Gamma(ika)}{\Gamma\left(-\lambda+\frac{ika}{2}\right)\Gamma\left(\lambda+\frac{1}{2}+\frac{ika}{2}\right)},\quad B_2=\frac{\Gamma\left(\frac{3}{2}\right)\Gamma(ika)}{\Gamma\left(-\lambda+\frac{1}{2}+ika\right)\Gamma\left(\lambda+1+\frac{ika}{2}\right)}.$$

The difference in sign in front of the coefficient c_2 in Eqs. (2) and (3) is explained by the fact that $\sinh x/a$ is an odd function and the second term of Eq. (1) changes sign during the transition from the positive to negative values of x.

The condition that at $+\infty$ only the transmitted wave is present leads to the following relation between the coefficients c_1 and c_2:

$$c_1A_1+c_2A_2=0.$$

The coefficient of transmission will then have the form

$$D=\frac{|c_1B_1+c_2B_2|^2}{|c_1A_1-c_2A_2|^2}.$$

Substituting in the last equation the values of the coefficients A_1, A_2, B_1 and B_2 and carrying out a simple transformation we get finally

$$D=\frac{\sinh^2\pi ka}{\sinh^2\pi ka+\cos^2\left(\frac{\pi}{2}\sqrt{1-8\mu V_0a^2\hbar^{-2}}\right)},\quad \text{if}\quad \frac{8\mu V_0a^2}{\hbar^2}<1$$

and

$$D=\frac{\sinh^2\pi ka}{\sinh^2\pi ka+\cosh^2\left(\frac{\pi}{2}\sqrt{8\mu V_0a^2\hbar^{-2}-1}\right)},\quad \text{if}\quad \frac{8\mu V_0a^2}{\hbar^2}>1.$$

6. The potential energy of the electron has the form shown in Fig. 10. The transmission coefficient D is

$$D\approx\exp\left[-\frac{2}{\hbar}\int_0^{x_0}\sqrt{2\mu(|E|-Fx)}\,dx\right],$$

where the points $x = 0$, $x = x_0 = \frac{|E|}{F}$ mark the limits of the allowable regions according to classical mechanics. Evaluating the integral in the exponent, we obtain

$$D \approx \exp\left(-\frac{4}{3}\frac{\sqrt{2\mu}}{\hbar F}|E|^{3/2}\right). \tag{1}$$

As regards the limits of applicability of this result, we note that the quasi-classical approach is not valid in the vicinity of the classical turning point x_0 inside the region $x - x_0 \leqslant \left(\frac{\hbar^2}{2\mu F}\right)^{1/3}$. Eq. (1) can be used if this region is smaller than the width of the barrier $x_0 = |E|F^{-1}$:

$$\left(\frac{\hbar^2}{2\mu F}\right)^{1/3} < \frac{|E|}{F} \quad \text{or} \quad \frac{\sqrt{2\mu}}{\hbar F}|E|^{3/2} \gg 1.$$

This requirement is therefore equivalent to the requirement that the transmission coefficient be small, i.e., $D \ll 1$.

The transmission coefficient D decreases rapidly with an increase in $|E|$ and increases with an increase in F (see Table 1).

Table 1 *Transmission coefficient*

F	10^6	5×10^6	10^7	2×10^7	3×10^7	5×10^7	$10^8 \frac{\text{V}}{\text{cm}}$
			Without the electrical image force				
$E = -2$V	10^{-84}	$1{\cdot}3\times10^{-17}$	$3{\cdot}5\times10^{-9}$	6×10^{-5}	$1{\cdot}5\times10^{-3}$	0·02	0·14
$E = -3$V	10^{-154}	$1{\cdot}3\times10^{-31}$	$3{\cdot}5\times10^{-16}$	19×10^{-8}	7×10^{-6}	8×10^{-4}	0·029
$E = -5$V	10^{-332}	4×10^{-67}	6×10^{-34}	$2{\cdot}5\times10^{-17}$	10^{-11}	$2{\cdot}5\times10^{-7}$	5×10^{-4}
			With the electrical image force				
$E = -2$V	10^{-80}	8×10^{-15}	$1{\cdot}3\times10^{-6}$	0·013	1*	1	1
$E = -3$V	10^{-150}	5×10^{-28}	7×10^{-14}	$2{\cdot}3\times10^{-6}$	7×10^{-4}	0·07	1
$E = -5$V	10^{-328}	8×10^{-65}	10^{-31}	2×10^{-15}	6×10^{-10}	10^{-5}	0·01

* If the transmission coefficient is equal to unity, then in the case of classica mechanics the electron can escape from the metal.

7. The total potential energy is

$$V = -Fx - \frac{e^2}{4x}.$$

Note that for small values of x (of the order of the interatomic distance) this equation ceases to be valid. However, for the calculation of the transmission coefficient the precise shape of the potential in the vicinity of this region is not important.

The transmission coefficient is

$$D \approx \exp\left\{-\frac{2}{\hbar}\int_{x_1}^{x_2}\sqrt{2\mu\left(|E| - Fx - \frac{e^2}{4x}\right)}\,dx\right\} = \exp\left\{-\frac{2}{\hbar}\int_{x_1}^{x_2} p\,dx\right\}.$$

In this case we determine the turning points x_1 and x_2 from the condition that the classical momentum of the particle vanishes:

$$p = \sqrt{2\mu\left(|E| - Fx - \frac{e^2}{4x}\right)} = 0,$$

$$x_{1,2} = \frac{|E| \pm \sqrt{E^2 - e^2F}}{2F};$$

the integral

$$\int_{x_1}^{x_2} p\,dx = \sqrt{2\mu}\int_{x_1}^{x_2}\sqrt{|E| - Fx - \frac{e^2}{4x}}\,dx$$

is a complete elliptic integral. When the independent variable is replaced by means of $F|E|^{-1}x = \xi$, the integral reduces to a function of one parameter

$$\int_{x_1}^{x_2} p\,dx = \frac{2}{3}\sqrt{2\mu}\,\frac{|E|^{3/2}}{F}\,\varphi(y), \quad y = \frac{\sqrt{e^2F}}{2|E|}.$$

Here $\varphi(y) = \frac{3}{2}\int_{\xi_1}^{\xi_2}\sqrt{1 - y^2\xi^{-1} - \xi}\,d\xi$; we find the limits of integration ξ_1, ξ_2 from the condition that the integrand vanishes. Introducing the notation $k_0 = \frac{4}{3}\sqrt{2\mu}\,|E|^{3/2}F^{-1}$, we get $D = \exp\left(-k_0\varphi(y)\right)$. We note that the transmission coefficient without allowance for the electrical image ($y = 0$) is $D = e^{-k_0}$ (see Prob. 6). The values of $\varphi(y)$ are given in Table 2. The influence of the electrical image force on the transmission coefficient through the potential barrier can be seen from Table 1.

Table 2

y	0	0·2	0·3	0·4	0·5	0·6	0·7	0·8	0·9	1·0
$\varphi(y)$	1·000	0·951	0·904	0·849	0·781	0·696	0·603	0·494	0·345	0·000

8. The wave function has the following form:

$$\psi_{\mathrm{I}} = A \sin kx, \qquad 0 < x < a,$$
$$\psi_{\mathrm{II}} = B_1 e^{\varkappa x} + B_2 e^{-\varkappa x}, \qquad a < x < a+b,$$
$$\psi_{\mathrm{III}} = C \sin k(2a+b-x), \qquad a+b < x < 2a+b,$$

where

$$k = \frac{\sqrt{2\mu E}}{\hbar}, \qquad \varkappa = \frac{\sqrt{2\mu(V_0 - E)}}{\hbar}.$$

The condition that the wave function and its derivative are continuous leads to the following relations:

$$A \sin ka = B_1 e^{\varkappa a} + B_2 e^{-\varkappa a},$$
$$Ak \cos ka = \varkappa(B_1 e^{\varkappa a} - B_2 e^{-\varkappa a}),$$
$$B_1 e^{\varkappa(a+b)} + B_2 e^{-\varkappa(a+b)} = C \sin ka,$$
$$\varkappa(B_1 e^{\varkappa(a+b)} - B_2 e^{-\varkappa(a+b)}) = -Ck \cos ka.$$

Eliminating B_1 and B_2 from these equations, we have

$$\left(\frac{\varkappa}{k}\tan ka + 1\right) A e^{\varkappa b} = \left(\frac{\varkappa}{k}\tan ka - 1\right) C,$$

$$\left(\frac{\varkappa}{k}\tan ka - 1\right) A e^{-\varkappa b} = \left(\frac{\varkappa}{k}\tan ka + 1\right) C.$$

From the condition that the determinant vanishes:

$$\begin{vmatrix} \left(\frac{\varkappa}{k}\tan ka + 1\right) e^{\varkappa b}, & -\left(\frac{\varkappa}{k}\tan ka - 1\right) \\ \left(\frac{\varkappa}{k}\tan ka - 1\right) e^{-\varkappa b}, & -\left(\frac{\varkappa}{k}\tan ka + 1\right) \end{vmatrix} = 0$$

we obtain

$$\left(\frac{\varkappa}{k}\tan ka + 1\right) e^{\varkappa b} = \pm\left(\frac{\varkappa}{k}\tan ka - 1\right).$$

This equation gives us the energy levels.

Making use of the inequality

$$\varkappa b \gg 1,$$

we can represent the above equation in the approximate form

$$\tan ka = -\frac{k}{\varkappa} \mp 2\frac{k}{\varkappa} e^{-\varkappa b}.$$

The right-hand side of this equation represents a small quantity. In the zero-order approximation we obtain ($k \ll \varkappa$):

$$k_0 = \frac{n\pi}{a}, \qquad E_n^{(0)} = \frac{n^2\pi^2\hbar^2}{2\mu a^2},$$

which is the energy of a particle in a potential box (see Prob. 1, §1). In the next approximation

$$k = \frac{n\pi}{a} - \frac{k_0}{a\varkappa_0} \mp 2\frac{k_0}{a\varkappa_0}e^{-\varkappa_0 b}, \quad (n = 1, 2, 3, \ldots),$$

$$E_n = E_n^{(0)} - \frac{2E_n^{(0)}}{a\varkappa_0} \mp 4\frac{E_n^{(0)}}{a\varkappa_0}e^{-\varkappa_0 b}, \qquad \varkappa_0 = \frac{1}{\hbar}\sqrt{2\mu(V_0 - E_n^{(0)})}.$$

The first two terms $E_n^{(1)} = E_n^{(0)} - \frac{2E_n^{(0)}}{a\varkappa_0}$ do not depend on b and give the approximate values of the energy levels of a particle in the potential well shown in Fig. 20 ($b \to \infty$).

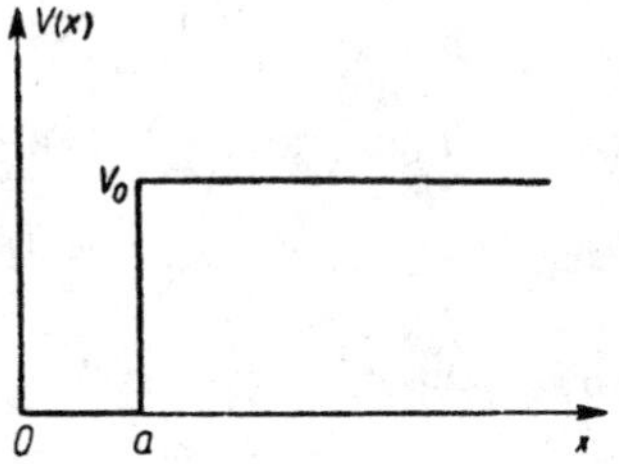

Fig. 20

In this approximation the levels are doubly degenerate; this corresponds to the possibility of finding the particle in either region I or III. The fact that b is finite, in other words, the possibility that the particle can penetrate the potential barrier, leads to the splitting of the levels. This splitting is exponentially small.

We shall find the coefficients A, B_1, B_2, and C in this approximation. For the lower level

$$E_n^- = E_n^{(1)} - 4\frac{E_n^{(0)}}{a\varkappa_0}e^{-\varkappa_0 b}$$

we have the coefficients

$$B_1 = (-1)^{n-1} \frac{k_0}{\varkappa_0} e^{-\varkappa_0 (b+a)} A,$$

$$C = A,$$

$$B_2 = (-1)^{n-1} \frac{k_0}{\varkappa_0} e^{\varkappa_0 a} A;$$

and for the upper level

$$E_n^+ = E_n^{(1)} + 4 \frac{E_n^{(0)}}{a\varkappa_0} e^{-\varkappa_0 b},$$

the coefficients

$$B_1 = -(-1)^{n-1} \frac{k_0}{\varkappa_0} e^{-\varkappa_0 (a+b)} A,$$

$$C = -A,$$

$$B_2 = (-1)^{n-1} \frac{k_0}{\varkappa_0} e^{\varkappa_0 a} A.$$

The value of A obtained from the normalization condition is equal to $\frac{1}{\sqrt{a}}$. (In evaluating the normalization integral we can neglect the contribution from region II.)

Thus the lower level is described by the wave function

$$\psi_{\mathrm{I}} = a^{-1/2} \sin kx,$$

$$\psi_{\mathrm{II}} = (-1)^{n-1} a^{-1/2} \frac{k_0}{\varkappa_0} \{e^{-\varkappa_0 (x-a)} + e^{-\varkappa_0 (a+b-x)}\},$$

$$\psi_{\mathrm{III}} = a^{-1/2} \sin k (2a + b - x).$$

Similarly, for the upper level we find

$$\psi_{\mathrm{I}} = a^{-1/2} \sin kx,$$

$$\psi_{\mathrm{II}} = (-1)^{n-1} a^{-1/2} \frac{k_0}{\varkappa_0} \{e^{-\varkappa_0 (x-a)} - e^{-\varkappa_0 (a+b-x)}\},$$

$$\psi_{\mathrm{III}} = -a^{-1/2} \sin k (2a + b - x).$$

Figure 21 shows the wave functions for $n = 1$ and $n = 4$.

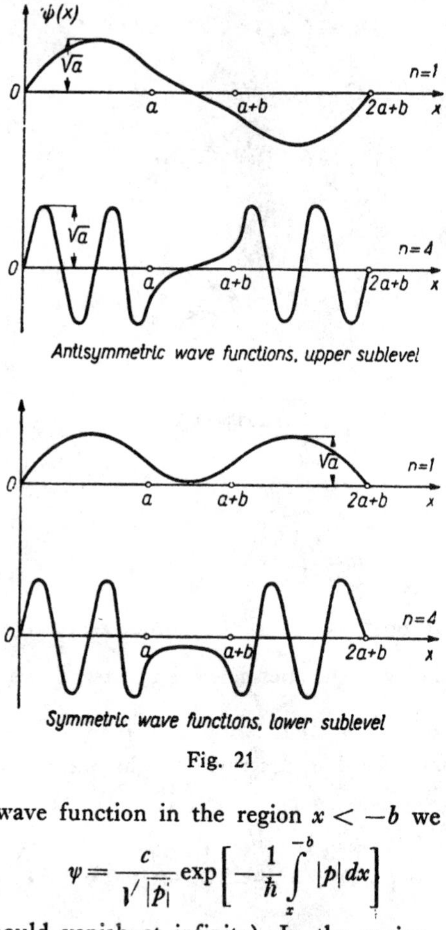

Fig. 21

9. For the wave function in the region $x < -b$ we have

$$\psi = \frac{c}{\sqrt{|p|}} \exp\left[-\frac{1}{\hbar}\int_x^{-b} |p|\, dx\right]$$

(the solution should vanish at infinity). In the region $-b<x<-a$

$$\psi = \frac{c}{\sqrt{p}} \exp\left(-i\frac{\pi}{4} + \frac{i}{\hbar}\int_{-b}^{x} p\, dx\right) + \frac{c}{\sqrt{p}} \exp\left(i\frac{\pi}{4} - \frac{i}{\hbar}\int_{-b}^{x} p\, dx\right)$$

$$= \frac{c}{\sqrt{p}} \exp\left(-i\frac{\pi}{4} + \frac{i}{\hbar}\int_{-b}^{-a} p\, dx - \frac{i}{\hbar}\int_{x}^{-a} p\, dx\right) +$$

$$+ \frac{c}{\sqrt{p}} \exp\left(i\frac{\pi}{4} - \frac{i}{\hbar}\int_{-b}^{-a} p\, dx + \frac{i}{\hbar}\int_{x}^{-a} p\, dx\right).$$

In the region $-a<x<+a$

$$\psi = c\exp\left(-i\frac{\pi}{4}+\frac{i}{\hbar}\int_{-b}^{-a} p\,dx\right)\left\{\frac{1}{2\sqrt{|p|}}\exp\left(-\frac{i\pi}{4}-\frac{1}{\hbar}\int_{-a}^{x}|p|\,dx\right)+\right.$$

$$\left.+\frac{1}{\sqrt{|p|}}\exp\left(i\frac{\pi}{4}+\frac{1}{\hbar}\int_{-a}^{x}|p|\,dx\right\}+c\exp\left(i\frac{\pi}{4}-\frac{i}{\hbar}\int_{-b}^{-a}p\,dx\right)\times$$

$$\times\left\{\frac{1}{2\sqrt{|p|}}\exp\left(i\frac{\pi}{4}-\frac{1}{\hbar}\int_{-a}^{x}|p|\,dx\right)+\frac{1}{\sqrt{|p|}}\exp\left(-i\frac{\pi}{4}+\frac{1}{\hbar}\int_{-a}^{x}|p|\,dx\right\}$$

$$=\frac{c}{\sqrt{|p|}}\sin\left(\frac{1}{\hbar}\int_{-b}^{-a}p\,dx\right)\exp\left(-\frac{1}{\hbar}\int_{-a}^{+a}|p|\,dx+\frac{1}{\hbar}\int_{x}^{+a}|p|\,dx\right)+$$

$$+\frac{2c}{\sqrt{|p|}}\cos\left(\frac{1}{\hbar}\int_{-b}^{-a}p\,dx\right)\exp\left(\frac{1}{\hbar}\int_{-a}^{+a}|p|\,dx-\frac{1}{\hbar}\int_{x}^{+a}|p|\,dx\right).$$

Similarly for $+a<x<+b$ we obtain

$$\psi = c\sin\left(\frac{1}{\hbar}\int_{-b}^{-a}p\,dx\right)\exp\left(-\frac{1}{\hbar}\int_{-a}^{+a}|p|\,dx\right)\left\{\frac{1}{2\sqrt{p}}\exp\left(-i\frac{\pi}{4}-\frac{i}{\hbar}\int_{+a}^{x}p\,dx\right)+\right.$$

$$\left.+\frac{1}{2\sqrt{p}}\exp\left(i\frac{\pi}{4}+\frac{i}{\hbar}\int_{+a}^{x}p\,dx\right)\right\}+2c\cos\left(\frac{1}{\hbar}\int_{-b}^{-a}p\,dx\right)\exp\left(\frac{1}{\hbar}\int_{-a}^{+a}|p|\,dx\right)\times$$

$$\times\left\{\frac{1}{\sqrt{p}}\exp\left(-i\frac{\pi}{4}+\frac{i}{\hbar}\int_{+a}^{x}p\,dx\right)+\frac{1}{\sqrt{p}}\exp\left(i\frac{\pi}{4}-\frac{i}{\hbar}\int_{a}^{x}p\,dx\right)\right\}$$

$$=\frac{c}{\sqrt{p}}\left\{\frac{1}{2}\sin\left(\frac{1}{\hbar}\int_{-b}^{-a}p\,dx\right)\exp\left(-\frac{1}{\hbar}\int_{-a}^{+a}|p|\,dx-i\frac{\pi}{4}\right)+\right.$$

$$\left.+2\cos\left(\frac{1}{\hbar}\int_{-a}^{+a}p\,dx\right)\exp\left(\frac{1}{\hbar}\int_{-a}^{+a}|p|\,dx+i\frac{\pi}{4}\right)\right\}\exp\left(-\frac{i}{\hbar}\int_{+a}^{+b}p\,dx+\right.$$

$$\left.+\frac{i}{\hbar}\int_{x}^{+b}p\,dx\right)+\frac{c}{\sqrt{p}}\left\{\frac{1}{2}\sin\left(\frac{1}{\hbar}\int_{-b}^{-a}p\,dx\right)\exp\left(-\frac{1}{\hbar}\int_{-a}^{+a}|p|\,dx+i\frac{\pi}{4}\right)+\right.$$

$$\left.+2\cos\left(\frac{1}{\hbar}\int_{-a}^{+a}p\,dx\right)\exp\left(\frac{1}{\hbar}\int_{-a}^{+a}|p|\,dx-i\frac{\pi}{4}\right)\right\}\exp\left(\frac{i}{\hbar}\int_{+a}^{+b}p\,dx-\frac{i}{\hbar}\int_{x}^{+b}p\,dx\right).$$

And, finally extending this solution to the region $x > +b$, we have

$$\psi = \frac{c}{\sqrt{|p|}}\left\{\frac{1}{2}\sin\left(\frac{1}{\hbar}\int_a^b p\,dx\right)\cos\left(\frac{1}{\hbar}\int_a^b p\,dx\right)\exp\left(-\frac{1}{\hbar}\int_{-a}^{+a}|p|\,dx\right)+\right.$$

$$\left.+2\cos\left(\frac{1}{\hbar}\int_a^b p\,dx\right)\sin\left(\frac{1}{\hbar}\int_a^b p\,dx\right)\exp\left(\frac{1}{\hbar}\int_{-a}^{+a}|p|\,dx\right)\right\}\exp\left(-\frac{1}{\hbar}\int_b^x|p|\,dx\right)+$$

$$+\frac{c}{\sqrt{|p|}}\left\{-\sin^2\left(\frac{1}{\hbar}\int_a^b p\,dx\right)\exp\left(-\frac{1}{\hbar}\int_{-a}^{+a}|p|\,dx\right)+\right.$$

$$\left.+4\cos^2\left(\frac{1}{\hbar}\int_a^b p\,dx\right)\exp\left(\frac{1}{\hbar}\int_{-a}^{+a}|p|\,dx\right)\right\}\exp\left(\frac{1}{\hbar}\int_b^x|p|\,dx\right).$$

In order that the solution should tend to zero as $x\to+\infty$ the coefficient of $\exp\left(\frac{1}{\hbar}\int_b^x|p|\,dx\right)$ must vanish, i.e.,

$$-\sin^2\left(\frac{1}{\hbar}\int_a^b p\,dx\right)\exp\left(-\frac{1}{\hbar}\int_{-a}^{+a}|p|\,dx\right)+4\cos^2\left(\frac{1}{\hbar}\int_a^b p\,dx\right)\exp\left(\frac{1}{\hbar}\int_{-a}^{+a}|p|\,dx\right)=0,$$

whence

$$\cot\left(\frac{1}{\hbar}\int_a^b p\,dx\right)=\pm\frac{1}{2}\exp\left(-\frac{1}{\hbar}\int_{-a}^{+a}|p|\,dx\right).$$

Assuming that the transparency of the barrier is not great, we get a condition determining the energy levels

$$\frac{1}{\hbar}\int_a^b p\,dx=\pi\left(n+\frac{1}{2}\right)\pm\frac{1}{2}\exp\left(-\frac{1}{\hbar}\int_{-a}^{+a}|p|\,dx\right).$$

We denote by $E_n^{(0)}$ the energy level of a single potential well

$$\frac{1}{\hbar}\int_a^b\sqrt{2\mu(E_n^{(0)}-V)}\,dx=\pi\left(n+\frac{1}{2}\right).$$

We determine the energy level in the double well $E_n = E_n^{(0)} + \Delta E_n$ from the quantization conditions found for this case. Expanding $\sqrt{2\mu(E_n-V)}$ into a series in ΔE_n and neglecting terms higher than the first power of ΔE_n, we obtain

$$\Delta E_n \frac{\mu}{\hbar} \int_a^b \frac{dx}{\sqrt{2\mu(E_n^{(0)} - V)}} = \pm \frac{1}{2} \exp\left(-\frac{1}{\hbar} \int_{-a}^{+a} |p|\, dx\right),$$

or

$$\Delta E_n = \pm \frac{\hbar\omega}{2\pi} \exp\left(-\frac{1}{\hbar} \int_{-a}^{+a} |p|\, dx\right),$$

where ω denotes the angular frequency of the classical motion in a single potential well: $\frac{2\pi}{\omega} = 2\mu \int_a^b \frac{dx}{p}$.

The splitting of the energy level E_n is $2\,|\Delta E_n|$.

10.

$$\tau = \frac{\pi^2}{\omega} \exp \frac{1}{\hbar} \int_{-a}^{+a} |p|\, dx.$$

11. In the region of the nth potential barrier $b_n < x < a_{n+1}$ the wave function can be written in the form

$$\psi = \frac{C_n}{\sqrt{|p|}} \exp\left(-\frac{1}{\hbar} \int_{b_n}^{x} |p|\, dx\right) + \frac{D_n}{\sqrt{|p|}} \exp\left(\frac{1}{\hbar} \int_{b_n}^{x} |p|\, dx\right)$$

$$= \frac{C_n}{\sqrt{|p|}} \exp\left(-\frac{1}{\hbar} \int_{b_n}^{a_{n+1}} |p|\, dx + \frac{1}{\hbar} \int_{x}^{a_{n+1}} |p|\, dx\right) +$$

$$+ \frac{D_n}{\sqrt{|p|}} \exp\left(\frac{1}{\hbar} \int_{b_n}^{a_{n+1}} |p|\, dx - \frac{1}{\hbar} \int_{x}^{a_{n+1}} |p|\, dx\right).$$

Extending this function to the region of the $(n+1)$th potential barrier $b_{n+1} < x < a_{n+2}$, we have

$$\psi = \frac{1}{\sqrt{|p|}} \exp\left(-\frac{1}{\hbar} \int_{b_{n+1}}^{x} |p|\, dx\right) \left\{ \frac{C_n}{2} \exp\left(-\frac{1}{\hbar} \int_{b_n}^{a_{n+1}} |p|\, dx\right) \cos\left(\frac{1}{\hbar} \int_{a_{n+1}}^{b_{n+1}} p\, dx\right) + \right.$$

$$\left. + D_n \exp\left(\frac{1}{\hbar} \int_{b_n}^{a_{n+1}} |p|\, dx\right) \sin\left(\frac{1}{\hbar} \int_{a_{n+1}}^{b_{n+1}} p\, dx\right) \right\} + \frac{1}{\sqrt{|p|}} \exp\left(\frac{1}{\hbar} \int_{b_{n+1}}^{x} |p|\, dx\right) \times$$

$$\times\left\{-C_n \exp\left(-\frac{1}{\hbar}\int_{b_n}^{a_{n+1}} |p|\,dx\right)\sin\left(\frac{1}{\hbar}\int_{a_{n+1}}^{b_{n+1}} p\,dx\right)+\right.$$

$$\left.+2D_n \exp\left(\frac{1}{\hbar}\int_{b_n}^{a_{n+1}} |p|\,dx\right)\cos\left(\frac{1}{\hbar}\int_{a_{n+1}}^{b_{n+1}} p\,dx\right)\right\}.$$

We introduce the notation

$$\frac{1}{\hbar}\int_{b_1}^{a_2} |p|\,dx = \frac{1}{\hbar}\int_{b_2}^{a_3} |p|\,dx = \ldots = \frac{1}{\hbar}\int_{b_{N-1}}^{a_N} |p|\,dx = \tau,$$

$$\frac{1}{\hbar}\int_{a_1}^{b_1} p\,dx = \frac{1}{\hbar}\int_{a_2}^{b_2} p\,dx = \ldots = \frac{1}{\hbar}\int_{a_N}^{b_N} p\,dx = \sigma.$$

Then the previous expression for ψ in the region of the $(n+1)$th barrier takes the form

$$\psi = \frac{1}{\sqrt{|p|}}\exp\left(-\frac{1}{\hbar}\int_{b_{n+1}}^{x} |p|\,dx\right)\left\{\frac{C_n}{2}e^{-\tau}\cos\sigma + D_n e^{\tau}\sin\sigma\right\}+$$

$$+\frac{1}{\sqrt{|p|}}\exp\left(\frac{1}{\hbar}\int_{b_{n+1}}^{x} |p|\,dx\right)\left\{-C_n e^{-\tau}\sin\sigma + 2D_n e^{\tau}\cos\sigma\right.$$

$$= \frac{C_{n+1}}{\sqrt{|p|}}\exp\left(-\frac{1}{\hbar}\int_{b_{n+1}}^{x} |p|\,dx\right) + \frac{D_{n+1}}{\sqrt{|p|}}\exp\left(\frac{1}{\hbar}\int_{b_{n+1}}^{x} |p|\,dx\right),$$

where

$$C_{n+1} = \frac{C_n}{2}e^{-\tau}\cos\sigma + D_n\, e^{\tau}\sin\sigma,$$

$$D_{n+1} = -C_n e^{-\tau}\sin\sigma + 2D_n e^{\tau}\cos\sigma.$$

The relation between the coefficients C_{n+1}, D_{n+1} and C_n, D_n can be expressed conveniently in matrix form:

$$\begin{pmatrix} C_{n+1} \\ D_{n+1} \end{pmatrix} = \begin{pmatrix} \frac{1}{2}e^{-\tau}\cos\sigma & e^{\tau}\sin\sigma \\ -e^{-\tau}\sin\sigma & 2e^{\tau}\cos\sigma \end{pmatrix}\begin{pmatrix} C_n \\ D_n \end{pmatrix} = A\begin{pmatrix} C_n \\ D_n \end{pmatrix}. \quad (1)$$

Using Eq. (1) N times we obtain the relation between C_N, D_N and C_0, D_0

$$\begin{pmatrix} C_N \\ D_N \end{pmatrix} = \begin{pmatrix} \frac{1}{2}e^{-\tau}\cos\sigma & e^{\tau}\sin\sigma \\ -e^{-\tau}\sin\sigma & 2e^{\tau}\cos\sigma \end{pmatrix}^N \begin{pmatrix} C_0 \\ D_0 \end{pmatrix} = A^N \begin{pmatrix} C_0 \\ D_0 \end{pmatrix}.$$

The wave function for the stationary state should decrease for $x<a_1$ and $x>b_N$, and therefore it is necessary that $C_0 = D_N = 0$. It is readily seen that this requires that the matrix element $(A^N)_{22}$ should vanish. The condition that $(A^N)_{22} = 0$ leads to the energy spectrum. To calculate this matrix element we shall consider the matrix

$$S = e^{At} = 1 + tA + \frac{t^2}{2!}A^2 + \ldots + \frac{t^N}{N!}A^N + \ldots$$

It is not difficult to show that the matrix S satisfies the equation

$$\frac{dS}{dt} = AS \tag{2}$$

with the initial condition

$$S(0) = 1.$$

We write Eq. (2) in the extended form

$$\frac{d}{dt}\begin{pmatrix} S_{11} & S_{12} \\ S_{21} & S_{22} \end{pmatrix} = \begin{pmatrix} \alpha & \beta \\ \gamma & \delta \end{pmatrix}\begin{pmatrix} S_{11} & S_{12} \\ S_{21} & S_{22} \end{pmatrix},$$

or

$$\frac{dS_{11}}{dt} = \alpha S_{11} + \beta S_{21}, \qquad \frac{dS_{12}}{dt} = \alpha S_{12} + \beta S_{22},$$

$$\frac{dS_{21}}{dt} = \gamma S_{11} + \delta S_{21}, \qquad \frac{dS_{22}}{dt} = \gamma S_{12} + \delta S_{22}.$$

Since the condition which leads to the energy spectrum may be written in the form

$$(A^N)_{22} = \left(\frac{d^N S_{22}}{dt^N}\right)_{t=0} = 0,$$

it suffices to consider the second pair of equations.

Setting $S_{12} = fe^{\lambda t}$, $S_{22} = ge^{\lambda t}$ we obtain

$$f\lambda = \alpha f + \beta g,$$
$$g\lambda = \gamma f + \delta g.$$

The values of λ are given by the equation

$$\begin{vmatrix} \alpha-\lambda & \beta \\ \gamma & \delta-\lambda \end{vmatrix} = 0, \qquad \lambda^2-\lambda\left(2e^\tau+\frac{1}{2}e^{-\tau}\right)\cos\sigma+1=0,$$

which has two roots, say, λ_1 and λ_2. Since $\lambda_1\lambda_2 = 1$, we can write $\lambda_{1,2}$ in the form

$$\lambda_{1,2} = \exp(\pm iu),$$

where

$$\cos u = \left(e^\tau+\frac{1}{4}e^{-\tau}\right)\cos\sigma.$$

If $\lambda_1 \neq \lambda_2$, then the solution satisfying the initial conditions $S_{12}(0) = 0$, $S_{22}(0) = 1$ has the form

$$S_{12} = \beta(e^{\lambda_1 t}-e^{\lambda_2 t}),$$

$$S_{22} = \frac{(\lambda_1-\alpha)e^{\lambda_1 t}-(\lambda_2-\alpha)e^{\lambda_2 t}}{\lambda_1-\lambda_2}.$$

The condition establishing the energy spectrum in this problem can now be written in the form

$$(A^N)_{22} = \left(\frac{d^N S_{22}}{dt^N}\right)_{t=0} = \frac{1}{\lambda_1-\lambda_2}\{(\lambda_1-\alpha)\lambda_1^N-(\lambda_2-\alpha)\lambda_2^N\} = 0.$$

We insert into this expression the values

$$\lambda_{1,2} = \exp(\pm iu)$$

and in the formula for $\cos u$ we neglect $e^{-\tau}$, which is equivalent to the assumption that the penetrability of the barrier is small:

$$\cos u \approx e^\tau \cos\sigma.$$

With this assumption the condition determining the energy levels takes the simple form

$$\frac{\sin(N+1)u}{\sin u} = 0.$$

This equation has the following roots:

$$u = \frac{n\pi}{N+1},$$

the values $u = 0, \pi, 2\pi$ being excluded.

Then $\cos u$ has N different values:

$$\cos u = \cos\frac{n\pi}{N+1} \approx e^\tau\cos\sigma \qquad (n = 1, 2, \ldots, N).$$

More precisely,

$$\cos\left(\frac{1}{\hbar}\int_{a_1}^{b_1} p\,dx\right) = \left[\exp\left(-\frac{1}{\hbar}\int_{b_1}^{a_2} |p|\,dx\right)\right]\cos\frac{\pi n}{N+1} \qquad (n = 1, 2, \ldots, N).$$

Since $\exp\left(-\frac{1}{\hbar}\int_{b_1}^{a_2} |p|\,dx\right)$ is a small quantity, then the preceding relation can be written in the form

$$\frac{1}{\hbar}\int_{a_1}^{b_1} p\,dx = \pi\left(m+\frac{1}{2}\right) + \left[\exp\left(-\frac{1}{\hbar}\int_{b_1}^{a_2} |p|\,dx\right)\right]\cos\frac{\pi n}{N+1} \tag{3}$$

$$(m = 0, 1, 2, \ldots) \qquad (n = 1, 2, \ldots, N).$$

This is the condition for determining the energy levels in the field $V(x)$. It is very similar to the quantization condition for the field of a single well. From (3) one may conclude that the energy spectrum in the field $V(x)$ can be represented roughly by the energy spectrum of one of the potential wells, all levels being split into N sublevels. We calculate the separation between the lines of one level ΔE_m:

$$\frac{1}{\hbar}\int_{a_1}^{b_1}\sqrt{2\mu(E_m^{(0)}-V)}\,dx + \frac{2\mu}{2\hbar}\int_{a_1}^{b_1}\frac{dx}{\sqrt{2\mu(E_m^{(0)}-V)}}\,\Delta E_m$$

$$= \pi\left(m+\frac{1}{2}\right) + \left[\exp\left(-\frac{1}{\hbar}\int_{b_1}^{a_2} |p|\,dx\right)\right]\cos\frac{\pi n}{N+1} \qquad (n = 1, 2, \ldots, N),$$

from which, after introducing the notation

$$\frac{\pi}{\omega} = \mu\int_{a_1}^{b_1}\frac{dx}{p} = \mu\int_{a_1}^{b_1}\frac{dx}{\sqrt{2\mu(E_m^{(0)}-V)}},$$

we obtain

$$\Delta E_m = \frac{\hbar\omega}{\pi}\left[\exp\left(-\frac{1}{\hbar}\int_{b_1}^{a_2} |p|\,dx\right)\right]\cos\frac{n\pi}{N+1} \qquad (n = 1, 2, \ldots, N).$$

The distance between the upper and lower sublevels is

$$\frac{2\hbar\omega}{\pi}\left[\exp\left(-\frac{1}{\hbar}\int_{b}^{a} |p|\,dx\right)\right]\cos\frac{\pi}{N+1}.$$

12. In the region $x<-b$, according to the conditions of the problem, there is only a wave travelling in the direction of $-\infty$, i.e.,

$$\psi=\frac{c}{\sqrt{p}}\exp\left(\frac{i}{\hbar}\int_x^{-b}p\,dx\right).$$

Extending this solution to the region $x>b$, we obtain the following relation for the wave function:

$$\psi=\frac{c}{\sqrt{p}}\exp\left(\frac{i}{\hbar}\int_b^x p\,dx\right)\left\{\frac{1}{8}\exp\left(-\frac{2}{\hbar}\int_a^b|p|\,dx+i\frac{\pi}{2}\right)\cos\left(\frac{1}{\hbar}\int_{-a}^{+a}p\,dx\right)+\right.$$

$$\left.+2\exp\left(\frac{2}{\hbar}\int_a^b|p|\,dx-i\frac{\pi}{2}\right)\cos\left(\frac{1}{\hbar}\int_{-a}^{+a}p\,dx\right)\right\}+\frac{c}{\sqrt{p}}\exp\left(-\frac{i}{\hbar}\int_b^x p\,dx\right)\times$$

$$\times\left\{\frac{1}{8}\exp\left(-\frac{2}{\hbar}\int_a^b|p|\,dx\right)\cos\left(\frac{1}{\hbar}\int_{-a}^{+a}p\,dx\right)-i\sin\left(\frac{1}{\hbar}\int_{-a}^{+a}p\,dx\right)+\right.$$

$$\left.+2\exp\left(\frac{2}{\hbar}\int_a^b|p|\,dx\right)\cos\left(\frac{1}{\hbar}\int_{-a}^{+a}p\,dx\right)\right\}.$$

We find the quasi-stationary levels from the condition that there is no travelling wave from $+\infty$.

Setting the second expression in the preceding relation equal to zero we get

$$\cot\left(\frac{1}{\hbar}\int_{-a}^{+a}p\,dx\right)=i\left\{\frac{1}{8}\exp\left(-\frac{2}{\hbar}\int_a^b|p|\,dx\right)+2\exp\left(\frac{2}{\hbar}\int_a^b|p|\,dx\right)\right\}^{-1}.$$

We regard $\exp\left(-\frac{2}{\hbar}\int_a^b|p|\,dx\right)$ as a small quantity and find that

$$\frac{1}{\hbar}\int_{-a}^{+a}p\,dx=\pi\left(n+\frac{1}{2}\right)-\frac{i}{2}\exp\left(-\frac{2}{\hbar}\int_a^b|p|\,dx\right).$$

We therefore obtain the condition for determining the quasi-stationary levels $E_n^{(0)}$ and their width Γ:

$$\frac{1}{\hbar}\int_{-a}^{+a}\sqrt{2\mu(E_n^{(0)}-V)}\,dx=\left(n+\frac{1}{2}\right)\pi\qquad(n=0,1,2,\ldots),$$

$$\frac{\hbar\omega}{2\pi}\exp\left(-\frac{2}{\hbar}\int_a^b|p|\,dx\right)=\Gamma,$$

where

$$\omega = \pi \left(\mu \int_{-a}^{+a} \frac{dx}{\sqrt{2\mu(E_n^{(0)} - V)}} \right)^{-1}.$$

We obtain the following value for the transmission coefficient:

$$D(E) = \left\{ 4 \exp \left(\frac{4}{\hbar} \int_a^b |p| \, dx \right) \cos^2 \left(\frac{1}{\hbar} \int_{-a}^{+a} p \, dx \right) + \sin^2 \left(\frac{1}{\hbar} \int_{-a}^{+a} p \, dx \right) \right\}^{-1}.$$

For the value of E which coincides with one of the quasi-levels we have $D(E_n^{(0)}) = 1$. For $|\Delta E| < |E_n^{(0)}|$ we have

$$D(E_n^{(0)} + \Delta E) = \frac{\Gamma^2}{\Gamma^2 + (\Delta E)^2}.$$

Fig. 22 illustrates the behaviour of $D(E)$ in the vicinity of a quasi-level.

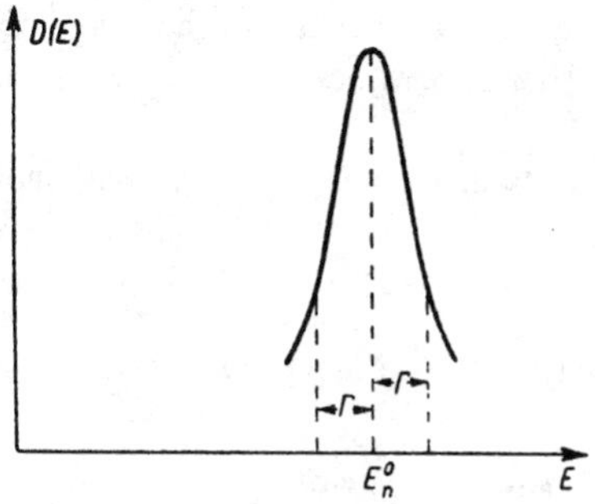

Fig. 22

13. On the left we have ($x < 0$)

$$\psi = e^{ikx} + Ae^{-ikx}, \quad k = \frac{\sqrt{2\mu E}}{\hbar},$$

and for $x > 0$

$$\psi = Be^{ikx}.$$

Integrating the Schrödinger equation in the neighbourhood about the point $x = 0$ we find that ψ' experiences a finite jump, and, consequently, ψ is continuous. The coefficients A and B are found from matching the solutions:

$$A = \frac{-iS}{1+iS}, \quad B = \frac{1}{1+iS}, \quad S = \frac{\mu q}{k\hbar^2}.$$

Thus, the transmission coefficient is

$$D = |B|^2 = \frac{1}{1+\frac{\mu^2 q^2}{k^2 \hbar^4}}.$$

This formula is valid both for $q>0$ and $q<0$. In the latter case, there is one discrete level:

$$E_0 = -\frac{\mu q^2}{2\hbar^2}.$$

Hence in this case

$$D = \frac{E}{E+|E_0|}.$$

Such a dependence of D on the energy is characteristic for energies when $\lambda \gg a$, where a is the characteristic dimension of the potential energy).

14. Since $V(x)$ in an even function, the wave function of the stationary state will be either even or odd. Therefore it is sufficient to consider the region $x>0$.

In this region the Schrödinger equation has the form

$$\psi'' + k^2\psi = \varkappa\delta(x-a)\psi, \quad k^2 = \frac{2\mu E}{\hbar^2}, \quad \varkappa = \frac{2\mu q}{\hbar^2} \tag{1}$$

Integrating this equation in the neighbourhood of the point $x = a$, we obtain

$$\psi'(a+0) - \psi'(a-0) = \varkappa\psi(a). \tag{2}$$

The odd wave function satisfying Eq. (1) has the form

$$\psi = \sin kx \qquad (x < a).$$
$$\psi = A\sin(kx+\delta) \qquad (x > 0),$$

The condition of continuity of the wave function and condition (2) give the two equations

$$\begin{aligned} \sin ka &= A\sin(ka+\delta), \\ Ak\cos(ka+\delta) - k\cos ka &= \varkappa \sin ka. \end{aligned} \tag{3}$$

From these equations it follows that the energy spectrum, as should have been expected, is continuous. Given the value of E, we can find A and δ.

We shall seek the values of E for which the coefficient A is a minimum. (Such values of E are called quasi-levels.)

From (3) it follows that

$$A^2 = 1 + \frac{\varkappa^2}{k^2}\sin^2 ka + \frac{2\varkappa}{k}\sin ka\cdot\cos ka.$$

In our case $\frac{\varkappa}{k} \gg 1$, and A will be a small quantity when sin ka is close to zero.

We set $\sin ka = \sin\,(n\pi + \varepsilon)$, where ε is a small quantity. If we discard terms above the second degree in ε, we find that

$$A^2 = 1 + \frac{\varkappa^2}{k^2}\varepsilon^2 + \frac{2\varkappa}{k}\varepsilon = \left(1 + \frac{\varkappa}{k}\varepsilon\right)^2.$$

We shall choose ε such that $A \approx 0$. To do so, we set

$$\varepsilon = -\frac{k}{\varkappa} \quad \left(\frac{k}{\varkappa} \ll 1\right).$$

Hence for energies equal to

$$E = \frac{n^2\pi^2\hbar^2}{2\mu\left(a + \frac{1}{\varkappa}\right)^2} \quad (n = 1, 2, \ldots)$$

the amplitude of the wave function in the region $|x| > a$ will be close to zero. Carrying out the calculation for the even wave function,

$$\begin{aligned} \psi &= \cos kx & x &< a, \\ \psi &= A\cos(kx + \delta) & x &> a, \end{aligned}$$

we obtain for E the following values:

$$E = \frac{\left(n + \frac{1}{2}\right)^2 \pi^2\hbar^2}{2\mu(a - {}^1/\varkappa)^2} \quad (n = 0, 1, 2, \ldots)$$

It is readily seen that for $\varkappa \to 0$ the spectrum is described by the single formula

$$E \to \frac{n^2\pi^2\hbar^2}{2\mu(2a)^2} \quad (n = 1, 2, \ldots)$$

i.e., the quasi-levels go over to the levels of a particle in a potential box (see Prob. 1, §1).

15.

$$D = \left\{1 + \frac{\varkappa^4}{4k^4}\sin^2 2ka + \frac{\varkappa^2}{k^2}\cos^2 2ka + \frac{\varkappa^3}{k^3}\sin 2ka\cdot\cos 2ka\right\}^{-1},$$

$$\varkappa = \frac{2\mu q}{\hbar^2}, \quad k^2 = \frac{2\mu E}{\hbar^2}.$$

We shall consider the case in which $\frac{\varkappa}{k} \gg 1$. We assume that the energy E of the particle is equal to the value of the quasi-level. Then

$$\sin 2ka = 2\varepsilon \qquad \left(\varepsilon = -\frac{k}{\varkappa}\right)$$

$$\cos 2ka = 1 \text{ (see Prob. 16, §2).}$$

Inserting the $\varkappa$ values in D, we obtain

$$D \approx 1 \qquad \text{(see Prob. 12, §2).}$$

16. Since the scattering particle can be in two states of energy ε_α ($\alpha = 1, 2$), then the wave function of the system can be written in the form

$$\psi(x) = \begin{pmatrix}\psi_1(x)\\ \psi_2(x)\end{pmatrix}.$$

The energy operator of the system consists of the kinetic energy of the electron $p^2/2\mu$, the energy of the scattering particle

$$\begin{pmatrix}\varepsilon_1 & 0\\ 0 & \varepsilon_2\end{pmatrix},$$

and the interaction energy, which can be set equal to the Hermitian operator

$$q\delta(x)\begin{pmatrix}0 & 1\\ 1 & 0\end{pmatrix}.$$

Upon interacting with the electron, the particle passes from one energy state to the other.

The Schrödinger equation is

$$\left\{\frac{p^2}{2\mu} + \begin{pmatrix}\varepsilon_1 & 0\\ 0 & \varepsilon_2\end{pmatrix} + q\delta(x)\begin{pmatrix}0 & 1\\ 1 & 0\end{pmatrix}\right\}\begin{pmatrix}\psi_1(x)\\ \psi_2(x)\end{pmatrix} = E\begin{pmatrix}\psi_1(x)\\ \psi_2(x)\end{pmatrix}.$$

Written for each component, this equation becomes

$$-\frac{\hbar^2}{2\mu}\psi_1'' + q\psi_2(0)\,\delta(x) = (E-\varepsilon_1)\,\psi_1;$$

$$-\frac{\hbar^2}{2\mu}\psi_2'' + q\psi_1(0)\delta(x) = (E-\varepsilon_2)\psi_2.$$

We assume that the electron moves from the left to the right and the particle is initially in the state ε_1. Then, for $x < 0$,

$$\psi_1 = e^{ik_1x} + Ae^{-ik_1x},$$

$$\psi_2 = Be^{-ik_2x}, \quad k_\alpha = \sqrt{2\mu(E-\varepsilon_\alpha)/\hbar^2};$$

for $x > 0$

$$\psi_1 = Ce^{ik_1x},$$

$$\psi^2 = De^{ik_2x}.$$

From the above equations, it follows that ψ_α is continuous when

$$D = B,$$

$$1 + A = C,$$

and ψ_α' undergoes a jump at $x = 0$:

$$\psi_1'(+0) - \psi_1'(-0) = \frac{2\mu}{\hbar^2}q\psi_2(0),$$

$$\psi_2'(+0) - \psi_2'(-0) = \frac{2\mu}{\hbar^2}q\psi_1(0).$$

Thus

$$ik_1(C + A - 1) = \frac{2\mu}{\hbar^2}qB,$$

$$2ik_2B = \frac{2\mu}{\hbar^2}qC,$$

and therefore

$$C = \left\{1 + \left(\frac{\mu q}{\hbar^2}\right)^2 \frac{1}{k_1k_2}\right\}^{-1}.$$

Hence, the probability of transmission without excitation is

$$W_t = C^2 = \left[1 + \left(\frac{\mu q}{\hbar^2}\right)^2 \frac{1}{k_1k_2}\right]^{-2}.$$

The probability of reflection without excitation is

$$W_r = |A|^2 = |C-1|^2 = \left[\frac{\left(\frac{\mu q}{\hbar^2}\right)^2 \frac{1}{k_1k_2}}{1 + \left(\frac{\mu q}{\hbar^2}\right)^2 \frac{1}{k_1k_2}}\right]^2.$$

Finally, in the case of excitation, transmission or reflection occurs with equal probability (the probability is proportional to the current and not to $|\psi|^2$).

$$W_e = 2|B|^2\frac{k_2}{k_1} = \frac{2\left(\frac{\mu q}{\hbar^2}\right)^2\frac{1}{k_1k_2}}{\left[1+\left(\frac{\mu q}{\hbar^2}\right)^2\frac{1}{k_1k_2}\right]^2}.$$

§3. Commutation relations. Uncertainty relation. Spreading of wave packets

1. At first, we shall consider the case of a discrete set of wave functions ψ_i. The mean values of the operators $\hat{A}$ and $\hat{B}$ in the states characterized by the function ψ ($\psi = \Sigma a_i\psi_i$) are given by

$$\bar{\hat{A}} = \sum_{i,k} a_i^* A_{ik} a_k, \qquad \bar{\hat{B}} = \sum_{i,k} a_i^* B_{ik} a_k.$$

We construct the non-negative quantity

$$J(\lambda) = \sum_i \{\sum_k (A_{ik}+i\lambda B_{ik})a_k\}^* \{\sum_l (A_{il}+i\lambda B_{il})a_l\} \geqslant 0$$

(here λ is a real parameter).

We collect together terms of the same power in λ. Since the operators $\hat{A}$ and $\hat{B}$ are Hermitian ($A_{ik} = A_{ki}^*$, $B_{ik} = B_{ki}^*$) we obtain

$$J(\lambda) = \sum_{i,k,l} \{a_k^* A_{ki} A_{il} a_l + i\lambda a_k^*(A_{ki}B_{il} - B_{ki}A_{il})a_l + \lambda^2 a_k^* B_{ki} B_{il} a_l\}$$
$$= \overline{\hat{A}^2} - \lambda\bar{\hat{C}} + \lambda^2\overline{\hat{B}^2}.$$

Here $\hat{C}$ is the Hermitian operator

$$\hat{C} = \frac{1}{i}(\hat{A}\hat{B} - \hat{B}\hat{A}).$$

The quadratic expression $J(\lambda)$ is non-negative and therefore $4\,\overline{\hat{A}^2}\,\overline{\hat{B}^2} \geqslant (\bar{\hat{C}})^2$. We note that the operators $\Delta\hat{A} = \hat{A} - \bar{\hat{A}}$ and $\Delta\hat{B} = \hat{B} - \bar{\hat{B}}$ satisfy the same commutation relation as $\hat{A}$ and $\hat{B}$:

$$\Delta\hat{A}\,\Delta\hat{B} - \Delta\hat{B}\,\Delta\hat{A} = i\hat{C}.$$

We thus obtain

$$\sqrt{\overline{(\Delta\hat{A})^2}\,\overline{(\Delta\hat{B})^2}} \geqslant \frac{|\bar{\hat{C}}|}{2}.$$

In the case of a continuous set of wave functions the proof can be made in a similar way. The expression

$$J(\lambda) = \int \{(\hat{A}+i\lambda\hat{B})\psi\}^* \{(\hat{A}+i\lambda\hat{B})\psi\}\, d\tau,$$

where λ is a real number, is non-negative, and can be written as follows:

$$J(\lambda) = \int \{(\hat{A}\psi)^* - i\lambda(\hat{B}\psi)^*\}\{\hat{A}\psi + i\lambda\hat{B}\psi\}\, d\tau$$
$$= \int \{\psi^*\hat{A}^2\psi + i\lambda\psi^*(\hat{A}\hat{B}-\hat{B}\hat{A})\psi + \lambda^2\psi^*\hat{B}^2\psi\}\, d\tau.$$

The operator $\hat{A}$ is Hermitian and therefore

$$\int (\hat{A}\psi^*)\varphi\, d\tau = \int \psi^*\hat{A}\varphi\, d\tau.$$

The rest of the proof is similar to the preceding one.

2. Proceeding in the same way as in the solution of the preceding problem, but taking λ to be complex, we obtain

$$\overline{(\hat{A})^2}\,\overline{(\hat{B})^2} \geqslant \frac{(\bar{C}\sin\varphi + \bar{D}\cos\varphi)^2}{4},$$

where φ is arbitrary.

We now let $\bar{C} = \varrho\sin\alpha$, $\bar{D} = \varrho\cos\alpha$, $\varrho = \sqrt{(\bar{C})^2+(\bar{D})^2}$.

Then

$$\overline{(\hat{A})^2}\,\overline{(\hat{B})^2} \geqslant \frac{\varrho^2}{4}\cos(\varphi-\alpha).$$

Since φ is arbitrary, then, setting $\varphi = \alpha$, we finally obtain

$$\overline{(\hat{A})^2}\,\overline{(\hat{B})^2} \geqslant \frac{(\bar{C})^2+(\bar{D})^2}{4}.$$

3.

$$\overline{(\Delta\hat{q})^2}\,\overline{(\Delta\hat{F})^2} \geqslant \frac{\hbar^2}{4}\left|\overline{\frac{\partial\hat{F}}{\partial p}}\right|^2.$$

4. The energy of an oscillator in the stationary state is

$$E = \int \psi(x)\left\{\frac{1}{2\mu}\hat{p}^2 + \frac{1}{2}kx^2\right\}\psi(x)\, dx = \frac{1}{2\mu}\overline{\hat{p}^2} + \frac{k}{2}\overline{\hat{x}^2};$$

since

$$\overline{\hat{p}^2} = \overline{(\hat{p}-\bar{p})^2} + (\bar{p})^2 = \overline{(\Delta\hat{p})^2} + (\bar{p})^2,$$
$$\overline{\hat{x}^2} = \overline{(\Delta\hat{x})^2} + (\bar{x})^2$$

and

$$\bar{p} = 0, \quad \bar{x} = 0,$$

then

$$E = \frac{1}{2\mu}\overline{(\Delta\hat{p})^2} + \frac{k}{2}\overline{(\Delta\hat{x})^2}.$$

From the uncertainty relation $\overline{(\Delta \hat{p})^2} \cdot \overline{(\Delta \hat{x})^2} \geqslant \frac{\hbar^2}{4}$ it follows that

$$E \geqslant \frac{\hbar^2}{8\mu \overline{(\Delta \hat{x})^2}} + \frac{k}{2}\overline{(\Delta \hat{x})^2}.$$

The expression on the right-hand side attains a minimum for

$$\overline{(\Delta \hat{x})^2} = \frac{\hbar}{2}\sqrt{\frac{1}{\mu k}};$$

then

$$E_{\min} \sim \frac{\hbar}{2}\sqrt{\frac{k}{\mu}} = \frac{\hbar\omega}{2},$$

where $\omega = \sqrt{\frac{k}{\mu}}$ is the frequency of the oscillator.

5. In this case we can neglect the screening of the field of the nucleus by the other electrons.

The energy of the K-electron is

$$E = \frac{p^2}{2\mu} - \frac{Ze^2}{r}.$$

Since $p \sim \frac{\hbar}{r}$, where r is the extent of the region of localization,

$$E \sim \frac{\hbar^2}{2\mu r^2} - \frac{Ze^2}{r}. \tag{1}$$

This expression has a minimum at $r = \frac{\hbar^2}{Ze^2\mu} = \frac{a}{Z}$ ($a = 0 \cdot 529 \times \times 10^{-8}$ cm — the radius of the first Bohr orbit).

At this energy

$$E \sim -\frac{Z^2}{2}\frac{\mu e^4}{\hbar^2} = -Z^2 \times 13 \cdot 5 \text{ eV}.$$

If we take into account the relativistic correction for the change of mass, then (1) takes the form

$$E \geqslant \{\mu_0^2 c^4 + c^2 p^2\}^{1/2} - \frac{Ze^2}{r} - \mu_0 c^2 \geqslant \left\{\mu_0^2 c^4 + \frac{c^2\hbar^2}{r^2}\right\}^{1/2} - \frac{Ze^2}{r} - \mu_0 c^2.$$

We thus obtain the energy

$$E \geqslant \mu_0 c^2\{(1-\alpha^2 Z^2)^{1/2} - 1\}, \quad \text{where } \alpha = \frac{e^2}{\hbar c}.$$

6. Let the dimensions of the region of localization for the first and second electrons be r_1 and r_2. Then the momenta of the electrons, according to the uncertainty relation, are

$$p_1 \sim \frac{\hbar}{r_1}, \quad p_2 \sim \frac{\hbar}{r_2},$$

and therefore the kinetic energy is of the order

$$\frac{\hbar^2}{2\mu}\left(\frac{1}{r_1^2}+\frac{1}{r_2^2}\right).$$

The potential energy of the interaction of the electrons with the nucleus with a charge Z is

$$-Ze^2\left(\frac{1}{r_1}+\frac{1}{r_2}\right),$$

and the interaction energy between the electrons is of the order $\dfrac{e^2}{r_1+r_2}$. To find the ground state energy we find the minimum total energy

$$E(r_1, r_2) = \frac{\hbar^2}{2\mu}\left(\frac{1}{r_1^2}+\frac{1}{r_2^2}\right) - Ze^2\left(\frac{1}{r_1}+\frac{1}{r_2}\right) + \frac{e^2}{r_1+r_2}.$$

This minimum exists for

$$r_1 = r_2 = \frac{\hbar^2}{\mu e^2}\frac{1}{Z-\frac{1}{4}}.$$

Therefore the ground state energy for an ion with two electrons and nuclear charge Z is

$$E \sim -\left(Z-\frac{1}{4}\right)^2\frac{\mu e^4}{\hbar^2} = -2\left(Z-\frac{1}{4}\right)^2 Ry,$$

$$Ry = \frac{1}{2}\frac{\mu e^4}{\hbar^2} = 13\cdot 5 \text{ eV}.$$

This coincides quite closely with the experimental data if one considers the exceptional simplicity of the solution.

	H^-	He	Li^+	Be^{++}	B^{+++}	C^{++++}
$E_{calc.}$ in Ry	$-1\cdot125$	$-6\cdot125$	$-15\cdot12$	$-28\cdot12$	$-45\cdot12$	$-66\cdot12$
$E_{exper.}$ in Ry	$-1\cdot05$	$-5\cdot807$	$-14\cdot56$	$-27\cdot31$	$-44\cdot06$	$-64\cdot8$

7. No.

8. The mean value of the particle momentum is p_0.

9. In the proof we make use of a relation which is valid if the Hamiltonian does not depend explicity on the time and if there is no magnetic field:

$$\hat{\mathbf{p}} = \frac{i\mu}{\hbar}(\hat{H}\mathbf{r} - \mathbf{r}\hat{H}).$$

The mean value of $\hat{\mathbf{p}}$ in the state ψ in the case of a discrete spectrum is

$$\overline{\hat{\mathbf{p}}} = \frac{i\mu}{\hbar}\int \psi^* (\hat{H}\mathbf{r} - \mathbf{r}\hat{H})\psi \, d\tau.$$

Since $\hat{H}$ is Hermitian,

$$\overline{\hat{\mathbf{p}}} = \frac{i\mu}{\hbar}\int \{\hat{H}^* \psi^* \cdot \mathbf{r}\psi - \psi^* \mathbf{r}\hat{H}\psi\} \, d\tau.$$

In the stationary state we have

$$\hat{H}\psi = E\psi, \qquad \hat{H}^*\psi^* = E\psi^*,$$

and we thus obtain

$$\overline{\hat{\mathbf{p}}} = 0.$$

10. The wave function $\psi(x, t)$ of a free particle is given by $\psi(x, 0)$ in the following manner:

$$\psi(x, t) = \frac{1}{(2\pi\hbar)^{1/2}} \int_{-\infty}^{+\infty} a(p) \exp\left\{\frac{i}{\hbar}\left(px - \frac{p^2}{2\mu}t\right)\right\} dp,$$

where

$$a(p) = \frac{1}{(2\pi\hbar)^{1/2}} \int_{-\infty}^{+\infty} \psi(x, 0) \exp\left(-\frac{i}{\hbar}px\right) dx$$

$$= \frac{1}{(2\pi\hbar)^{1/2}} \int_{-\infty}^{+\infty} \varphi(x) \exp\left(\frac{i}{\hbar}(p_0 - p)x\right) dx.$$

The function $a(p)$ differs significantly from zero only for values of p which satisfy the condition

$$\frac{|p_0 - p|}{\hbar}\delta \leqslant 1.$$

If this condition is satisfied, the oscillating factor $\exp \frac{i}{\hbar}(p_0 - p)x$ changes very little for values of x lying in the interval

$$-\delta < x < +\delta,$$

and, therefore, we can represent $\psi(x, t)$ approximately in the form

$$\psi(x, t) \approx \frac{1}{(2\pi\hbar)^{1/2}} \int_{p_0-\frac{\hbar}{\delta}}^{p_0+\frac{\hbar}{\delta}} a(p) \exp\left\{\frac{i}{\hbar}\left(px - \frac{p^2}{2\mu}t\right)\right\} dp$$

or

$$\psi(x, t) \approx \frac{1}{(2\pi\hbar)^{1/2}} \exp\left\{\frac{i}{\hbar}\left(p_0 x - \frac{p_0^2}{2\mu} t\right)\right\} \int_{-\frac{\hbar}{\delta}}^{+\frac{\hbar}{\delta}} a(p+p_0) \times$$

$$\times \exp\left\{+\frac{i}{\hbar}\left[p\left(x - \frac{p_0}{\mu} t\right) - \frac{p^2}{2\mu} t\right]\right\} dp.$$

From the last relation it follows that the wave function $\psi(x, t)$ differs appreciably from zero only when the oscillating factor

$$\exp\left\{\frac{i}{\hbar}\left[p\left(x - \frac{p_0}{\mu}t\right) - \frac{p^2}{2\mu}t\right]\right\}$$

varies little for changes in p lying in the interval $-\frac{\hbar}{\delta} < p < +\frac{\hbar}{\delta}$. Consequently, the dimension of the wave packet at time t is of the order

$$\delta_t \sim \delta + \frac{\hbar t}{2\mu\delta}.$$

11. For the solution of the problem it is necessary to find the wave function $\psi(x, t)$ satisfying the Schrödinger equation

$$i\hbar \frac{\partial\psi}{\partial t} = \hat{H}\psi \tag{1}$$

and having the given value $\psi(x, 0)$ at $t = 0$. If $\hat{H}$ does not depend explicitly on the time, Eq. (1) has the solution

$$\psi_n(x, t) = \psi_n(x) \exp\left(-\frac{i}{\hbar} E_n t\right), \tag{2}$$

where $\psi_n(x)$ are time-independent eigenfunctions of the operator $\hat{H}$

$$\hat{H}\psi_n(x) = E_n \psi_n(x).$$

We find the coefficients of the expansion of $\psi(x, 0)$ into a series in $\psi_n(x)$:

$$\psi(x, 0) = \sum_n a_n \psi_n(x), \qquad a_n = \int \psi_n^*(x)\, \psi(x, 0)\, dx.$$

The function $\sum_n a_n \psi_n(x)\, e^{-\frac{i}{\hbar}E_n t}$ satisfies Eq. (1), and for $t=0$ is equal to $\psi(x,\ 0)$.
Thus

$$\psi(x,t)=\sum_n a_n\psi_n(x)\exp\left(-i\frac{E_n}{\hbar}t\right)$$

or

$$\psi(x,t)=\int G_t(\xi,x)\psi(\xi,0)\,d\xi, \tag{3}$$

where

$$G_t(\xi,x)=\sum_n \psi_n^*(\xi)\psi_n(x)\exp\left(-\frac{i}{\hbar}E_n t\right).$$

Hence to solve the problem it is sufficient to calculate the Green's function $G_t\,(\xi,\ x)$ and use Eq. 3.

(a) In the case of a free particle the eingenfunctions are

$$\psi_{\mathbf{p}}(\mathbf{r})=\frac{1}{(2\pi\hbar)^{3/2}}\exp\left(i\frac{\mathbf{pr}}{\hbar}\right),\quad E_p=\frac{\mathbf{p}^2}{2\mu}$$

and the corresponding Green's function is

$$G_t(\boldsymbol{\rho},\mathbf{r})=\iiint\frac{d\mathbf{p}}{(2\pi\hbar)^{3/2}}\exp\left\{\frac{i}{\hbar}\left[\mathbf{p}(\mathbf{r}-\boldsymbol{\rho})-\frac{p^2t}{2\mu}\right]\right\}$$
$$=\left(\frac{\mu}{2\pi i\hbar t}\right)^{3/2}\exp\left[\frac{i\mu}{2\hbar t}(\mathbf{r}-\boldsymbol{\rho})^2\right].$$

According to the conditions of the problem we have

$$\psi(\mathbf{r},0)=\frac{1}{(\pi\delta^2)^{3/4}}\exp\left\{\frac{i\mathbf{p}_0\mathbf{r}}{\hbar}-\frac{r^2}{2\delta^2}\right\},$$

and therefore

$$\psi(\mathbf{r},t)=\iiint\left(\frac{\mu}{2\pi i\hbar t}\right)^{3/2}\frac{1}{(\pi\delta^2)^{3/4}}\exp\left\{-\frac{\varrho^2}{2\delta^2}+\frac{i\mathbf{p}_0\boldsymbol{\rho}}{\hbar}+\frac{i\mu}{2\hbar t}(\mathbf{r}-\boldsymbol{\rho})^2\right\}d\boldsymbol{\rho},$$

from which we obtain $\psi(\mathbf{r},t)=\dfrac{1}{(\pi\delta^2)^{3/4}\left(1+\dfrac{\hbar^2t^2}{\mu^2\delta^4}\right)^{3/4}}\times$

$$\times\exp\left\{-\frac{\left(\mathbf{r}-\dfrac{\mathbf{p}_0t}{\mu}\right)^2}{2\delta^2\left(1+\dfrac{\hbar^2t^2}{\mu^2\delta^4}\right)}\left(1-\frac{i\hbar t}{\mu\delta^2}\right)-\frac{ip_0^2t}{2\mu\hbar}+\frac{i\mathbf{p}_0\mathbf{r}}{\hbar}\right\}.$$

For the probability density we obtain

$$\psi^*(\mathbf{r}, t)\psi(\mathbf{r}, t) = \left[\pi\delta^2\left(1+\frac{\hbar^2t^2}{\mu^2\delta^4}\right)\right]^{-3/2} \exp\left\{-\frac{\left(\mathbf{r}-\frac{\mathbf{p}_0 t}{\mu}\right)^2}{\delta^2\left(1+\frac{\hbar^2t^2}{\mu^2\delta^4}\right)}\right\}.$$

From this expression it is seen that centre of gravity of the wave packet moves with the velocity $\mathbf{p}_0/\mu$. The size of the packet δ_t is at first of the same order as δ, and increases with time according to the law

$$\delta_t = \delta\sqrt{1+\frac{\hbar^2t^2}{\mu^2\delta^4}},$$

but its distribution over r remains, as before, a Gaussian one. The time τ for the wave packet to change its initial dimensions by one order of magnitude is

$$\tau \sim \frac{\delta^2\mu}{\hbar}.$$

For $t \gg \tau$ the linear dimensions of the packet increase proportionally to the time

$$\delta_t \sim \frac{\hbar}{\mu\delta}t.$$

We shall consider some specific examples.

For an electron localized initially in a region of $\delta \sim 10^{-8}$ cm the value of τ is of the order 10^{-16} sec. For a "classical" particle $\mu = 1$ g, $\delta = 10^{-5}$ cm we find that $\tau = 10^{17}$ sec $\sim$ 3000 million years.

(b) The wave functions for the one-dimensional motion of a particle in a uniform field $V = -Fx$ has the form (see Prob. 14, §1)

$$\psi_E(x) = A\int_{-\infty}^{+\infty} \exp\left[i\left(\frac{1}{3}u^3 - uq\right)\right] du, \quad q = \left(x+\frac{E}{F}\right)a,$$

where

$$a = \left(\frac{2\mu F}{\hbar}\right)^{1/2}, \quad A = \frac{a}{2\pi\sqrt{F}}.$$

We calculate the Green's function

$$G_t(\xi, x) = \int_{-\infty}^{+\infty} dE \exp\left(-\frac{i}{\hbar}Et\right)\psi_E^*(\xi)\psi_E(x)$$

$$= A^2\int_{-\infty}^{+\infty} dE \exp\left(-\frac{i}{\hbar}Et\right)\iint_{-\infty}^{+\infty} du\, dv \exp\left[-i\left(\frac{v^3}{3}-v\eta\right)+i\left(\frac{u^3}{3}-uq\right)\right],$$

where

$$\eta = \left(\xi + \frac{E}{F}\right) a.$$

We first integrate over E

$$G_t(\xi, x) = A^2 \int\int_{-\infty}^{+\infty} du\, dv \exp\left[-i\left(\frac{v^3}{3} - \frac{u^3}{3} - v\xi a + uxa\right)\right] \int_{-\infty}^{+\infty} dE \exp\left[-i\frac{Ea}{F}\left(u + \right.\right.$$

$$\left.\left. + \frac{Ft}{a\hbar} - v\right)\right]$$

$$= A^2 \int\int_{-\infty}^{+\infty} du\, dv \exp\left[-i\left(\frac{v^3}{3} - \frac{u^3}{3} - v\xi a + uxa\right)\right] \frac{2\pi F}{a} \delta\left(u + \frac{Ft}{a\hbar} - v\right).$$

Using the properties of the δ-function we integrate over v and rearrange the expression in the exponent in a form convenient for the subsequent integration over u:

$$G_t(\xi, x) = \frac{2\pi F}{a} A^2 \int_{-\infty}^{+\infty} du \exp\left\{-i\frac{Ft}{a\hbar}\left[u + \frac{Ft}{2a\hbar} + \frac{a^2\hbar}{2Ft}(x - \xi)\right]^2 - \frac{i}{12}\left(\frac{Ft}{a\hbar}\right)^3 + \right.$$

$$\left. + \frac{i}{2}\frac{Ft}{\hbar}(x + \xi) + \frac{ia^3\hbar}{4Ft}(x - \xi)^2\right\}.$$

Finally, for the Green's function we obtain

$$G_t(\xi, x) = \left(\frac{\mu}{2\pi i\hbar t}\right)^{1/2} \exp\left\{-\frac{i}{12}\left(\frac{Ft}{\hbar a}\right)^3 + \frac{iFt}{2\hbar}(x + \xi) + \frac{i\mu}{2\hbar t}(x - \xi)^2\right\}.$$

As should be expected, for $F \to 0$ this expression goes over into the Green's function for the one-dimensional motion of a free particle. With the help of (3) we can determine the time-variation of the wave function $\psi(x, 0) = (\pi\delta^2)^{-1/4} \exp\left(-\frac{x^2}{2\delta^2} + \frac{ip_0 x}{\hbar}\right)$ given for $t = 0$.

We thus obtain

$$\psi(x, t) = \frac{1}{\left\{\pi\delta^2\left(1 + \frac{\hbar^2 t^2}{\mu^2\delta^4}\right)\right\}^{1/4}} \exp\left\{-\frac{\left(x - \frac{p_0 t}{\mu} - \frac{Ft^2}{2\mu}\right)^2}{2\delta^2\left(1 + \frac{\hbar^2 t^2}{\mu^2\delta^4}\right)}\left(1 - \frac{i\hbar t}{\mu\delta^2}\right) + \right.$$

$$\left. + \frac{i}{\hbar}(p_0 + Ft)x - \frac{i}{\hbar}\int_0^t \frac{1}{2\mu}(p_0 + Ft)^2\, dt\right\}.$$

In the general case of three-dimensional motion in a uniform field with the initial wave function

$$\psi(\mathbf{r}, 0) = (\pi\delta^2)^{-3/4} \exp\left[-\frac{r^2}{2\delta^2} + \frac{i\mathbf{p}_0\mathbf{r}}{\hbar}\right]$$

we obtain

$$\psi(\mathbf{r}, t) = \left\{\pi\delta^2\left(1+\frac{\hbar^2t^2}{\mu^2\delta^4}\right)\right\}^{-3/4} \exp\left\{-\frac{\left(\mathbf{r}-\frac{\mathbf{p}_0 t}{\mu}-\frac{\mathbf{F}t^2}{2\mu}\right)^2}{2\delta^2\left(1+\frac{\hbar^2t^2}{\mu^2\delta^4}\right)}\left(1-\frac{i\hbar t}{\mu\delta^2}\right)+\right.$$

$$\left.+\frac{i}{\hbar}(\mathbf{p}_0+\mathbf{F}t)\mathbf{r}-\frac{i}{\hbar}\int_0^t \frac{1}{2\mu}(\mathbf{p}_0+\mathbf{F}t)^2 dt\right\}.$$

From this expression it follows that the distribution of the probability density preserves its Gaussian form, and the centre of gravity of the wave packet moves in accordance with the laws of classical mechanics for uniform acceleration. The change in the dimensions of the packet with time takes place in the same way as in the absence of the field (case (a) above).

(c) The eigenfunctions of the Schrödinger equation

$$-\frac{\hbar^2}{2\mu}\psi_n'' + \frac{1}{2}\mu\omega^2x^2\psi_n = E_n\psi_n$$

have the following form:

$$\psi_n(x) = c_n \exp\left(-\frac{1}{2}\alpha^2x^2\right)H_n(\alpha x),$$

where

$$\alpha = \left(\frac{\mu\omega}{\hbar}\right)^{1/2}, \quad c_n^2 = \frac{1}{2^n n!}\frac{\alpha}{\sqrt{\pi}}, \quad E_n = \hbar\omega\left(n+\frac{1}{2}\right).$$

The wave function we are seeking is, according to (2),

$$\psi(x, t) = \sum_n a_n\psi_n(x)\exp\left(-i\frac{E_n}{\hbar}t\right), \tag{4}$$

where

$$a_n = c_n c \int_{-\infty}^{+\infty} H_n(\alpha x)\exp\left\{-\alpha^2\frac{(x-x_0)^2}{2}+\frac{i}{\hbar}p_0x-\frac{\alpha^2x^2}{2}\right\}dx.$$

To calculate a_n we make use of the generating function for Tchebycheff-Hermite polynomials

$$e^{-\lambda^2+2\lambda\eta} = \sum_{n=0}^{\infty} \frac{\lambda^n}{n!} H_n(\eta). \tag{5}$$

It is readily seen that $\frac{a_n}{c_n c}$ is the coefficient for the term $\frac{\lambda^n}{n!}$ in the expansion of

$$\int_{-\infty}^{+\infty} \exp\left\{-\lambda^2 + 2\lambda a x - \frac{1}{2}a^2(x-x_0)^2 + \frac{i}{\hbar} p_0 x - \frac{1}{2}a^2 x^2\right\} dx$$

into a series in λ. We thus obtain

$$a_n = c_n c \sqrt{\pi} \left(a x_0 + \frac{i p_0}{a\hbar}\right)^n \exp\left\{-\frac{1}{2}a^2 x_0^2 + \frac{1}{4}\left(a x_0 + \frac{i p_0}{a\hbar}\right)^2\right\}.$$

After substituting this expression into (4) we can sum over n, again using relation (5). We introduce the notation

$$x_0 + \frac{i p_0}{\hbar a^2} = Q e^{-i\delta}$$

and after some calculations we obtain

$$\psi(x, t) = c \exp\left\{-\frac{a^2}{2}[x - Q\cos(\omega t + \delta)]^2 - i x Q a^2 \sin(\omega t + \delta) - \frac{i\omega t}{2} + \right.$$
$$\left. + \frac{a^2 Q^2}{4} i[\sin 2(\omega t + \delta) - \sin 2\delta]\right\}.$$

In this case the motion does not produce a spreading of the wave packet. The centre of gravity of the motion obeys, as before, the laws of classical mechanics, undergoing harmonic oscillations with an amplitude Q and a frequency ω. From the expression obtained for ψ it also follows that the mean momentum at the time t is given by

$$P(t) = \hbar Q a^2 \sin(\omega t + \delta),$$

i.e., the classical momentum of an oscillating particle.

The expression

$$\exp\left\{-\frac{1}{2} i\omega t + \frac{a^2 Q^2}{4} i[\sin 2(\omega t + \delta) - \sin 2\delta]\right\}$$

can be written as

$$\exp\left\{-\frac{i}{\hbar}\int_0^t \frac{1}{2\mu} P^2(t)\,dt\right\}.$$

12. We consider the operator

$$\hat{a}(s) = e^{s\hat{L}}\,\hat{a}\,e^{-s\hat{L}},$$

where s is an auxiliary parameter, and find the differential equation which $\hat{a}(s)$ satisfies. This operator is

$$\frac{d\hat{a}(s)}{ds} = \hat{L}e^{s\hat{L}}\,\hat{a}\,e^{-s\hat{L}} - e^{s\hat{L}}\,\hat{a}\,e^{-s\hat{L}}\,\hat{L} = [\hat{L}\hat{a}(s)].$$

We differentiate this equation once again:

$$\frac{d^2\hat{a}(s)}{ds^2} = \left[\hat{L}, \frac{d\hat{a}(s)}{ds}\right] = [\hat{L}[\hat{L}\hat{a}(s)]].$$

It is readily seen that the derivative $\dfrac{d^n\hat{a}(s)}{ds^n}$ is equal to the expression obtained as the result of n successive permutations of the operator $\hat{L}$ with the operator $\hat{a}(s)$.

We represent the operator

$$e^{\hat{L}}\,\hat{a}e^{-\hat{L}} = \hat{a}(1)$$

in the form of a Taylor series

$$\hat{a}(1) = \hat{a}(0) + \frac{d\hat{a}(0)}{ds} + \frac{1}{2!}\frac{d^2\hat{a}(0)}{ds^2} + \ldots$$

and express the derivative with respect to s at the point $s = 0$ through another permutation of the operator $\hat{L}$ with $\hat{a}(0) = \hat{a}$; as a result, we obtain the proof of the relation given in the problem.

13. We express the Hamiltonian

$$\hat{H} = \frac{1}{2\mu}\hat{p}^2 + \frac{1}{2}\mu\omega^2\hat{x}^2 - f(t)\hat{x}$$

in terms of operators $\hat{a}$ and $\hat{a}^+$ (see Prob. 6, §1)

$$\hat{H} = \hbar\omega\left(\hat{a}^+\hat{a} + \frac{1}{2}\right) - f(t)\sqrt{\frac{\hbar}{2\mu\omega}}(\hat{a} + \hat{a}^+)$$

and look for the solution of the Schrödinger equation

$$i\hbar\dot{\psi}(t) = \hat{H}\psi(t)$$

in the form

$$\psi(t) = c(t)\exp\left(\alpha(t)\hat{a}^+\right)\exp\left[\beta(t)\hat{a}\right]\exp\left[\gamma(t)\hat{a}^+\hat{a}\right]\psi(-\infty).$$

In differentiating the operator acting on $\psi(-\infty)$ with respect to t we keep in mind the fact that $\hat{a}$ and $\hat{a}^+$ do not commute:

$$\hat{a}\hat{a}^+ - \hat{a}^+\hat{a} = [\hat{a},\ \hat{a}^+] = 1.$$

We transform the expression for the time derivative

$$\dot{\psi} = [\dot{c}\exp(\alpha\hat{a}^+)\exp(\beta\hat{a})\exp(\gamma\hat{a}^+\hat{a}) + c\dot{\alpha}\hat{a}^+\exp(\alpha\hat{a}^+)\exp(\beta\hat{a})\exp(\gamma\hat{a}^+\hat{a}) + c\dot{\beta}\exp(\alpha\hat{a}^+)\hat{a}\exp(\beta\hat{a})\exp(\gamma\hat{a}^+\hat{a}) + c\dot{\gamma}\exp(\alpha\hat{a}^+)\exp(\beta\hat{a})\hat{a}^+\hat{a}\exp(\gamma\hat{a}^+\hat{a})]\psi(-\infty)$$

so that it takes the form

$$\dot{\psi} = \hat{G}e^{\alpha\hat{a}^+}e^{\beta\hat{a}}e^{\gamma\hat{a}^+\hat{a}}\psi(-\infty),$$

where $\hat{G}$ is some operator. We first consider the third term on the right-hand side of the equation for $\dot{\psi}$. Noting that

$$[(\hat{a}^+)^n\hat{a}] = -n(\hat{a}^+)^{n-1},$$

we see that

$$e^{\alpha\hat{a}^+}\hat{a} = \left(1 + \alpha\hat{a}^+ + \frac{1}{2}(\alpha\hat{a}^+)^2 + \dots\right)\hat{a}$$

$$= (\hat{a} - \alpha)\left(1 + \alpha\hat{a}^+ + \frac{1}{2}(\alpha\hat{a}^+)^2 + \dots\right) = (\hat{a} - \alpha)e^{\alpha\hat{a}^+}.$$

In view of this, the third term may be written in the form

$$c\dot{\beta}(\hat{a} - \alpha)e^{\alpha\hat{a}^+}e^{\beta\hat{a}}e^{\gamma\hat{a}^+\hat{a}}\psi(-\infty).$$

Similarly, we rearrange the last term

$$c\dot{\gamma}e^{\alpha\hat{a}^+}e^{\beta\hat{a}}\hat{a}^+\hat{a}e^{\gamma\hat{a}^+\hat{a}} = c\dot{\gamma}e^{\alpha\hat{a}^+}(\hat{a}^+\hat{a} + \beta\hat{a})e^{\beta\hat{a}}e^{\gamma\hat{a}^+\hat{a}}$$

$$= c\dot{\gamma}(\hat{a}^+\hat{a} - \alpha\hat{a}^+ + \beta\hat{a} - \alpha\beta)e^{\alpha\hat{a}^+}e^{\beta\hat{a}}e^{\gamma\hat{a}^+\hat{a}}.$$

Therefore,

$$\hat{G} = c\dot{\gamma}\hat{a}^+\hat{a} + c(\dot{\alpha} - \alpha\dot{\gamma})\hat{a}^+ + c(\dot{\beta} + \beta\dot{\gamma})\hat{a} + \dot{c} - c\dot{\gamma}\alpha\beta - c\dot{\beta}\alpha;$$

and to satisfy the Schrödinger equation we should set

$$i\hbar\hat{G} = \hat{H}.$$

If we compare the coefficients of the operators $\hat{a}^+\hat{a}$, $\hat{a}$, $\hat{a}^+$ and the free terms, we get the set of equations

$$\dot\gamma = -i\omega,$$
$$\dot\alpha + i\omega\alpha = i(2\hbar\mu\omega)^{-1/2} f(t),$$
$$\dot\beta - i\omega\beta = i(2\hbar\mu\omega)^{-1/2} f(t),$$
$$\frac{\dot c}{c} = -i\frac{\omega}{2} + \alpha(\dot\beta - i\omega\beta).$$

Solving these equations for the initial conditions $\alpha(-\infty) = 0$, $\beta(-\infty) = 0$, $|c(-\infty)| = 1$, we find

$$\alpha(t) = \frac{i\exp(-i\omega t)}{\sqrt{2\hbar\mu\omega}} \int_{-\infty}^{t} f(t')\, e^{i\omega t'}\, dt',$$

$$\beta(t) = \frac{i\exp(+i\omega t)}{\sqrt{2\hbar\mu\omega}} \int_{-\infty}^{t} f(t')\, e^{-i\omega t'}\, dt = -\alpha^*(t),$$

$$\gamma(t) = -i\omega t,$$

$$c(t) = \exp\left(-\frac{i}{2}\omega t\right)\exp\left\{-\frac{1}{2\hbar\mu\omega}\int_{-\infty}^{t} dt'\, f(t')\, e^{-i\omega t'}\int_{-\infty}^{t'} dt''\, f(t'')\, e^{i\omega t''}\right\}.$$

The probability of transition from the state $\psi(-\infty)$ to the nth excited state for $t = +\infty$ is

$$W_n = \lim_{t\to\infty}\left|\int \psi_n^* \psi(t)\, dx\right|^2 = \lim_{t\to\infty}\left| c \int \psi_n^*\, e^{\alpha\hat a^+}\, e^{\beta\hat a}\, e^{\gamma\hat a^+\hat a}\, \psi(-\infty)\, dx\right|^2.$$

If the initial state was a ground state $\psi(-\infty) = \psi_0$, then, since $\hat a\psi_0 = 0$, we obtain, without difficulty:

$$e^{\beta\hat a}\, e^{\gamma\hat a^+\hat a}\, \psi_0 = \psi_0.$$

Furthermore, since the normalized wave functions have the form

$$\psi_n = \frac{1}{\sqrt{n!}}(\hat a^+)^n \psi_0 \qquad \text{(see Prob. 6c, §1)},$$

then

$$e^{\alpha\hat a^+}\psi_0 = \sum_{m=0}^{\infty} \frac{\alpha^m}{\sqrt{m!}}\psi_m.$$

On the basis of the orthogonality relations $\int \psi_n^* \psi_m\, dx = \delta_{nm}$ we obtain

$$W_{n0} = \lim_{t\to\infty} |c(t)|^2 \frac{|\alpha(t)|^{2n}}{n!}.$$

From the formulae for $a(t)$ and $c(t)$ we find

$$\lim_{t\to\infty} |a(t)|^2 = \frac{1}{2\hbar\mu\omega}\left|\int_{-\infty}^{+\infty} f(t) e^{i\omega t}\, dt\right|^2 \equiv \nu,$$

$$\lim_{t\to\infty} |c(t)|^2 = \exp\left\{-\frac{1}{2\hbar\mu\omega}\left|\int_{-\infty}^{+\infty} f(t) e^{i\omega t}\, dt\right|^2\right\} = e^{-\nu}.$$

Finally, we obtain for W_{n0} the Poisson distribution

$$W_{n0} = e^{-\nu}\frac{\nu^n}{n!},$$

where

$$\nu = \frac{1}{2\hbar\mu\omega}\left|\int_{-\infty}^{+\infty} f(t) e^{i\omega t}\, dt\right|^2.$$

(a) For $f(t) = f_0 \exp\left(-\frac{t^2}{\tau^2}\right)$ we find

$$\nu = \frac{\pi f_0^2 \tau^2}{2\hbar\mu\omega} \exp\left(-\frac{1}{2}\omega^2\tau^2\right).$$

(b) For $f(t) = f_0[1+(t/\tau)^2]^{-1}$ we find

$$\nu = \frac{\pi^2 f_0^2 \tau^2}{2\hbar\mu\omega} e^{-2\omega\tau}.$$

14. The Schrödinger equation for an oscillator with a perturbing force has the form

$$i\hbar\frac{\partial\psi}{\partial t} = -\frac{\hbar^2}{2\mu}\frac{\partial^2\psi}{\partial x^2} + \frac{\mu\omega^2 x^2}{2}\psi - f(t)\, x\psi.$$

We introduce the new coordinates $x_1 = x - \xi(t)$; then

$$i\hbar\frac{\partial\psi}{\partial t} = i\hbar\dot{\xi}\frac{\partial\psi}{\partial x_1} - \frac{\hbar^2}{2\mu}\frac{\partial^2\psi}{\partial x_1^2} + \frac{1}{2}\mu\omega^2(x_1+\xi)^2\psi - f(t)(x_1+\xi)\psi.$$

If we set

$$\psi = \exp\left(\frac{i}{\hbar}\mu\dot{\xi}x_1\right)\varphi(x_1, t),$$

then for φ we obtain the equation

$$i\hbar\frac{\partial\varphi}{\partial t} = -\frac{h^2}{2\mu}\frac{\partial^2\varphi}{\partial x_1^2} + \frac{1}{2}\mu\omega^2 x_1^2\varphi + (\mu\ddot{\xi} + \mu\omega^2\xi - f)\, x_1\varphi - L\varphi, \qquad (1)$$

where L is the Lagrangian

$$L=\frac{1}{2}\mu\dot{\xi}^2-\frac{1}{2}\mu\omega^2\xi^2+f(t)\xi.$$

In Eq. (1) the term $(\mu\ddot{\xi}+\mu\omega^2\xi-f)x_1\varphi$ is equal to zero, since ξ, as a function of time, satisfies the classical equation of motion of an oscillator with a perturbing force

$$\mu\ddot{\xi}+\mu\omega^2\xi=f(t).$$

If we introduce another new function χ defined by the relation $\varphi=\chi\exp\left(\frac{i}{\hbar}\int_0^t L\,dt\right)$, we obtain for the function χ an equation of the same form as the equation of motion of a free oscillator:

$$i\hbar\frac{\partial\chi}{\partial t}=-\frac{\hbar^2}{2\mu}\frac{\partial^2\chi}{\partial x_1^2}+\frac{1}{2}\mu\omega^2x_1^2\chi.$$

Thus the wave function of the oscillator with a perturbing force can be represented in the form

$$\psi(x,t)=\chi[x-\xi(t),t]\exp\left\{\frac{i\mu}{\hbar}\dot{\xi}(x-\xi)+\frac{i}{\hbar}\int_0^t L\,dt\right\}.$$

15. We shall seek the solution to the Schrödinger equation

$$i\hbar\frac{\partial\psi}{\partial t}=-\frac{\hbar^2}{2\mu}\frac{\partial^2\psi}{\partial x^2}+\frac{1}{2}\mu\omega^2(t)x^2\psi \tag{1}$$

in the form

$$\psi(x,t)=\int G(x,t;x',\tau)\psi(x',\tau)\,dx'.$$

We can readily show that the Green's function $G(x,t;\ x',\tau)$ should satisfy Eq. (1) and the initial condition

$$\lim_{t\to\tau+0}G(x,t;x',\tau)=\delta(x-x'). \tag{2}$$

Let us try to satisfy these two conditions by setting

$$G(x,t;x',\tau)\sim\exp\left\{\frac{i}{2\hbar}[a(t)x^2+2b(t)x+c(t)]\right\}. \tag{3}$$

Substituting (3) in (1), we obtain equations for a, b, c:

$$\left.\begin{aligned} \frac{1}{\mu}\frac{da}{dt} &= -\frac{a^2}{\mu^2} - \omega^2(t), \\ \frac{db}{dt} &= -\frac{a}{\mu}b, \\ \frac{dc}{dt} &= i\hbar\frac{a}{\mu} - \frac{b^2}{\mu}. \end{aligned}\right\} \tag{4}$$

The solution to the system of equations (4) has the form

$$a = \mu\frac{\dot{Z}}{Z}, \quad b = \frac{\text{const}}{Z}, \quad c = i\hbar \ln Z - \frac{1}{\mu}\int^t b^2\,dt, \tag{5}$$

where Z is the solution of the equation

$$\ddot{Z} = -\omega^2(t)Z.$$

Let us now try to select the constants of integration so as to satisfy the initial condition (2). For this purpose let us take one of the possible expressions for the δ-function, viz.,

$$\delta(x-x') = \lim_{t\to\tau}\sqrt{\frac{\mu}{2\pi i\hbar(t-\tau)}}\exp\left\{\frac{i\mu}{2\hbar(t-\tau)}(x-x')^2\right\} \tag{6}$$

(see Prob. 11a, §3).

In order for expression (3) to go over into (6) for $t \to \tau$ it is necessary and sufficient that

$$Z = 0, \quad \dot{Z} = 1 \quad \text{for} \quad t = \tau,$$

$$b = -\frac{\mu x'}{Z},$$

$$c = i\hbar \ln Y + \mu x'^2 \frac{Y}{Z},$$

where Y is the solution of the equation $\ddot{Y} = -\omega^2(t)Y$ satisfying the initial condition $Y = 1$, $\dot{Y} = 0$ for $t = \tau$.

We note that since $\dot{Z}Y - \dot{Y}Z = 1$,

$$\frac{Y}{Z} = -\int\frac{dt}{Z^2}.$$

Accordingly, for the Green's function in our problem we get an expression of the following form:

$$G(x, t; x', \tau) = \sqrt{\frac{\mu}{2\pi i\hbar Z}}\exp\left\{\frac{i\mu}{2\hbar Z}(\dot{Z}x^2 - 2xx' + Yx'^2)\right\}.$$

When ω is constant we have

$$Z = \frac{1}{\omega}\sin\omega(t-\tau), \qquad Y = \cos\omega(t-\tau),$$

and in this case the Green's function is equal to

$$G(x, t; x', \tau) = \sqrt{\frac{\mu\omega}{2\pi\hbar i \sin\omega(t-\tau)}} \exp\left\{\frac{i\mu\omega}{2\hbar\sin\omega(t-\tau)}(\cos\omega(t-\tau)x^2 - \right.$$
$$\left. - 2xx' + \cos\omega(t-\tau)x'^2)\right\}.$$

16.

$$|\psi(x, t)|^2 = \frac{\alpha}{\sqrt{2\pi}}\left(\cos^2\omega t + \frac{\alpha^4\hbar^2}{\mu^2\omega^2}\sin^2\omega t\right)^{-1/2} \times$$
$$\times \exp\left\{-\frac{\alpha^2(x - Q\cos(\omega t+\delta))^2}{\cos^2\omega t + \frac{\alpha^4\hbar^2}{\mu^2\omega^2}\sin^2\omega t}\right\},$$

where

$$x_0 + i\frac{p_0}{\mu\omega} = Q e^{-i\delta}.$$

If $\alpha = \sqrt{\frac{\mu\omega}{\hbar}}$, we obtain the results of Prob. 11c, §3.

17.

$$G(x, t; x', \tau) = \sqrt{\frac{\mu}{2\pi i\hbar Z}} \exp\left\{\frac{i\mu}{2\hbar Z}[\dot{Z}(x-\xi)^2 - 2(x-\xi)x' + Yx'^2] + \right.$$
$$\left. + \frac{i}{\hbar}\mu\dot{\xi}(t)(x-\xi) + \frac{i}{\hbar}\int_\tau^t L\,dt\right\}.$$

In this expression, ξ satisfies the equation $\mu\ddot{\xi} = -\mu\omega^2\xi + f(t)$ and the initial conditions $\xi(\tau) = 0$, $\dot{\xi}(\tau) = 0$; L is the Lagrangian: $L = \frac{1}{2}\mu\dot{\xi}^2 - \frac{1}{2}\mu\omega^2\xi^2 + f\xi$.

18. The probability of the transition from state n to state m is given by the relation (we employ the system of units in which $\hbar = 1$, $\mu = 1$, $\omega = 1$)

$$P_{mn}(t, 0) = |G_{mn}(t, 0)|^2, \tag{1}$$

where

$$G_{mn}(t, 0) = \int\int \psi_m^*(x) G(x, t; x', 0)\psi_n(x')\,dx\,dx'.$$

By means of the generating function

$$\exp\left\{-\frac{1}{2}(2z^2+x^2-4zx)\right\}=\sum_{n=0}^{\infty}\sqrt{\frac{\sqrt{\pi}\,2^n}{n!}}\,z^n\psi_n(x)$$

we construct the function $G(u, v)$

$$G(u,v)=\iint\exp\left\{-\frac{1}{2}(2v^2+x^2-4vx)-\frac{1}{2}(2u^2+x'^2-4ux')\right\}\times$$

$$\times\, G(x,t;x',0)\,dx\,dx'=\sum_{m=0}^{\infty}\sum_{n=0}^{\infty}\sqrt{\frac{\pi 2^{n+m}}{n!\,m!}}\,v^m u^n G_{mn}(t,0). \qquad (2)$$

From (2) it follows that, apart from the factor $\sqrt{\frac{\pi 2^{i+j}}{i!\,j!}}$, the quantities $G_{ij}(t, 0)$ are the coefficients of the individual terms in the expansion of $G(u, v)$ in powers of u, v, while $|G_{ij\,(t,\,0)}|^2$, by Eq. (1), give the transition, probabilities. We now calculate $G(u, v)$. We insert in (2) the expression for the Green's function G (see Prob. 17, §3) and obtain

$$G(u,v)=(2\pi iZ)^{-1/2}\exp\left\{i\int_0^t L\,dt-i\xi\dot{\xi}+\frac{i\dot{Z}}{2Z}\xi^2-u^2-v^2\right\}\times$$

$$\times\iint\exp\left\{-\frac{1}{2}\left[\left(1-\frac{i\dot{Z}}{Z}\right)x^2+\frac{2i}{Z}xx'+\left(1-\frac{iY}{Z}\right)x'^2-\right.\right.$$

$$\left.\left.-2\left(2v-\frac{i\dot{Z}}{Z}\xi+i\dot{\xi}\right)x-2\left(2u+\frac{i\xi}{Z}\right)x'\right]\right\}dx\,dx'.$$

If ω is constant then ξ, Z, Y have the form

$$\xi(t)=\int_0^t\sin(t-t')\,f(t')\,dt',\qquad Z=\sin t,\qquad Y=\cos t.$$

Evaluation of the integral, gives us

$$\iint_{-\infty}^{+\infty}dx\,dy\exp\left\{-\frac{1}{2}(ax^2+2bxy+cy^2-2px-2qy)\right\}$$

$$=\frac{2\pi}{\sqrt{ac-b^2}}\exp\left\{\frac{aq^2-2bpq+cp^2}{2(ac-b^2)}\right\}.$$

After some simple calculations we obtain for the function $G(u,v)$:

$$G(u,v)=\sqrt{\pi}\,e^{iF(t)}e^{-\frac{w}{2}}\exp\left\{-\frac{AB}{w}uv+Au+Bv\right\}.$$

Here

$$A = i\int_0^t e^{-it'} f(t')\,dt', \quad B = e^{-it}A, \quad 2w = |A|\cdot|B| = \dot{\xi}^2 + \xi^2,$$

and $F(t)$ is some real function of time.

To expand $G(u, v)$ into a power series we use the relation

$$\exp\left\{\alpha+\beta-\frac{\alpha\beta}{w}\right\} = \sum_{m,\,n=0}^{\infty} c(m, n|w)\frac{\alpha^m}{m!}\frac{\beta^n}{n!},$$

where

$$c(m, n|w) = \sum_{l=0}^{\min(m,n)} \frac{m!\,n!}{l!(m-l)!(n-l)!}(-w)^{-l}.$$

Expanding into a series, we obtain

$$G(u, v) = \sqrt{\pi}\,e^{iF(t)}e^{-\frac{w}{2}}\sum_{m,\,n=0}^{\infty} c(m, n|w)\,\frac{(Bv)^m}{m!}\frac{(Au)^m}{n!}. \tag{3}$$

From (2) and (3) it follows that

$$G_{mn}(t, 0) = \frac{\exp\left(-\frac{1}{2}w\right)A^n B^m}{\sqrt{2^{n+m}\,m!\,n!}}\,c(m, n|w)\,e^{iF(t)},$$

and the transition probability is

$$P_{mn}(t, 0) = \frac{e^{-w}w^{m+n}}{m!\,n!}\{c(m, n|w)\}^2.$$

In the special $n = 0$ the transition probability has the form

$$P_{m0}(t, 0) = \frac{e^{-w}w^m}{m!}, \quad \text{since} \quad c(m, 0|w) = 1.$$

After the calculation of the transition probability we can find the mean value of the energy and the square of the energy of the oscillator at the time t.

The mean values are given by

$$\overline{E} = \sum_{m=0}^{\infty} P_{mn}(t, 0)\left(m+\frac{1}{2}\right),$$

$$\overline{E^2} = \sum_{m=0}^{\infty} P_{mn}(t, 0)\left(m+\frac{1}{2}\right)^2.$$

We shall first investigate an expression of a similar type:

$$\left(1-\frac{\alpha}{w}\right)^m e^{\alpha} = \Phi(m, \alpha \,|\, w).$$

It is readily shown that

$$\Phi(m, \alpha \,|\, w) = \sum_{n=0}^{\infty} \frac{\alpha^n}{n!} c(m, n \,|\, w).$$

From

$$\sum_{m=0}^{\infty} \frac{e^{-w} w^m}{m!} \Phi(m, \alpha \,|\, w)\, \Phi(m, \beta \,|\, w) = e^{\alpha\beta w^{-1}}$$

it follows that

$$\sum_{m=0}^{\infty} \frac{e^{-w} w^m}{m!} c(m, n \,|\, w)\, c(m, n' \,|\, w) = \delta_{nn'} n!\, w^{-n},$$

from which we at once obtain the physically obvious relation

$$\sum_{m=0}^{\infty} P_{mn} = 1.$$

Let us consider the relation

$$\sum_{m=0}^{\infty} m \frac{e^{-w} w^m}{m!} \Phi(m, \alpha \,|\, w)\, \Phi(m, \beta \,|\, w) = \left\{w - \alpha - \beta + \frac{\alpha\beta}{w}\right\} e^{\alpha\beta w^{-1}}.$$

We differentiate the left- and right-hand sides n times with respect to α and m times with respect to β; we then set $\alpha = 0$, $\beta = 0$ and obtain

$$\sum_{m=0}^{\infty} m P_{mn} = n + w.$$

Thus the mean energy of the oscillator at the time t is $\bar{E} = E_n + w$. Here w is the work done by the force $f(t)$ in the interval of time t:

$$w = \int_0^t f(t)\dot{\xi}\, dt = \int_0^t (\ddot{\xi} + \xi)\dot{\xi}\, dt = \frac{1}{2}(\dot{\xi}^2 + \xi^2)_{t=t} - \frac{1}{2}(\dot{\xi}^2 + \xi^2)_{t=0}.$$

Similarly, we obtain

$$\overline{E^2} = 2wE_n.$$

21. (a) $$\hat{x}(t) = x - \frac{i\hbar}{\mu} t \frac{\partial}{\partial x},$$

(b) $$\hat{x}(t) = x \cos \omega t - \frac{i\hbar}{\mu\omega} \sin \omega t \frac{\partial}{\partial x}.$$

22. $$\overline{(\Delta x)_t^2} = \overline{(\Delta x)_0^2} + \frac{t}{\mu} \overline{[(\Delta p)(\Delta x) + (\Delta x)(\Delta p)]_0} + \frac{t^2}{\mu^2} \overline{(\Delta p)_0^2}.$$

Note: From the above relation we readily find the time τ at which the quantity $\overline{(\Delta x)^2}$ is a minimum. The function $\overline{(\Delta x)_t^2}$ is symmetric with respect to τ. If at the time $t = 0$ the wave function has the form $\psi(x) = \varphi(x) \exp\left(\frac{ip_0 x}{\hbar}\right)$ ($\varphi(x)$ being a real function), then $\tau = 0$ (see Prob. 11a, §3).

23. (a) $$\overline{(\Delta x)_t^2} = \overline{(\Delta x)_{t=0}^2} + \frac{\hbar^2 t^2}{\mu^2} \int_{-\infty}^{+\infty} \left(\frac{\partial \varphi}{\partial x}\right)^2 dx,$$

(b) $$\overline{(\Delta x)_t^2} = \overline{(\Delta x)_{t=0}^2} \times \cos^2 \omega t + \frac{\hbar^2}{\mu^2 \omega^2} \int_{-\infty}^{+\infty} \left(\frac{\partial \varphi}{\partial x}\right)^2 \sin^2 \omega t \, dx.$$

§ 4. Angular momentum. Spin

1. For an infinitesimal rotation of the coordinate system, a wave function is transformed in the following manner:

$$\psi'(\mathbf{r}) = \{1 + i\,\mathbf{d\alpha}\hat{\mathbf{l}}\} \psi(\mathbf{r}). \tag{1}$$

Here $\mathbf{d\alpha}$ is a vector along the axis of rotation and is equal in magnitude to the angle of rotation, while $\mathbf{l}$ is the operator of the orbital angular momentum.

Let us first consider a rotation about the z axis by the angle $d\alpha$. For such a rotation we have

$$\psi'(r, \theta, \varphi) = \psi(r, \theta, \varphi + d\alpha) = \psi(r, \theta, \varphi) + \frac{\partial \psi}{\partial \varphi} d\alpha. \tag{2}$$

Comparing (2) with (1), we get

$$\hat{l}_z = -i \frac{\partial}{\partial \varphi}.$$

In order to obtain the operator $\hat{l}_x$ in spherical coordinates we make a rotation about the x axis. Then

$$\psi'(r,\theta,\varphi)=\psi(r,\theta+d\theta,\varphi+d\varphi)=\left\{1+\left(\frac{d\theta}{da}\frac{\partial}{\partial\theta}+\frac{d\varphi}{da}\frac{\partial}{\partial\varphi}\right)da\right\}\psi(r,\theta,\varphi),$$

from which we have

$$\hat{l}_x=-i\left(\frac{d\theta}{da}\frac{\partial}{\partial\theta}+\frac{d\varphi}{da}\frac{\partial}{\partial\varphi}\right).$$

Let us compute $\frac{d\theta}{da}$ and $\frac{d\varphi}{da}$. It is readily seen that

$$z'-z=-y\,da,$$
$$y'-y=z\,da,$$

and since $z'=r\cos\theta$, $y'=r\sin\theta\sin\varphi$, we have

$$\frac{d\theta}{da}=-\sin\varphi,\qquad \frac{d\varphi}{da}=-\cot\theta\cos\varphi.$$

Therefore,

$$\hat{l}_x=i\left(\sin\varphi\frac{\partial}{\partial\theta}+\cot\theta\cos\varphi\frac{\partial}{\partial\varphi}\right).$$

Similarly, we obtain

$$\hat{l}_y=-i\left(\cos\varphi\frac{\partial}{\partial\theta}-\cot\theta\sin\varphi\frac{\partial}{\partial\varphi}\right).$$

5.

$$\hat{l}_{z'}=\hat{l}_x\cos(xz')+\hat{l}_y\cos(yz')+\hat{l}_z\cos(zz').$$

7. The eigenfunctions are the spherical functions $Y_l^m\ (\theta,\varphi)$. Here θ and φ are the polar angle components of the momentum vector with respect to the same axes as in the coordinate representation.

8. The transformation can be written in the matrix form

$$\begin{pmatrix} e^{i(\psi+\varphi)}\cos^2\frac{\theta}{2}, & \frac{i}{\sqrt{2}}e^{i\psi}\sin\theta, & -e^{-i(\psi-\varphi)}\sin^2\frac{\theta}{2} \\ \frac{i}{\sqrt{2}}e^{i\varphi}\sin\theta, & \cos\theta, & \frac{i}{\sqrt{2}}e^{-i\varphi}\sin\theta \\ -e^{i(\psi-\varphi)}\sin^2\frac{\theta}{2}, & \frac{i}{\sqrt{2}}e^{-i\psi}\sin\theta, & e^{-i(\psi+\varphi)}\cos^2\frac{\theta}{2} \end{pmatrix}\begin{pmatrix} Y_{11} \\ Y_{10} \\ Y_{1,-1} \end{pmatrix}=\begin{pmatrix} Y'_{11} \\ Y'_{10} \\ Y'_{1,-1} \end{pmatrix}.$$

9. Using the result obtained in the preceding problem for $M = 1$, we have

$$w(+1) = \cos^4\frac{\theta}{2}, \quad w(0) = \frac{1}{2}\sin^2\theta, \quad w(-1) = \sin^4\frac{\theta}{2};$$

for $M = 0$

$$w(+1) = \frac{1}{2}\sin^2\theta, \quad w(0) = \cos^2\theta, \quad w(-1) = \frac{1}{2}\sin^2\theta;$$

finally, for $M = -1$

$$w(+1) = \sin^4\frac{\theta}{2}, \quad w(0) = \frac{1}{2}\sin^2\theta, \quad w(-1) = \cos^4\frac{\theta}{2}.$$

10.

$$w\left(+\frac{1}{2}\right) = \cos^2\frac{\theta}{2}, \quad w\left(-\frac{1}{2}\right) = \sin^2\frac{\theta}{2}.$$

The mean value of the spin component is $\frac{1}{2}\cos\theta$.

11. We make use of the matrix transformation of the components of the spin function for a rotation of the coordinate axes. This matrix has the form

$$\begin{pmatrix} e^{\frac{1}{2}i(\varphi+\psi)}\cos\frac{\theta}{2}, & ie^{\frac{1}{2}i(\varphi-\psi)}\sin\frac{\theta}{2} \\ ie^{-\frac{1}{2}i(\varphi-\psi)}\sin\frac{\theta}{2}, & e^{-\frac{1}{2}i(\varphi+\psi)}\cos\frac{\theta}{2} \end{pmatrix}.$$

With the aid of this matrix we find the spin function in the new coordinate system

$$\psi_1' = e^{\frac{1}{2}i(\varphi+\psi)+i\alpha}\cos\frac{\theta}{2}\cos\delta + ie^{\frac{1}{2}i(\varphi-\psi)+i\beta}\sin\frac{\theta}{2}\sin\delta,$$

$$\psi_2' = ie^{-\frac{1}{2}i(\varphi-\psi)+i\alpha}\sin\frac{\theta}{2}\cos\delta + e^{-\frac{1}{2}i(\varphi+\psi)+i\beta}\cos\frac{\theta}{2}\sin\delta.$$

We find the probability that the spin is directed along axis z':

$$w_1 = \psi_1'^*\psi_1' = \cos^2\frac{\theta}{2}\cos^2\delta + \sin^2\frac{\theta}{2}\sin^2\delta + \frac{1}{2}\sin\theta\sin 2\delta\cdot\sin(\psi+\alpha-\beta).$$

From this formula it follows that the probability of a given value of the spin component in an arbitrary direction depends only on the difference $\alpha - \beta$ and not on α and β separately.

12. The direction of the spin is given by the angles

$$\theta = 2\delta, \qquad \Phi = \frac{\pi}{2} + \beta - \alpha.$$

13. Yes, we can. In the case of a mixed ensemble, irrespective of the direction of the inhomogeneous magnetic field, the beam would always split into two parts. In the case of a pure ensemble one of the two beams can be made to vanish by an adjustment of the orientation of the instrument.

17. For an infitesimal rotation about the x axis by an angle $d\alpha$ the components of the spin function vary in accordance with the relation

$$(1+i\,d\alpha\,\hat{s}_x)\begin{pmatrix}\psi_1\\ \psi_0\\ \psi_{-1}\end{pmatrix} = \begin{pmatrix}\psi_1'\\ \psi_0'\\ \psi_{-1}'\end{pmatrix}, \tag{1}$$

where

$$\hat{s}_x = \frac{1}{\sqrt{2}}\begin{pmatrix}0 & 1 & 0\\ 1 & 0 & 1\\ 0 & 1 & 0\end{pmatrix}.$$

Equation (1) is equivalent to the three differential equations

$$\frac{d\psi_1}{d\alpha} = \frac{i}{\sqrt{2}}\,\psi_0,$$

$$\frac{d\psi_0}{d\alpha} = \frac{i}{2}\,(\psi_1+\psi_{-1}),$$

$$\frac{d\psi_{-1}}{d\alpha} = \frac{i}{\sqrt{2}}\,\psi_0.$$

The solution of these equations readily gives us a transformation matrix which has the form

$$\begin{pmatrix} \cos^2\dfrac{\alpha}{2} & \dfrac{i}{\sqrt{2}}\sin\alpha & -\sin^2\dfrac{\alpha}{2}\\[2ex] \dfrac{i}{\sqrt{2}}\sin\alpha & \cos\alpha & \dfrac{i}{\sqrt{2}}\sin\alpha\\[2ex] -\sin^2\dfrac{\alpha}{2} & \dfrac{i}{\sqrt{2}}\sin\alpha & \cos^2\dfrac{\alpha}{2} \end{pmatrix}$$

Similarly, we obtain the transformation matrix for the rotation about the z axis by an angle α:

$$\begin{pmatrix} e^{i\alpha} & 0 & 0 \\ 0 & 1 & 0 \\ 0 & 0 & e^{-i\alpha} \end{pmatrix}.$$

We take the Euler angles ψ, θ and φ to characterize the rotation. Then, in order to find the matrix for the transformation being sought, we must multiply the three matrices together. Computing the product, we have

$$\begin{pmatrix} e^{i(\psi+\varphi)}\cos^2\dfrac{\theta}{2} & \dfrac{i}{\sqrt{2}}e^{i\varphi}\sin\theta & -e^{-i(\psi-\varphi)}\sin^2\dfrac{\theta}{2} \\ \dfrac{i}{\sqrt{2}}e^{i\psi}\sin\theta & \cos\theta & \dfrac{i}{\sqrt{2}}e^{-i\psi}\sin\theta \\ -e^{i(\psi-\varphi)}\sin^2\dfrac{\theta}{2} & \dfrac{i}{\sqrt{2}}e^{-i\varphi}\sin\theta & e^{-i(\psi+\varphi)}\cos^2\dfrac{\theta}{2} \end{pmatrix}. \tag{2}$$

Note that this same result may be obtained by considering the transformation of the symmetric spinor of rank two. The components of the spin function and the components of the symmetric spinor are related by

$$\psi^{11} = \psi_1, \qquad \psi^{12} = \psi^{21} = \frac{1}{\sqrt{2}}\psi_0, \qquad \psi^{22} = \psi_{-1}. \tag{3}$$

A spinor of rank two transforms like the product of two spinors of rank one, i.e.,

$$\begin{aligned} \psi'^{11} &= \alpha^2\psi^{11} + 2\alpha\beta\psi^{12} + \beta^2\psi^{22}, \\ \psi'^{12} &= \alpha\gamma\psi^{11} + (\alpha\delta + \beta\gamma)\psi^{12} + \beta\delta\psi^{22}, \\ \psi'^{22} &= \gamma^2\psi^{11} + 2\gamma\delta\psi^{12} + \delta^2\psi^{22}. \end{aligned}$$

Or, by replacing the spinor components by the components of the spin function according to (3), we get

$$\begin{aligned} \psi_1' &= \alpha^2\psi_1 + \sqrt{2}\alpha\beta\psi_0 + \beta^2\psi_{-1}, \\ \psi_0' &= \sqrt{2}\alpha\gamma\psi_1 + (\alpha\delta + \beta\gamma)\psi_0 + \sqrt{2}\beta\delta\psi_{-1}, \\ \psi_{-1}' &= \gamma^2\psi_1 + \sqrt{2}\gamma\delta\psi_0 + \delta^2\psi_{-1}. \end{aligned}$$

Inserting in these relations the values of the coefficients

$$\alpha = e^{\frac{1}{2}i(\varphi+\psi)}\cos\frac{\theta}{2}, \qquad \beta = i\sin\frac{\theta}{2}e^{\frac{1}{2}i(\varphi-\psi)},$$

$$\gamma = i\sin\frac{\theta}{2}e^{-\frac{1}{2}i(\varphi-\psi)}, \qquad \delta = e^{-\frac{1}{2}i(\varphi+\psi)}\cos\frac{\theta}{2},$$

we obtain (2).

18. In order to find the probabilities we employ a formal method in which, instead of particles of angular momentum j, we consider a system consisting of $2j$ particles of spin $\frac{1}{2}$. In accordance with the conditions of the problem, the component of the angular momentum of the particle is equal to j; therefore, in an equivalent system of $2j$ particles all of the particles have a spin component of $+\frac{1}{2}$ along the z axis. The probability of a spin component of $+\frac{1}{2}$ (or $-\frac{1}{2}$) along the z' axis for each such particle is equal to $\cos^2\frac{\theta}{2}$ $\left(\text{or } \sin^2\frac{\theta}{2}\right)$ (see Prob. 10, §4). In order for the component of the total angular momentum of these particles along the z' axis to be equal to m, the $j+m$ particles must have a component of $+\frac{1}{2}$ along the z axis and the remaining j-m particles, $-\frac{1}{2}$. We obtain the probability we are seeking $w(m)$ by multiplying $\left(\cos^2\frac{\theta}{2}\right)^{j+m}\left(\sin^2\frac{\theta}{2}\right)^{j-m}$ by the number of ways in which $2j$ particles can be divided into two such groups, i.e., by $\frac{(2j)!}{(j+m)!\,(j-m)!}$. Therefore

$$w(m) = \frac{(2j)!}{(j+m)!\,(j-m)!}\left(\cos^2\frac{\theta}{2}\right)^{j+m}\left(\sin^2\frac{\theta}{2}\right)^{j-m}.$$

It is readily seen that $\sum_{-j}^{+j} w(m) = 1$.

19. The state of a system with angular momentum J will be described by a symmetric spinor of rank $2J$. To solve the problem we should establish the relation between the components

$$\psi^{\overbrace{11\ldots1}^{J+M}\overbrace{22\ldots2}^{J-M}}, \quad \psi'^{\overbrace{11\ldots1}^{J+M'}\overbrace{22\ldots2}^{J-M'}}.$$

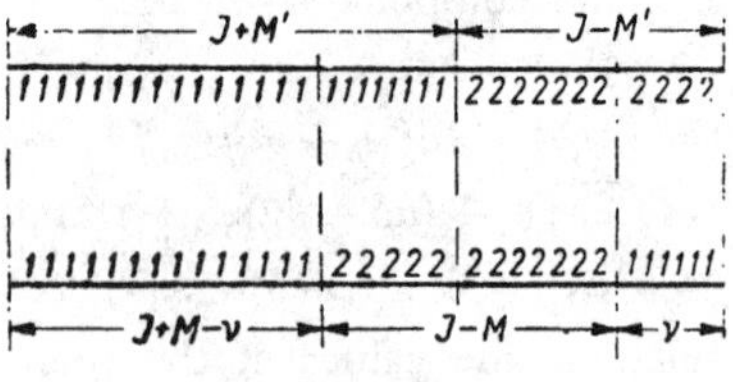

Fig. 23

From Fig. 23 it is readily established that

$$\psi'^{\overbrace{11\ldots1}^{J+M'}\overbrace{22\ldots2}^{J-M'}} = \frac{(J+M')!\,(J-M')!}{(2J)!}\times$$

$$\times \sum_{\nu=0} (2J)! \frac{(\gamma)^{\nu}(\beta)^{M'-M+\nu}(\alpha)^{J+M-\nu}(\delta)^{J-M'-\nu}}{\nu!\,(M'-M+\nu)!\,(J+M-\nu)!\,(J-M'-\nu)!}\,\psi_{\overbrace{11\ldots1}^{J+M}\overbrace{22\ldots2}^{J-M}},$$

where α, β, γ, δ are the Cayley-Klein parameters.

Since

$$\psi_{\overbrace{11\ldots1}^{J+M}\overbrace{22\ldots2}^{J-M}} = \sqrt{\frac{(J+M)!\,(J-M)!}{(2J)!}}\,\psi(M),$$

$$\psi'_{\overbrace{11\ldots1}^{J+M'}\overbrace{22\ldots2}^{J-M'}} = \sqrt{\frac{(J+M')!\,(J-M')!}{(2J)!}}\,\psi'(M'),$$

and since we were given the condition that $\psi(M) = 1$, then

$$\psi'(M') = \sqrt{(J+M')!\,(J-M')!\,(J+M)!\,(J-M)!}\times$$

$$\times \sum_{\nu=0} \frac{(\gamma)^{\nu}(\beta)^{M'-M+\nu}(\alpha)^{J+M-\nu}(\delta)^{J-M'-\nu}}{\nu!\,(M'-M+\nu)!\,(J+M-\nu)!\,(J-M'-\nu)!}.$$

It thus follows that

$$P(M,M') = (J+M')!\,(J-M')!\,(J+M)!\,(J-M)!\left(\cos\frac{\vartheta}{2}\right)^{4J}\times$$

$$\times\left\{\sum_{\nu=0} \frac{(-1)^{\nu}\left[\tan\left(\frac{1}{2}\vartheta\right)\right]^{2\nu-M+M'}}{\nu!\,(M'-M+\nu)!\,(J+M-\nu)!\,(J-M'-\nu)!}\right\}^2.$$

When the value unity divided by a negative factorial occurs in the sum, the term should be set equal to zero, i.e., the summation over ν should include only those terms satisfying the inequalities

$$\nu \geqslant M-M',$$
$$\nu \leqslant J+M,$$
$$\nu \leqslant J-M'.$$

21. Let us first find the eigenfunctions of the operator $\hat{j}_z$. We write the operator $\hat{j}_z$ in matrix form:

$$\hat{j}_z = \begin{pmatrix} \hat{l}_z + \frac{1}{2} & 0 \\ 0 & \hat{l}_z - \frac{1}{2} \end{pmatrix}.$$

Since $\hat{l}_z = -i\frac{\partial}{\partial\varphi}$, the equation determining the eigenfunctions and the eigenvalues of $\hat{j}_z$ has the form

$$\begin{pmatrix} -i\frac{\partial}{\partial\varphi}+\frac{1}{2}, & 0 \\ 0 & -i\frac{\partial}{\partial\varphi}-\frac{1}{2} \end{pmatrix}\begin{pmatrix}\psi_1\\ \psi_2\end{pmatrix}=m\begin{pmatrix}\psi_1\\ \psi_2\end{pmatrix},$$

or

$$-i\frac{\partial\psi_1}{\partial\varphi}+\frac{1}{2}\psi_1=m\psi_1,$$

$$-i\frac{\partial\psi_2}{\partial\varphi}-\frac{1}{2}\psi_2=m\psi_2.$$

It follows from this that

$$\psi_1=f_1(r,\vartheta)\exp\left[i\left(m-\frac{1}{2}\right)\varphi\right],\qquad \psi_2=f_2(r,\vartheta)\exp\left[i\left(m+\frac{1}{2}\right)\varphi\right],$$

where f_1 and f_2 are any functions of r and ϑ, m being a half-integrer.

From all the possible functions of the form

$$\begin{pmatrix} f_1(r,\vartheta)\exp\left[i\left(m-\frac{1}{2}\right)\varphi\right] \\ f_2(r,\vartheta)\exp\left[i\left(m+\frac{1}{2}\right)\varphi\right]\end{pmatrix}$$

we should select those which are at the same time the eigenfunctions of the operator $\hat{\mathbf{l}}^2$. Such eigenfunctions are

$$\begin{pmatrix}\psi_1\\ \psi_2\end{pmatrix}=\begin{pmatrix}R_1(r)\,Y_{l,\,m-\frac{1}{2}}(\vartheta,\varphi)\\ R_2(r)\,Y_{l,\,m+\frac{1}{2}}(\vartheta,\varphi)\end{pmatrix}.$$

The last step consists of choosing R_1 and R_2 so that the function is an eigenfunction of the operator of the square of the total angular momentum. Therefore let us write the equation $\hat{\mathbf{j}}^2\psi=j(j+1)\psi$ in the matrix form

$$\begin{pmatrix}\hat{\mathbf{l}}^2+\frac{3}{4}+\hat{l}_z & \hat{l}_x-i\hat{l}_y \\ \hat{l}_x+i\hat{l}_y & \hat{\mathbf{l}}^2+\frac{3}{4}-\hat{l}_z\end{pmatrix}\begin{pmatrix}R_1(r)\,Y_{l,\,m-\frac{1}{2}}(\vartheta,\varphi)\\ R_2(r)\,Y_{l,\,m+\frac{1}{2}}(\vartheta,\varphi)\end{pmatrix}$$

$$=j(j+1)\begin{pmatrix}R_1(r)\,Y_{l,\,m-\frac{1}{2}}(\vartheta,\varphi)\\ R_2(r)\,Y_{l,\,m+\frac{1}{2}}(\vartheta,\varphi)\end{pmatrix}.$$

Let us take into account the properties of spherical functions:

$$(\hat{l}_x+i\hat{l}_y)\,Y_{lm}=\sqrt{(l+m+1)(l-m)}\,Y_{l,m+1},$$

$$(\hat{l}_x-i\hat{l}_y)\,Y_{lm}=\sqrt{(l-m+1)(l+m)}\,Y_{l,m-1}.$$

Then, from the matrix relation it appears that R_1 and R_2 should satisfy two homogeneous equations

$$\left[l(l+1)-j(j+1)+m+\frac{1}{4}\right]R_1+\sqrt{\left(l+\frac{1}{2}\right)^2-m^2}\,R_2=0,$$

$$\sqrt{\left(l+\frac{1}{2}\right)^2-m^2}\,R_1+\left[l(l+1)-j(j+1)-m+\frac{1}{4}\right]R_2=0.$$

From the condition of the compatability of these equations, it follows that j can be equal to either $l+\frac{1}{2}$ or $j=l-\frac{1}{2}$. Setting $j=l+\frac{1}{2}$, we get

$$R_1=\sqrt{l+\frac{1}{2}+m}\,R(r),\qquad R_2=\sqrt{l-m+\frac{1}{2}}\,R(r).$$

Therefore,

$$\psi\left(l,j=l+\frac{1}{2},\,m\right)=R(r)\begin{Bmatrix}\sqrt{\dfrac{l+m+\frac{1}{2}}{2l+1}}\,Y_{l,\,m-\frac{1}{2}}\\[2ex]\sqrt{\dfrac{l-m+\frac{1}{2}}{2l+1}}\,Y_{l,\,m+\frac{1}{2}}\end{Bmatrix},$$

$$(l=0,1,2,\ldots);$$

similarly, for $j=l-\frac{1}{2}$ we have

$$\psi\left(l,j=l-\frac{1}{2},\,m\right)=R(r)\begin{Bmatrix}\sqrt{\dfrac{l-m+\frac{1}{2}}{2l+1}}\,Y_{l,\,m-\frac{1}{2}}\\[2ex]-\sqrt{\dfrac{l+m+\frac{1}{2}}{2l+1}}\,Y_{l,\,m+\frac{1}{2}}\end{Bmatrix},$$

$$(l=0,\ 1,\ 2,\ldots).$$

The factor $\frac{1}{\sqrt{2l+1}}$ was added to normalize the function.

22.

	Probability	
	$j=l+\frac{1}{2}$	$j=l-\frac{1}{2}$
Orbital angular momentum component $m-\frac{1}{2}$ Spin component $\frac{1}{2}$	$\frac{l+m+\frac{1}{2}}{2l+1}$	$\frac{l-m+\frac{1}{2}}{2l+1}$
Orbital angular momentum component $m+\frac{1}{2}$ Spin component $-\frac{1}{2}$	$\frac{l-m+\frac{1}{2}}{2l+1}$	$\frac{l+m+\frac{1}{2}}{2l+1}$

$$\bar{l}_z\left(j=l+\frac{1}{2}\right)=\frac{2ml}{2l+1},\quad \bar{s}_z\left(j=l+\frac{1}{2}\right)=\frac{m}{2l+1},$$

$$\bar{l}_z\left(j=l-\frac{1}{2}\right)=\frac{2m(l+1)}{2l+1},\quad \bar{s}_z\left(j=l-\frac{1}{2}\right)=-\frac{m}{2l+1}.$$

23. The eigenfunctions of the operator of the spin component along the direction Θ, Φ can be found from the relation

$$(\sigma_x \sin\Theta\cos\Phi+\sigma_y\sin\Theta\sin\Phi+\sigma_z\cos\Theta)\begin{pmatrix}\alpha\\ \beta\end{pmatrix}=\begin{pmatrix}\alpha\\ \beta\end{pmatrix},$$

from which it follows that

$$\alpha\sin\Theta e^{i\Phi}-\beta\cos\Theta=\beta. \tag{1}$$

From (1) we find the ratio α/β:

$$\frac{\alpha}{\beta}=\cot\frac{\Theta}{2}e^{-i\Phi}. \tag{2}$$

On the other hand, from the explicit form of the functions $\psi(l, j=l+1/2, m)$ and $\psi(l, j=l-1/2, m)$ we find

$$\frac{\alpha}{\beta}=c_j\frac{Y_{l,m-\frac{1}{2}}(\vartheta,\varphi)}{Y_{l,m+\frac{1}{2}}(\vartheta,\varphi)}=c_j\frac{P_l^{m-\frac{1}{2}}(\cos\vartheta)}{P_l^{m+\frac{1}{2}}(\cos\vartheta)}e^{-i\varphi}, \qquad (3)$$

where

$$c_j=\sqrt{\frac{l+m+\frac{1}{2}}{l-m+\frac{1}{2}}} \quad \text{for} \quad j=l+\frac{1}{2},$$

$$c_j=-\sqrt{\frac{l-m+\frac{1}{2}}{l+m+\frac{1}{2}}} \quad \text{for} \quad j=l-\frac{1}{2}.$$

Setting (2) equal to (3) we find that $\Phi=\varphi$, i.e., the direction of the spin at a given point in space lies in a plane passing through the z axis and the given point. The angle Θ is determined from the condition

$$\cot\frac{\Theta}{2}=c_j\frac{P_l^{m-\frac{1}{2}}(\cos\vartheta)}{P_l^{m+\frac{1}{2}}(\cos\vartheta)}.$$

24. The operator of the square of the total spin is

$$\hat{\mathbf{S}}^2=\hat{\mathbf{s}}_1^2+\hat{\mathbf{s}}_2^2+2\hat{\mathbf{s}}_1\hat{\mathbf{s}}_2=\frac{3}{2}\begin{pmatrix}1&0\\0&1\end{pmatrix}_1\begin{pmatrix}1&0\\0&1\end{pmatrix}_2+$$

$$+\frac{1}{2}\left\{\begin{pmatrix}0&1\\1&0\end{pmatrix}_1\begin{pmatrix}0&1\\1&0\end{pmatrix}_2+\begin{pmatrix}0&-i\\i&0\end{pmatrix}_1\begin{pmatrix}0&-i\\i&0\end{pmatrix}_2+\begin{pmatrix}1&0\\0&-1\end{pmatrix}_1\begin{pmatrix}1&0\\0&-1\end{pmatrix}_2\right\};$$

the indices 1, 2 indicate the number of the particle.

We shall find the eigenfunctions of this operator starting with the case for which the projection of the total spin is zero:

$$\hat{\mathbf{S}}^2\left\{a\begin{pmatrix}1\\0\end{pmatrix}_1\begin{pmatrix}0\\1\end{pmatrix}_2+b\begin{pmatrix}0\\1\end{pmatrix}_1\begin{pmatrix}1\\0\end{pmatrix}_2\right\}=\lambda\left\{a\begin{pmatrix}1\\0\end{pmatrix}_1\begin{pmatrix}0\\1\end{pmatrix}_2+b\begin{pmatrix}0\\1\end{pmatrix}_1\begin{pmatrix}1\\0\end{pmatrix}_2\right\}.$$

Hence we have

$$(\lambda-1)a-b=0, \qquad -a+(\lambda-1)b=0.$$

Two values of λ are thus obtained: $\lambda=2$ and $\lambda=0$.

For $\lambda = 2$, we have $a = b$, and for $\lambda = 0$, we have $a = -b$. Taking into account the normalization condition $a^2 + b^2 = 1$, we obtain the eigenfunctions in the form

$$\frac{1}{\sqrt{2}}\left\{\begin{pmatrix}1\\0\end{pmatrix}_1\begin{pmatrix}0\\1\end{pmatrix}_2+\begin{pmatrix}0\\1\end{pmatrix}_1\begin{pmatrix}1\\0\end{pmatrix}_2\right\} \quad \lambda = 2 \quad (S = 1),$$

$$\frac{1}{\sqrt{2}}\left\{\begin{pmatrix}1\\0\end{pmatrix}_1\begin{pmatrix}0\\1\end{pmatrix}_2-\begin{pmatrix}0\\1\end{pmatrix}_1\begin{pmatrix}1\\0\end{pmatrix}_2\right\} \quad \lambda = 0 \quad (S = 0).$$

It may readily be shown that the functions $\begin{pmatrix}1\\0\end{pmatrix}_1\begin{pmatrix}1\\0\end{pmatrix}_2$ and $\begin{pmatrix}0\\1\end{pmatrix}_1\begin{pmatrix}0\\1\end{pmatrix}_2$ will also be eigenfunctions of the operator $\hat{\mathbf{S}}^2$ with spin components along the z axis of 1 and -1, respectively. These functions will also be eigenfunctions of the operator $\hat{\mathbf{s}}_1\hat{\mathbf{s}}_2$.

25. The wave function of the system $\Psi(J, M)$ is the sum of the products of the wave functions of the individual particles

$$\Psi(J, M) = c_1\psi_1^{(1)}\psi_{M-1}^{(2)} + c_0\psi_0^{(1)}\psi_M^{(2)} + c_{-1}\psi_{-1}^{(1)}\psi_{M+1}^{(2)}.$$

Here, the lower index for the wave functions denotes the value of the magnetic quantum numbers.

The coefficients c_i should be determined from the condition

$$\hat{\mathbf{J}}^2\Psi(J, M) = J(J+1)\Psi(J, M). \tag{1}$$

It is more convenient to write the wave functions for the first particle in the form

$$\psi_1^{(1)} = \begin{pmatrix}1\\0\\0\end{pmatrix}, \quad \psi_0^{(1)} = \begin{pmatrix}0\\1\\0\end{pmatrix}, \quad \psi_{-1}^{(1)} = \begin{pmatrix}0\\0\\1\end{pmatrix}.$$

For this form of the wave functions the operator $\hat{\mathbf{l}}_1$ will be a matrix with three rows and three columns

$$\hat{l}_{1x} = \frac{1}{\sqrt{2}}\begin{pmatrix}0&1&0\\1&0&1\\0&1&0\end{pmatrix}, \quad \hat{l}_{1y} = \frac{1}{\sqrt{2}}\begin{pmatrix}0&-i&0\\i&0&-i\\0&i&0\end{pmatrix}, \quad \hat{l}_{1z} = \begin{pmatrix}1&0&0\\0&0&0\\0&0&-1\end{pmatrix}.$$

And for the operator $\hat{\mathbf{J}}^2$ we obtain

$$\hat{\mathbf{J}}^2 = \hat{\mathbf{l}}_1^2 + \hat{\mathbf{l}}_2^2 + 2\hat{\mathbf{l}}_1\hat{\mathbf{l}}_2$$

$$= \begin{pmatrix} l(l+1)+2+2\hat{l}_{2z} & \sqrt{2}\,\hat{l}_{2-} & 0 \\ \sqrt{2}\,\hat{l}_{2+} & l(l+1)+2 & \sqrt{2}\,\hat{l}_{2-} \\ 0 & \sqrt{2}\,\hat{l}_{2+} & l(l+1)+2-2\hat{l}_{2z} \end{pmatrix},$$

where

$$\hat{l}_+ = \hat{l}_x + i\hat{l}_y, \quad \hat{l}_- = \hat{l}_x - i\hat{l}_y.$$

Making use of the properties of the operators $\hat{l}_+$ $\hat{l}_-$

$$\hat{l}_+\psi_m = \sqrt{(l+m+1)(l-m)}\,\psi_{m+1},$$
$$\hat{l}_-\psi_m = \sqrt{(l+m)(l-m+1)}\,\psi_{m-1},$$

we find that condition (1) gives two equations

$$[J(J+1)-l(l+1)-2M]\,c_1 = \sqrt{2}\sqrt{(l+M)(l-M+1)}\,c_0,$$
$$[J(J+1)-l(l+1)+2M]\,c_{-1} = \sqrt{2}\sqrt{(l+M+1)(l-M)}\,c_0$$

(the third equation is satisfied identically).

Solving these equations we get for $c(J, M)$

$$\begin{pmatrix} c_1(l+1, M) & c_0(l+1, M) & c_{-1}(l+1, M) \\ c_1(l, M) & c_0(l, M) & c_{-1}(l, M) \\ c_1(l-1, M) & c_0(l-1, M) & c_{-1}(l-1, M) \end{pmatrix}$$

$$= \begin{vmatrix} \sqrt{\frac{(l+M)(l+M+1)}{2(2l+1)(l+1)}} & \sqrt{\frac{(l+M+1)(l-M+1)}{(2l+1)(l+1)}} & \sqrt{\frac{(l-M)(l-M+1)}{2(2l+1)(l+1)}} \\ -\sqrt{\frac{(l+M)(l-M+1)}{2l(l+1)}} & \frac{M}{\sqrt{l(l+1)}} & \sqrt{\frac{(l+M+1)(l-M)}{2l(l+1)}} \\ \sqrt{\frac{(l-M)(l-M+1)}{2l(2l+1)}} & -\sqrt{\frac{(l+M)(l-M)}{l(2l+1)}} & \sqrt{\frac{(l+M)(l+M+1)}{2l(2l+1)}} \end{vmatrix}.$$

Owing to the orthogonality of this matrix its inverse matrix is identical with its transposed matrix; therefore each of the functions $\psi_1^{(1)}\psi_{M-1}^{(2)}$, $\psi_0^{(1)}\psi_M^{(2)}$, $\psi_{-1}^{(1)}\psi_{M+1}^{(2)}$ is expressed as a linear combination of the functions $\Psi(l-1, M)$, $\Psi(l, M)$, $\Psi(l-1, M)$ with the coefficients appearing in the columns of this matrix.

26. The wave function of the system corresponding to the angular momentum $J = j_1 + j_2$ and projection $M = J$ has the form

$$\Psi^{M=J}_{J=j_1+j_2} = \psi^{j_1}_{j_1}(1)\,\psi^{j_2}_{j_2}(2).$$

(The superscripts indicate the value of the projection of the angular momentum.)

Applying the operator $\hat{J}_-$, we obtain

$$\hat{J}\Psi^{M=J}_{J=j_1+j_2} = \sqrt{J(J+1)-M(M-1)}\,\Psi^{J-1}_J$$
$$= \sqrt{j_1(j_1+1)-j_1(j_1-1)}\,\psi^{j_1-1}_{j_1}(1)\psi^{j_2}_{j_2}(2) +$$
$$+ \sqrt{j_2(j_2+1)-j_2(j_2-1)}\,\psi^{j_1}_{j_1}(1)\psi^{j_2-1}_{j_2}(2).$$

Finally we have

$$\Psi_J^{J-1} = \frac{1}{\sqrt{j_1+j_2}}\{\sqrt{j_1}\,\psi_{j_1}^{j_1-1}(1)\,\psi_{j_2}^{j_2}(2) + \sqrt{j_2}\,\psi_{j_1}^{j_1}(1)\,\psi_{j_2}^{j_2-1}(2)\}.$$

From the condition of orthonormality of the functions ψ_J^{J-1} and ψ_{J-1}^{J-1}, we find that

$$\Psi_{J-1}^{J-1} = \frac{1}{\sqrt{j_1+j_2}}\{\sqrt{j_1}\,\psi_{j_1}^{j_1-1}(1)\,\psi_{j_2}^{j_2}(2) - \sqrt{j_2}\,\psi_{j_1}^{j_1}(1)\,\psi_{j_2}^{j_2-1}(2)\}.$$

32. The operator of the magnetic dipole-dipole interaction has the form

$$\hat{V} = \mu^2 r^{-3}\{(\hat{\sigma}_1\hat{\sigma}_2) - 3r^{-2}(\hat{\sigma}_1\mathbf{r})(\hat{\sigma}_2\mathbf{r})\}.$$

Using the result of Prob. 26, §4, we obtain:
for the singlet state

$$\psi_0^0 = \frac{1}{\sqrt{2}}\left\{\begin{pmatrix}1\\0\end{pmatrix}_1\begin{pmatrix}0\\1\end{pmatrix}_2 - \begin{pmatrix}0\\1\end{pmatrix}_1\begin{pmatrix}1\\0\end{pmatrix}_2\right\},$$

$$E_{00} = 0$$

for the triplet state

$$\psi_1^1 = \begin{pmatrix}1\\0\end{pmatrix}_1\begin{pmatrix}1\\0\end{pmatrix}_2 \quad \text{or} \quad \psi_1^{-1} = \begin{pmatrix}0\\1\end{pmatrix}_1\begin{pmatrix}0\\1\end{pmatrix}_2,$$

$$E_{1,1} = E_{1,-1} = -2\mu^2 a^{-3},$$

$$\psi_1^0 = \frac{1}{\sqrt{2}}\left\{\begin{pmatrix}1\\0\end{pmatrix}_1\begin{pmatrix}0\\1\end{pmatrix}_2 + \begin{pmatrix}0\\1\end{pmatrix}_1\begin{pmatrix}1\\0\end{pmatrix}_2\right\} \qquad E_{1,0} = 4\mu^2 a^{-3}.$$

Here μ is the magnetic moment of the proton and the z axis is the line joining the protons.

33. $$\tau = \frac{\pi\hbar}{4\mu^2}a^3.$$

35. $\bar{\mu}_x = \bar{\mu}_y = 0,$

$$\bar{\mu}_z = M_J\left\{\frac{1}{2}(g_1+g_2) + \frac{1}{2}(g_1-g_2)\frac{J_1(J_1+1) - J_2(J_2+1)}{J(J+1)}\right\}.$$

36. $-0{\cdot}24$.
37. $-1{\cdot}91$.
38. (a) 0·879, (b) 0·5, (c) 0·689, (d) 0·310.
39. The weight of the D wave is 0·04.

40. (a) $-\frac{1}{10}\overline{r^2}$; (b) $\frac{1}{20}\overline{r^2}$. Note the difference in the sign of the quadrupole moment in the $^1P_1$ and $^3P_1$ states.

41. The quadrupole moment is defined by the following expression:

$$Q_0 = Q_j^{m_j=j} = \int (3\cos^2\theta - 1) r^2 |\Psi_j^{m_j=j}|^2 d\tau.$$

In the case $j = l + \frac{1}{2}$ the wave function has the form

$$\Psi_{j=l+\frac{1}{2}}^{m_j=j} = R(r)\, Y_l^l(\theta,\varphi) \begin{pmatrix} 1 \\ 0 \end{pmatrix}$$

and in the case $j = l - \frac{1}{2}$

$$\Psi_{j=l-\frac{1}{2}}^{m_j=j} = R(r) \frac{1}{\sqrt{2l+1}} \begin{pmatrix} Y_l^{l-1}(\theta,\varphi) \\ -\sqrt{2l}\, Y_l^l(\theta,\varphi) \end{pmatrix}$$

Taking the average, we obtain

$$Q_0 = -<r^2> \frac{2j-1}{2j+2}$$

The quadrupole moment depends only on j, and not on l. If the magnetic quantum number $m_j \neq j$, then

$$Q_j^{m_j} = Q_0 \frac{3m_j^2 - j(j+1)}{j(2j-1)}$$

42. To determine the quadrupole moment it is necessary to take the average of the operator

$$\hat{Q} = \frac{3Q_0}{j(2j-1)} \{\hat{j}_{1z}^2 + \hat{j}_{2z}^2 + \hat{j}_{3z}^2 - j(j+1)\};$$

here $\hat{j}_{1z}, \hat{j}_{2z}, \hat{j}_{3z}$ are the operators of the projection of the angular momenta of the first, second, and third particles, and Q is the quadrupole moment produced by the individual protons.

The wave function was found during the solution of Prob. 28.

Taking the averages, we obtain

$$<\hat{Q}> = \frac{3Q_0}{j(2j-1)} \left\{ -j(j+1) + j^2 + \frac{2}{2j-1} \cdot 2 \sum_{\frac{1}{2}}^{j-1} m_j^2 \right\}.$$

Since

$$\sum_{\frac{1}{2}}^{j-1} m_j^2 = \frac{j(2j-1)(j-1)}{6}$$

then we finally obtain

$$<\hat{Q}> = Q_0\left\{1-\frac{4}{2j-1}\right\}.$$

43. Let us take a coordinate system in which the mixed derivatives of the electrostatic potential φ at the position of the nucleus is zero $\left(\frac{\partial^2\varphi}{\partial x\partial y}=\frac{\partial^2\varphi}{\partial y\partial z}=\frac{\partial^2\varphi}{\partial z\partial x}=0\right)$. This can always be done by a suitable rotation of the coordinate axes.

With such a choice for the coordinate system the operator of the quadrupole energy $\hat{H}$ has the form

$$\hat{H} = \frac{Q_0}{2I(2I-1)}\left\{\frac{\partial^2\varphi}{\partial x^2}\hat{I}_x^2+\frac{\partial^2\varphi}{\partial y^2}\hat{I}_y^2+\frac{\partial^2\varphi}{\partial z^2}\hat{I}_z^2\right\}$$

(This is the same form as the Hamilton operator for an asymmetric top; see Prob. 21, §8.)

The operator $\hat{H}$ can be put in a more convenient form for calculating the matrix elements if in place of $\hat{I}_x$ and $\hat{I}_y$ we introduce the operators $\hat{I}_+ = \hat{I}_x + i\hat{I}_y$, $\hat{I}_- = \hat{I}_x - i\hat{I}_y$. Replacing $\hat{I}_x$ and $\hat{I}_y$ by $\hat{I}_+$, and $\hat{I}_-$, we obtain

$$\hat{H} = \frac{Q_0}{2I(2I-1)}\left\{\frac{1}{2}\frac{\partial^2\varphi}{\partial z^2}\left(3\hat{I}_z^2-\mathbf{I}^2\right)+\frac{1}{4}\left(\frac{\partial^2\varphi}{\partial x^2}-\frac{\partial^2\varphi}{\partial y^2}\right)\left(\hat{I}_+^2+\hat{I}_-^2\right)\right\}.$$

The matrix elements of $\hat{H}$ different from zero have the form

$$\left(m\,|\hat{H}|\,m\right)=\frac{A}{2}\frac{\partial^2\varphi}{\partial z^2}[3m^2-I(I+1)]\Big),$$

$$(m\,|\hat{H}|\,m\pm2)=\frac{A}{4}\left(\frac{\partial^2\varphi}{\partial x^2}-\frac{\partial^2\varphi}{\partial y^2}\right)[(I\mp m)(I\mp m-1)(I\pm m+1)(I\pm m+2)]^{\frac{1}{2}}.$$

Here

$$A=\frac{Q_0}{2I(2I-1)}.$$

To determine the energy the secular equation of degree $(2I+1)$ should be solved. Since the matrix elements are different from zero only for transitions between states with the same parity of m, then the secular equation separates into two equations.

When I is an integer, one of these equations is of degree I and the other of degree $I+1$. When I is a half-integer, both equations, of the degree $I+\frac{1}{2}$, are identical.

Solving these secular equations, we obtain

$$I=1 \quad E_1=-\frac{Q_0}{2}\frac{\partial^2\varphi}{\partial x^2};\; E_2=-\frac{Q_0}{2}\frac{\partial^2\varphi}{\partial y^2};\; E_3=-\frac{Q_0}{2}\frac{\partial^2\varphi}{\partial z^2};$$

$$I=\frac{3}{2} \quad E_{1,2}=\frac{Q_0}{\sqrt{12}}\left\{\left(\frac{\partial^2\varphi}{\partial x^2}\right)^2+\left(\frac{\partial^2\varphi}{\partial y^2}\right)^2+\left(\frac{\partial^2\varphi}{\partial x^2}\right)\left(\frac{\partial^2\varphi}{\partial y^2}\right)\right\}^{\frac{1}{2}};$$

$$E_{3,4}=-\frac{Q_0}{\sqrt{12}}\left\{\left(\frac{\partial^2\varphi}{\partial x^2}\right)^2+\left(\frac{\partial^2\varphi}{\partial y^2}\right)^2+\left(\frac{\partial^2\varphi}{\partial x^2}\right)\left(\frac{\partial^2\varphi}{\partial y^2}\right)\right\}^{\frac{1}{2}}$$

$$I=2 \quad E_1=\frac{Q_0}{4}\left(\frac{\partial^2\varphi}{\partial x^2}\right);\; E_2=\frac{Q_0}{4}\left(\frac{\partial^2\varphi}{\partial y^2}\right);\; E_3=\frac{Q_0}{4}\left(\frac{\partial^2\varphi}{\partial z^2}\right);$$

$$E_{4,5}=\pm\frac{Q_0}{2\sqrt{3}}\left\{\left(\frac{\partial^2\varphi}{\partial x^2}\right)^2+\left(\frac{\partial^2\varphi}{\partial y^2}\right)^2+\left(\frac{\partial^2\varphi}{\partial x^2}\right)\left(\frac{\partial^2\varphi}{\partial y^2}\right)\right\}^{\frac{1}{2}}.$$

48. The state with the angular momentum J ($\hat{\mathbf{J}}^2$ is a constant of motion) can be built up from states with $L=J-\frac{1}{2}$ and $L=J+\frac{1}{2}$. The reflection ($x\to -x$, $y\to -y$, $z\to -z$) does not change the Hamiltonian of an isolated system (parity is a constant of motion). The states with $L=J-\frac{1}{2}$ and $L=J+\frac{1}{2}$ have different parity. From this it follows that in states with a given value of J the orbital angular momentum L corresponding to the relative motion of the particles has a definite value.

49. The spin of the α-particle and the spin of the nucleus B are zero, and therefore the orbital angular momentum L of the relative motion of the system α-particle-plus-reaction-product will be unity. Consequently, this system must be in an odd state (the α-particle has even parity), while the initial state was assumed to be even.

51. No.

53. $J=0$, $J=2$.

The wave functions for $J=1$ and $J=3$ are symmetric with respect to the particles, and therefore states with such J cannot be realized.

§ 5. Centrally symmetric field

1. If we replace R_{nl} by $\frac{\chi_{nl}}{r}$, the equation for R_{nl} takes the form

$$-\frac{\hbar^2}{2\mu}\chi''_{nl}-\left(E_{nl}-U(r)-\frac{\hbar^2 l(l+1)}{2\mu r^2}\right)\chi_{nl}=0.$$

This expression is the same as the Schrödinger equation for one-dimensional motion in the region $0 \leqslant r < \infty$ with an effective potential

$$U_{\text{eff}}(r) = U(r) + \frac{\hbar^2 l(l+1)}{2\mu r^2}.$$

Since $\chi_{nl} = rR_{nl}$ vanishes for $r = 0$, then we can take $U = +\infty$ for $r < 0$ in this one-dimensional problem.

2. In the equation for $\chi = Rr$

$$\chi'' + \left\{ \frac{2\mu}{\hbar^2}[E - U(r)] - \frac{l(l+1)}{r^2} \right\} \chi = 0$$

we make the substitution

$$\chi = A \exp\left(i \frac{S}{\hbar} \right),$$

where A and S are real functions.

Equating to zero the real and imaginary parts separately, we obtain

$$2A'S' + S''A = 0, \tag{1}$$

$$S'^2 - \frac{\hbar^2 A''}{A} = 2\mu [E - U(r)] - \frac{\hbar^2 l(l+1)}{r^2}. \tag{2}$$

From the first equation we have

$$A = \frac{\text{const}}{\sqrt{S'}}.$$

We shall find an approximate solution of the second equation by considering $\hbar^2$ to be a small quantity. Here it is necessary, however, to remember that in passing over to classical mechanics ($\hbar \to 0$) $\hbar l$ should be regarded as a finite quantity, since $\hbar l$ represents the angular momentum in classical mechanics. Hence only the term $\hbar^2 A''/A$ in Eq. (2) can be considered to be a small quantity. For small r, when the dominating term on the right-hand side of (2) becomes $\hbar^2 l(l+1)/r^2$, we have $S' \approx i\hbar \sqrt{l(l+1)}\, r^{-1}$, $A \sim \sqrt{r}$, from which we obtain the approximate expression $\hbar^2 A'' A^{-1} \approx -\frac{\hbar^2}{4} r^{-2}$. Therefore, we obtain a better approximation of S if we take this term into account and insert in (2) this approximate relation (for large r the correction is not important). Hence we obtain

$$S = \int \sqrt{2\mu [E - U(r)] - \frac{\hbar^2 \left(l + \frac{1}{2} \right)^2}{r^2}}\, dr,$$

$$A = \text{const}\left[2\mu\,[E-U(r)] - \frac{\hbar^2\left(l+\frac{1}{2}\right)^2}{r^2}\right]^{-1/4}$$

3. We represent the Hamiltonian in the form

$$\hat{H} = \hat{H}_0 + \frac{\hbar^2}{2\mu}\frac{l(l+1)}{r^2}, \quad \text{where } \hat{H}_0 = -\frac{\hbar^2}{2\mu r^2}\frac{\partial}{\partial r}\left(r^2\frac{\partial}{\partial r}\right) + U(r).$$

Then the minimum values of energy and the eigenfunctions corresponding to them are related in the following way:

$$E_l^{\min} = \int \psi_l^*\left\{\hat{H}_0 + \frac{\hbar^2}{2\mu}\frac{l(l+1)}{r^2}\right\}\psi_l\, d\tau,$$

$$E_{l+1}^{\min} = \int \psi_{l+1}^*\left\{\hat{H}_0 + \frac{\hbar^2}{2\mu}\frac{(l+1)(l+2)}{r^2}\right\}\psi_{l+1}\, d\tau.$$

The last expression can be rearranged to give

$$E_{l+1}^{\min} = \int \psi_{l+1}^*\left\{\hat{H}_0 + \frac{\hbar^2 l(l+1)}{2\mu r^2}\right\}\psi_{l+1}\, d\tau + \int \frac{\hbar^2}{\mu}\frac{l+1}{r^2}\psi_{l+1}^*\psi_{l+1}\, d\tau.$$

Let us compare the first term of this expression with $E_l^{\min}$. Since ψ_l corresponds to the minimum eigenvalue of the operator $\hat{H}_0 + \frac{\hbar^2}{2\mu}\frac{l(l+1)}{r^2}$, then

$$\int \psi_{l+1}^*\left\{\hat{H}_0 + \frac{\hbar^2}{2\mu}\frac{l(l+1)}{r^2}\right\}\psi_{l+1}\, d\tau > \int \psi_l^*\left\{\hat{H}_0 + \frac{\hbar^2}{2\mu}\frac{l(l+1)}{r^2}\right\}\psi_l\, d\tau.$$

The integral $\int \frac{\hbar^2}{\mu}\frac{(l+1)}{r^2}\psi_{l+1}^*\psi_{l+1}\, d\tau$, is always greater than zero. Hence we have $E_l^{\min} < E_{l+1}^{\min}$, which was to be proved.

4. $\hat{\mathbf{p}}_1 + \hat{\mathbf{p}}_2 \equiv \hat{\mathbf{P}} = -i\hbar\nabla_R$; $\hat{\mathbf{l}}_1 + \hat{\mathbf{l}}_2 \equiv \hat{\mathbf{L}} = [\mathbf{R}\hat{\mathbf{P}}] + [\mathbf{r}\hat{\mathbf{p}}]$, where $\hat{\mathbf{p}} = -i\hbar\nabla_r$.

5. The potential energy is $U(r) = \frac{1}{2}\mu\omega^2 r^2$.

The radial part R of the wave function satisfies the equation

$$R'' + \frac{2}{r}R' + \left\{\frac{2\mu E}{\hbar^2} - \frac{\mu^2\omega^2 r^2}{\hbar^2} - \frac{l(l+1)}{r^2}\right\}R = 0.$$

Substituting $\chi = Rr$ and introducing the notation

$$k = \frac{1}{\hbar}\sqrt{2\mu E}, \qquad \frac{\mu\omega}{\hbar} = \lambda,$$

we have

$$\chi'' + \left\{k^2 - \lambda^2 r^2 - \frac{l(l+1)}{r^2}\right\}\chi = 0. \tag{1}$$

Taking into account the asymptotic behaviour of χ for $r \to 0$ and for $r \to \infty$, we look for a solution χ in the form

$$\chi = r^{l+1} \exp\left(-\frac{\lambda}{2} r^2\right) u(r). \tag{2}$$

Substituting (2) into (1), we get the following equation for the function $u(r)$:

$$u'' + 2\left\{\frac{l+1}{r} - \lambda r\right\} u' - \left\{2\lambda\left(l+\frac{3}{2}\right) - k^2\right\} u = 0. \tag{3}$$

Introducing a new independent variable $\xi = \lambda r^2$, we obtain from (3) the following differential equation:

$$\xi \frac{d^2 u}{d\xi^2} + \left\{\left(l+\frac{3}{2}\right) - \xi\right\} \frac{du}{d\xi} + \left\{\frac{1}{2}\left(l+\frac{3}{2}\right) - \frac{1}{2} s\right\} u = 0,$$

where

$$s = \frac{k^2}{2\lambda} = \frac{E}{\hbar\omega}.$$

The solution of this equation is a confluent hypergeometric function

$$u = F\left\{\frac{1}{2}\left(l+\frac{3}{2}-s\right), l+\frac{3}{2};\ \xi\right\}.$$

The condition that R decrease for $r \to \infty$ gives us

$$\frac{1}{2}\left(l+\frac{3}{2}-s\right) = -n_r \quad (n_r = 0, 1, 2, \ldots),$$

and consequently the energy levels are given by $E_{n_r l} = \hbar\omega\left(l + 2n_r + \frac{3}{2}\right)$, and the wave functions are

$$\psi_{n_r lm} = r^l \exp\left(-\frac{\lambda}{2} r^2\right) F\left\{-n_r, l+\frac{3}{2}, \lambda r^2\right\} Y_{lm}(\vartheta, \varphi).$$

6. The wave functions are

$$\Phi_{n_1 n_2 n_3}(x, y, z) = \varphi_{n_1}(x)\varphi_{n_2}(y)\varphi_{n_3}(z),$$

where

$$\varphi_n(x) = (2^n \lambda^{n-\frac{1}{2}} n!)^{-1/2} \pi^{-1/4} \left(\lambda x - \frac{\partial}{\partial x}\right)^n \exp\left(-\frac{1}{2}\lambda x^2\right).$$

The corresponding energy levels are

$$E_{n_1 n_2 n_3} = \hbar\omega\left(n_1+n_2+n_3+\frac{3}{2}\right) \quad \text{(see Prob. 6, § 1)}.$$

The relation between $\psi_{n_r lm}$ and $\Phi_{n_1 n_2 n_3}$ for $n_r = 0$, $l = 1$ has the form

$$\psi_{011} = \frac{1}{\sqrt{2}}(\Phi_{100} + i\,\Phi_{010}),$$

$$\psi_{010} = \Phi_{001},$$

$$\psi_{01,-1} = \frac{1}{\sqrt{2}}(\Phi_{100} - i\,\Phi_{010}).$$

7. $$Z_n = (n+1)(n+2),$$

where

$$n = 2n_r + l.$$

8. For $_2He^4$

$$\varrho(\mathbf{r}) = \frac{4}{(r_0\sqrt{2\pi})^3}\exp\left(-\frac{1}{2}\frac{r^2}{r_0^2}\right),$$

where

$$r_0 = \sqrt{\frac{\hbar}{2\mu\omega}}; \qquad R = r_0.$$

For $_8O^{16}$

$$\varrho(\mathbf{r}) = \frac{4}{(r_0\sqrt{2\pi})^3}\left(1+\frac{r^2}{r_0^2}\right)\exp\left[-\frac{1}{2}\left(\frac{r}{r_0}\right)^2\right],$$

$$R = 3{\cdot}73\, r_0.$$

9. The equation for the radial wave function has the form

$$-\frac{\hbar^2}{2\mu}\frac{1}{r^2}\frac{d}{dr}\left(r^2\frac{dR}{dr}\right) + \frac{\hbar^2}{2\mu}\frac{l(l+1)}{r^2}R + \left\{U(r) - E\right\}R = 0,$$

where μ is the reduced mass; $\mu = \dfrac{M_p M_n}{M_p + M_n} \approx \dfrac{M}{2}$, since $M_p \approx M_n = M$.

Setting $l = 0$ and $R = \chi(r)/r$, we get

$$\frac{d^2\chi}{dr^2} + \frac{2\mu}{\hbar^2}\left[E + A\exp\left(-\frac{r}{a}\right)\right]\chi = 0.$$

Introducing the change of variables

$$\xi = \exp\left(-\frac{r}{2a}\right),$$

we obtain

$$\frac{d^2\chi}{d\xi^2}+\frac{1}{\xi}\frac{d\chi}{d\xi}+\left(c^2-\frac{k^2}{\xi^2}\right)\chi=0,$$

where

$$c^2=\frac{8\mu}{\hbar^2}Aa^2,$$

$$k^2=-\frac{8\mu}{\hbar^2}Ea^2>0.$$

The general solution of this equation is

$$\chi=B_1 J_k(c\xi)+B_2 J_{-k}(c\xi).$$

For $r\to\infty$ ($\xi=0$) the wave function of the stationary state should vanish; hence $B_2=0$, and consequently

$$R=\frac{B_1}{r}J_k\left[c\exp\left(-\frac{r}{2a}\right)\right].$$

For R to be finite when $r=0$ it is necessary that

$$J_k(c)=0.$$

This equation gires the relation between a and A. To obtain the values of a and A for the ground state one must take for c the first root of the Bessel function (the radial wave function cannot have any nodes).

$a\cdot 10^{13}$	k	c	A MeV
1	0.45	3.1	100
2	0.91	3.7	36
4.4	2.02	5.1	14

10. The mean value of the energy E in the state described by the wave function $\psi(\mathbf{r})$ is given by the following relation:

$$E=\frac{\hbar^2}{2\mu}\int(\nabla\psi)^2\,d\tau+\int U\psi^2\,d\tau.$$

In accordance with the variational principle the magnitude of E assumes the value of the ground state energy if ψ is the exact function of that state. If some other functions dependent on one or more parameters $\alpha,\beta,\ldots$ are taken, then the energy E will be a function of these parameters $E(\alpha,\beta,\ldots)$. The best approximation of the energy and the function ψ for the ground state will be obtained for the values $\alpha=\alpha_0$, $\beta=\beta_0,\ldots,$

satisfying the conditions

$$\left(\frac{\partial E(\alpha, \beta, \ldots)}{\partial \alpha}\right)_{\substack{\alpha=\alpha_0 \\ \beta=\beta_0 \\ \ldots}} = 0, \quad \left(\frac{\partial E(\alpha, \beta, \ldots)}{\partial \beta}\right)_{\substack{\alpha=\alpha_0 \\ \beta=\beta_0 \\ \ldots}} = 0, \ldots$$

Here $E(\alpha_0,\beta_0,\ldots)$ is always greater than the ground state energy and the closer it is to that value, the more meaningful is the choice of the class of trial functions.

In our case $\psi = \frac{1}{\sqrt{4\pi}} R(r)$, where $R(r) = c \exp\ (-\alpha r/2a)$. From the normalization condition it follows that $c^2 = \frac{1}{2}(\alpha/a)^3$, so that

$$E(\alpha) = c^2 \frac{\hbar^2}{2\mu} \int_0^\infty \left(\frac{\alpha}{2a}\right)^2 \exp\left[-\frac{\alpha r}{a}\right] r^2\, dr - c^2 A \int_0^\infty \exp\left(-\frac{\alpha r}{a} - \frac{r}{a}\right) r^2\, dr$$

$$= \frac{\hbar^2}{2\mu}\left(\frac{\alpha}{2a}\right)^2 - A\left(\frac{\alpha}{\alpha+1}\right)^3.$$

We find the minimum of $E(\alpha)$

$$\frac{dE(\alpha)}{d\alpha} = \frac{\hbar^2 \alpha}{4\mu a^2} - \frac{3A\alpha^2}{(\alpha+1)^4} = 0,$$

from which we have

$$\frac{(\alpha_0+1)^4}{\alpha_0} = \frac{12 A\mu a^2}{\hbar^2} = 22 \cdot 3; \quad \alpha_0 = 1 \cdot 34.$$

The energy for this value of the parameter is

$$E = -2 \cdot 14 \text{ MeV}.$$

The exact solution of this problem leads for the given values of A and a to $E = -2 \cdot 2$ MeV (see preceding problem).

11. The equation for the radial part of the wave function for $r < a$ has the form

$$\frac{1}{r^2}\frac{d}{dr}\left(r^2 \frac{dR}{dr}\right) - \frac{l(l+1)}{r^2} R + k^2 R = 0, \tag{1}$$

where

$$k^2 = \frac{2\mu E}{\hbar^2}$$

for $r = a$, $R = 0$.

In place of R we introduce a new function $\chi(r)$ defined by the relation

$$\chi(r)=\sqrt{r}\,R(r).$$

Substituting into Eq. (1), we obtain the equation

$$\chi''+\frac{1}{r}\chi'+\left\{k^2-\frac{\left(l+\frac{1}{2}\right)^2}{r^2}\right\}\chi=0$$

for the determination of $\chi(r)$. This equation is satisfied by Bessel functions of half-integer order

$$\chi(r)=J_{l+1/2}(kr),$$
$$R(r)=cr^{-1/2}J_{l+1/2}(kr).$$

The values of the energy $E=\dfrac{\hbar^2k^2}{2\mu}$ of the stationary states are determined from the condition that the Bessel functions vanish for $r=a$

$$J_{l+1/2}(ka)=0,$$

and c is found from the normalization condition.

The energy level for a particle with angular momentum $l=0$ is most easily determined. In this case

$$J_{1/2}(kr)=\sqrt{\frac{2}{\pi kr}}\sin kr$$

and the energy is

$$E_{n0}=\frac{\hbar^2}{2\mu}\frac{n^2\pi^2}{a^2}.$$

12. The problem reduces to the solution of the one-dimensional problem with the potential

$$U(r)=\begin{cases}-U_0 & 0<r<a,\\ 0 & r>a,\\ \infty & r<0.\end{cases}$$

In Prob. 4, §1, we set $U_1=\infty$, $U_2=U_0$, and obtain the equation giving the energy levels for the discrete spectrum

$$ka=n\pi-\arcsin\frac{\hbar k}{\sqrt{2mU_0}},\qquad k=\frac{\sqrt{2mE}}{\hbar}.$$

The energy levels are easily found by graphical construction (see Fig. 24).

The depth of the well at which the first discrete level occurs is

$$U_0 = \frac{\pi^2 \hbar^2}{8ma^2}.$$

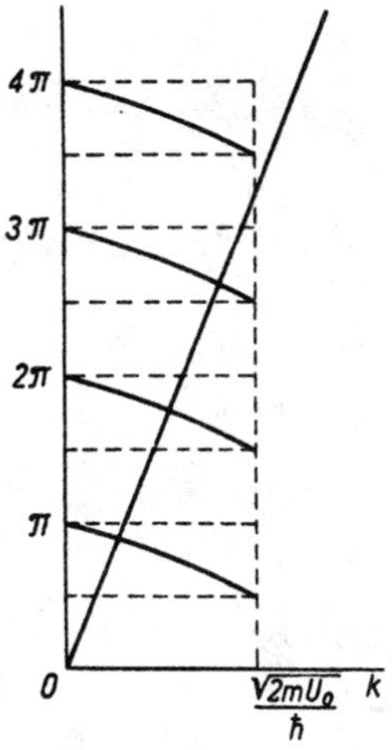

Fig. 24

13. If the edges of the well are rounded off, all energy levels are displaced upwards, i.e., $\Delta E > 0$. The levels of states with large l are displaced to a greater extent, since particles in states of higher angular momentum spend relatively more time near the edge of the well.

14. The radial wave function satisfies the equation

$$\chi'' + \frac{2\mu}{\hbar^2}(E-U)\chi = 0.$$

In the region I, where $U = 0$, the solution vanishing for $r = 0$ has the form

$$\chi = A \sin kr, \quad k^2 = \frac{2\mu E}{\hbar^2}.$$

In the region II ($U = U_0$) the general solution is

$$\chi = B_+ \exp\left[\varkappa(r-r_1)\right] + B_- \exp\left[-\varkappa(r-r_1)\right], \quad \varkappa^2 = \frac{2\mu(U_0-E)}{\hbar^2}.$$

The coefficients B_+ and B_- are found from the condition of the continuity of χ and χ' at the boundary of regions I and II:

$$A \sin kr_1 = B_+ + B_-,$$

$$Ak \cos kr_1 = \varkappa(B_+ - B_-),$$

whence

$$\left.\begin{aligned} B_+ &= \frac{1}{2} A\left(\sin kr_1 + \frac{k}{\varkappa}\cos kr_1\right), \\ B_- &= \frac{1}{2} A\left(\sin kr_1 - \frac{k}{\varkappa}\cos kr_1\right). \end{aligned}\right\} \tag{1}$$

The solution in the region III, where once again $U = 0$, is

$$\chi = C_+ e^{ik(r-r_2)} + C_- e^{-ik(r-r_2)}.$$

The condition of continuity at the boundary of regions II and III gives

$$B_+ e^{\varkappa(r_2-r_1)} + B_- e^{-\varkappa(r_2-r_1)} = C_+ + C_-,$$

$$\varkappa(B_+ e^{\varkappa(r_2-r_1)} - B_- e^{-\varkappa(r_2-r_1)}) = ik(C_+ - C_-).$$

From this we find

$$C_+ = \frac{1}{2} B_+\left(1 + \frac{\varkappa}{ik}\right) e^{\varkappa(r_2-r_1)} + \frac{1}{2} B_-\left(1 - \frac{\varkappa}{ik}\right) e^{-\varkappa(r_2-r_1)},$$

$$C_- = \frac{1}{2} B_+\left(1 - \frac{\varkappa}{ik}\right) e^{\varkappa(r_2-r_1)} + \frac{1}{2} B_-\left(1 + \frac{\varkappa}{ik}\right) e^{-\varkappa(r_2-r_1)}.$$

Let us express C_+ (and C_-) in terms of A by means of (1)

$$C_+ = \frac{1}{4} A \sin kr_1\left(1 + \frac{\varkappa}{ik}\right) e^{\varkappa(r_2-r_1)} \left\{ 1 + \frac{1 - \frac{\varkappa}{ik}}{1 + \frac{\varkappa}{ik}} e^{-2\varkappa(r_2-r_1)} + \right.$$

$$\left. + \frac{k}{\varkappa} \cot kr_1 \left[1 - \frac{1 - \frac{\varkappa}{ik}}{1 + \frac{\varkappa}{ik}} e^{-2\varkappa(r_2-r_1)} \right] \right\}. \tag{2}$$

Expressions (1) and (2) constitute the stationary wave function for the particle. The behaviour of the wave function depends to a great extent on the energy of the particle. Let us consider the relation between

C_+ and C_- and the energy. We shall assume that $\varkappa(r_2 - r_1) \gg 1$. Then all of the terms containing the factor $\exp[-2\varkappa(r_2 - r_1)]$ can be neglected and we have

$$C_+ \approx \frac{1}{4} A \sin kr_1 \left(1 + \frac{\varkappa}{ik}\right) e^{\varkappa(r_2 - r_1)} \left\{1 + \frac{k}{\varkappa} \cot kr_1\right\}, \qquad C_- = C_+^*.$$

Therefore if the quantity in the braces is not too small the coefficients C_+ and C_- are considerably greater than A, i. e., the wave function differs significantly from zero only in the region III (Fig. 25 *a*).

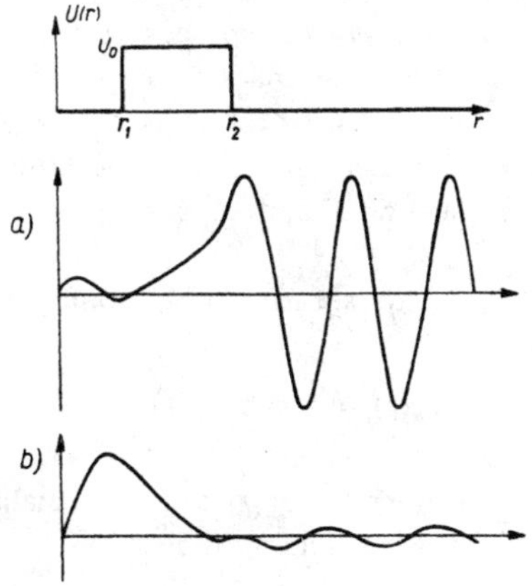

Fig. 25

For certain values of the energy, when the expression in the braces of (2) is small, C_+ and C_- can assume abnormally small values. Such energies lie in the vicinity of the roots of the transcendental equation

$$1 + \sqrt{\frac{E_n}{U_0 - E_n}} \cot \sqrt{\frac{2\mu E_n}{\hbar^2}}\, r_1 = 0,$$

and are called the *quasi-stationary* levels.

It can readily be seen that the values of E_n are the levels of the discrete spectrum for a particle in the potential shown in Fig. 26 ($r_2 \to \infty$).

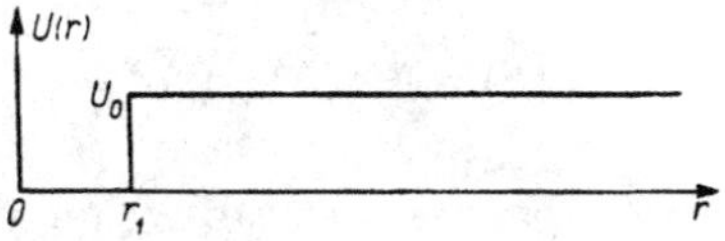

Fig. 26

Therefore the wave functions corresponding to the values of the energies in the narrow band in the vicinity of the quasi-levels are negligibly small in region III (Fig. 25 b.)

The probability that a particle of a given energy be found in the region I is equal to zero.

Actually, the wave function of a particle with a given energy belongs to a continuous spectrum, and the integral of $|\psi(r, E)|^2$ over region III is divergent, while the integral over region I is finite. This also holds for states in the vicinity of quasi-levels. Therefore, if we wish to find the probability of the particles leaving region I, we must investigate the state consisting of a superposition of a series of stationary states of close-lying energies, i.e., a "wave packet" localized in region I, and we must investigate its spreading with time. As a wave function for $t = 0$ we take the function χ_0 which, practically speaking, vanishes in region III, and coincides in regions I and II with the wave function of the quasi-stationary level.

Let us expand $\chi_0(r)$ into a series of stationary wave functions

$$\chi_0(r) = \int_0^\infty \varphi(E)\chi_E(r)\,dE. \tag{3}$$

We shall assume the functions $\chi_E(r)$ to be normalized over the energy scale. The state of the particle at time t is

$$\chi_0(r, t) = \int \varphi(E)\chi_E(r)\exp\left[-\frac{i}{\hbar}Et\right]dE.$$

Let us consider the probability that the particle is in the initial state $\chi_0(r)$ at the time t. This probability is

$$W(t) = \left|\int_0^\infty \chi_0(r)\chi_0(r, t)\,dr\right|^2 = \left|\int_0^\infty |\varphi(E)|^2\exp\left[-\frac{i}{\hbar}Et\right]dE\right|^2. \tag{4}$$

Therefore, the problem reduces to one of finding the splitting of the energy of the initial state $|\varphi(E)|^2$.

From equation (3) it follows that

$$\varphi(E) = \int_0^\infty \chi_0(r)\chi_E(r)\,dr. \tag{5}$$

In accordance with what we said earlier, for the function $\chi_0(r)$ we can take the eigenfunction of an auxiliary problem with a potential illustrated graphically in Fig. 26 for $r < r_1$:

$$\chi_0(r) = a \sin k_0 r \quad \left(k_0^2 = \frac{2\mu E_0}{\hbar^2}\right)$$

and for $r > r_1$

$$\chi_0(r) = -\frac{k_0}{\varkappa_0} a e^{-2\varkappa(r-r_1)}, \quad \varkappa^2 = \varkappa_0^2 - k_0^2.$$

We determine the value of k_0 from the condition

$$\sin k_0 r_1 = -\frac{k_0}{\varkappa_0}, \quad \cos k_0 r_1 = \frac{\varkappa}{\varkappa_0},$$

and the normalization constant is $a = \sqrt{\dfrac{2\varkappa}{1+\varkappa r_1}}$.

The functions $\chi_E(r)$ in the regions I, II and III have now been determined except for a common factor A. It is now necessary to choose A so that $\chi_E(r)$ is normalized in energy space. The asymptotic form $\chi_E(r)$ is determined by the value of the coefficients C_+ and C_-. The normalization

$$\int_0^\infty \chi_E(r) \chi_{E'}^*(r)\, dr = \delta(E-E')$$

gives

$$|C_+| = |C_-| = \frac{1}{\hbar}\sqrt{\frac{\mu}{2\pi k}}.$$

From this, with the help of (2), we can find the relation between A and the energy. Previously, we said that the ratio $\dfrac{|C_\pm|}{A(E)} = \dfrac{\frac{1}{\hbar}\sqrt{\mu/2\pi k}}{A(E)}$ is large for practically all values of energy and is small only when E is close to one of the quasi-levels. Therefore $\varphi(E)$ has one sharp maximum near $E_0 = \dfrac{\hbar^2 k_0^2}{2\mu}$. In the regions of other quasi-levels, although $A(E)$ increases again, the integral in (5) will be almost, equal to zero, since the function $\chi_0(r)$ is almost orthogonal to the eigenfunctions $\chi_0(r)$ belonging to other quasi-levels. In view of this, the values of E near E_0 are important in the expression (4) for the probability $W(t)$.

After these preliminary considerations let us proceed to calculate $\varphi(E)$. First of all, let us find the relation between A and E. From the expression for C_+ and C_- we have

$$C_+ = C_-^* = \frac{1}{2} B_+ \left(1 + \frac{\varkappa}{ik}\right) e^{\varkappa(r_2 - r_1)} + \frac{1}{2} B_- \left(1 - \frac{\varkappa}{ik}\right) e^{-\varkappa(r_2 - r_1)}.$$

Moreover, B_+ and B_- are given in terms of A by

$$B_+ = \frac{A}{2} \left(\sin kr_1 + \frac{k}{\varkappa} \cos kr_1\right),$$

$$B_- = \frac{A}{2} \left(\sin kr_1 - \frac{k}{\varkappa} \cos kr_1\right).$$

In the vicinity of the quasi-levels we can set $k - k_0 = \Delta k$ and assume that the condition

$$|\Delta k| \ll k_0 \text{ and } |\Delta k| \ll \varkappa. \tag{6}$$

is satisfied. Then the main terms in B_+ and B_- are

$$B_+ = \frac{A}{2} \frac{\varkappa_0}{\varkappa^2} (1 + \varkappa r_1) \Delta k,$$

$$B_- = -A \frac{k_0}{\varkappa_0}.$$

Assuming that $e^{-\varkappa(r_2 - r_1)} \ll 1$, we get

$$|C_\pm| = \frac{A}{4} \frac{\varkappa_0^2}{k\varkappa^2} (1 + \varkappa r_1) e^{\varkappa(r_2 - r_1)} \times \sqrt{(\Delta k)^2 + \left(\frac{4\varkappa^3}{\varkappa_0^4} \cdot \frac{k^2}{1 + \varkappa r_1} e^{-2\varkappa(r_2 - r_1)}\right)^2},$$

and since $|C_+| = \frac{1}{\hbar} \sqrt{\frac{\mu}{2\pi k}}$, it follows that

$$A(E) = \frac{\frac{1}{\hbar} \sqrt{\frac{\mu}{2\pi k}} \frac{4k\varkappa^2}{\varkappa_0^2 (1 + \varkappa r_1)} e^{-\varkappa(r_2 - r_1)}}{\sqrt{(\Delta k)^2 + \left(\frac{4\varkappa^3}{\varkappa_0^4} \cdot \frac{k^2}{1 + \varkappa r_1} \cdot e^{-2\varkappa(r_2 - r_1)}\right)^2}}.$$

Now the integral (5) determining $\varphi(E)$ can be calculated without difficulty under the assumption that the inequalities (6) are still valid. The function $\chi_E(r)$ in regions I and II does not differ much from $A(E) a^{-1} \chi_0$, while in region III it does not play a significant role in the determination of $\varphi(E)$, as $\chi_0(r)$ decreases exponentially for $r > r_1$.

We therefore have

$$\varphi(E) = \frac{A(E)}{a}\int_0^\infty \chi_0^2(r)\,dr = \frac{A(E)}{a}.$$

After simple transformations, we obtain

$$\varphi^2(E) = \frac{\hbar}{2\pi\tau}\frac{1}{(E-E_0)^2+\frac{\hbar^2}{4\tau^2}},$$

where

$$E-E_0 = \frac{\hbar^2}{\mu}k_0\,\Delta k$$

and

$$\tau = \frac{\mu}{8\hbar}\frac{\varkappa_0^4}{\varkappa^3 k^3}\,e^{2\varkappa(r_2-r_1)}(1+\varkappa r_1).$$

Integrating in (4), we obtain the decay law

$$W(t) = \exp\left[-\frac{t}{\tau}\right].$$

The probability of a particle remaining in the initial state inside the barrier $W(t)$ diminishes to $1/e$ in the time

$$\tau = \frac{1}{16}\frac{\hbar}{U_0}\left(\frac{U_0^2}{E(U_0-E)}\right)^{3/2} e^{2\varkappa(r_2-r_1)}(1+\varkappa r_1).$$

§6. Motion of a particle in a magnetic field

1. According to classical mechanics, an electron moving in a uniform magnetic field traces a helix whose axis is in the direction of the magnetic field. The motion in a plane perpendicular to the magnetic field has a frequency double the Larmor frequency $\left(\omega = \frac{e\mathcal{H}}{2\mu c}\right)$. We shall consider the motion of a wave packet on the basis of quantum mechanics. The Schrödinger equation for a particle in a magnetic field can be written

$$i\hbar\frac{\partial\Psi}{\partial t} = -\frac{\hbar^2}{2\mu}\Delta\Psi + \frac{\hbar}{i}\,\omega\left(x\frac{\partial\Psi}{\partial y} - y\frac{\partial\Psi}{\partial x}\right) + \frac{1}{2}\mu\omega^2(x^2+y^2)\Psi.$$

To solve the problem it will be convenient to make use of a rotating coordinate system:

$$\left.\begin{aligned} x &= x'\cos\omega t' - y'\sin\omega t', \\ y &= x'\sin\omega t' + y'\cos\omega t', \\ z &= z', \\ t &= t', \end{aligned}\right\} \qquad (1)$$

$$\Psi(x, y, z, t) = \Psi'(x', y', z', t').$$

Then

$$\frac{\partial}{\partial t'} = \frac{\partial}{\partial t} + \omega\left(x\frac{\partial}{\partial y} - y\frac{\partial}{\partial x}\right), \qquad \Delta' = \Delta.$$

The Schrödinger equation in the new variables takes the following form:

$$i\hbar\frac{\partial\Psi'}{\partial t'} = -\frac{\hbar^2}{2\mu}\Delta'\Psi' + \frac{1}{2}\mu\omega^2(x'^2 + y'^2)\Psi'.$$

This equation can be solved by separating the variables x', y', z'. The equation for the function $\varphi(z')$ describes the motion of a free particle along the z axis. The solution of the equation giving the function $\psi(x, y, t)$ has the form

$$\psi(x, y, t) = \sum A_{nm}\chi_n\left(x'\sqrt{\frac{\mu\omega}{\hbar}}\right)\chi_m\left(y'\sqrt{\frac{\mu\omega}{\hbar}}\right)e^{-i\omega t(n+m+1)}.$$

Here x' and y' are functions of the coordinates x, y and of the time t as given by (1): χ_n is the eigenfunction of a harmonic oscillator; A_{nm} are coefficients chosen to satisfy the initial conditions. This expression for $\psi(x, y, t)$ changes sign only if t increases by the period of classical motion $T = \frac{\pi}{\omega}$. Then x' and y' change sign. Recalling that for the oscillator eigenfunction

$$\chi_n(-\xi) = (-1)^n\chi_n(\xi),$$

we then have

$$\psi\left(x, y, t + \frac{\pi}{\omega}\right) = \sum A_{nm}(-1)^n\chi_n\left(x'\sqrt{\frac{\mu\omega}{\hbar}}\right)(-1)^m\chi_m\left(y'\sqrt{\frac{\mu\omega}{\hbar}}\right)\times$$

$$\times\exp\left[-i\omega t(n+m+1) - i\pi(n+m+1)\right] = -\psi(x, y, t).$$

Therefore, in the plane perpendicular to the magnetic field the wave packet will undergo cyclical changes with a period equal to that of the classical motion in a magnetic field. In the direction of the magnetic field the wave packet will spread as in the case of the wave packet for a free particle. The wave function $\psi(x, y, t)$ can be found in explicit form if the initial function is given in the form

$$\psi(x, y, 0) = \exp\left\{-\frac{a^2}{2}[(x-x_0)^2+(y-y_0)^2]+\frac{ip_{0x}x}{\hbar}+\frac{ip_{0y}y}{\hbar}\right\}.$$

Assuming that all $A_{nm} = 0$, except for $A_{00} = 1$, we find that such a wave packet does not spread in the x, y plane and that its centre of gravity describes the classical trajectory.

2. To find the operator $\hat{\mathbf{v}}$ it is necessary to employ the commutation relation for $\mathbf{r}$ and the Hamiltonian:

$$\hat{\mathbf{v}} = \frac{i}{\hbar}(\hat{H}\mathbf{r}-\mathbf{r}\hat{H}).$$

Since

$$\hat{H} = \frac{1}{2\mu}\left(\hat{\mathbf{p}}-\frac{e}{c}\mathbf{A}\right)^2+U(\mathbf{r}),$$

we obtain

$$\mu\hat{\mathbf{v}} = \hat{\mathbf{p}}-\frac{e}{c}\mathbf{A}.$$

We now find the commutation rules for these operators

$$\hat{v}_x\hat{v}_y-\hat{v}_y\hat{v}_x = \frac{e}{\mu^2 c}[-(\hat{p}_xA_y-A_y\hat{p}_x)+(\hat{p}_yA_x-A_x\hat{p}_y)]$$

$$= \frac{ie\hbar}{\mu^2 c}\left(\frac{\partial A_y}{\partial x}-\frac{\partial A_x}{\partial y}\right) = \frac{ie\hbar}{\mu^2 c}\mathcal{H}_z.$$

By cyclical permutation we obtain the remaining two relations.

3. We take the z axis in the direction of magnetic field, whose intensity we denote by $\mathcal{H}$. The particle's velocity components satisfy the following commutation rules (see Prob. 2, §6):

$$\hat{v}_x\hat{v}_y-\hat{v}_y\hat{v}_x = \frac{ie\hbar}{\mu^2 c}\mathcal{H}, \quad \hat{v}_y\hat{v}_z-\hat{v}_z\hat{v}_y = 0, \quad \hat{v}_z\hat{v}_x-\hat{v}_x\hat{v}_z = 0.$$

The energy operator is equal to

$$\hat{H} = \frac{1}{2}\mu[\hat{v}_x^2+\hat{v}_y^2+\hat{v}_z^2].$$

We represent H in the form of two commuting operators

$$\hat{H}_1 = \frac{1}{2}\mu[\hat{v}_x^2+\hat{v}_y^2], \quad \hat{H}_2 = \frac{1}{2}\mu\hat{v}_z^2.$$

The eigenvalues of $\hat{H}$ are equal to the sum of the eigenvalues of $\hat{H}_1$ and $\hat{H}_2$. Let us find the eigenvalues of $\hat{H}_1$. For this purpose we introduce the notation $\hat{v}_x = a\hat{Q}$, $\hat{v}_y = a\hat{P}$, where $a = \sqrt{\frac{e\hbar\mathcal{H}}{\mu^2 c}}$. In terms of the variables $\hat{P}, \hat{Q}$ the commutation rule has the form $\hat{P}\hat{Q} - \hat{Q}\hat{P} = -i$ and the operator $\hat{H}_1$, the form $\hbar\frac{e\mathcal{H}}{2\mu c}(\hat{P}^2 + \hat{Q}^2)$. On the basis of Prob. 6, §1 we can write the eigenvalues of $\hat{H}_1$ in the form:

$$E_{1n} = \hbar\frac{e\mathcal{H}}{\mu c}\left(n + \frac{1}{2}\right) \qquad (n = 0, 1, 2, \ldots).$$

The eigenvalues of $\hat{H}_2$ form a continuous spectrum. Thus the energy of a charged particle moving in a magnetic field is

$$E_{nv_z} = \hbar\frac{e\mathcal{H}}{\mu c}\left(n + \frac{1}{2}\right) + \frac{1}{2}\mu v_z^2.$$

4. We take the z axis in the direction of the magnetic field, and the x axis in the direction of the electric field. We take the vector potential of the magnetic field in the form $A_y = \mathcal{H}x$, $A_x = A_z = 0$.

The Hamiltonian in this case can be written in the form

$$\hat{H} = \frac{1}{2\mu}\hat{p}_x^2 + \frac{1}{2\mu}\left(\hat{p}_y - \frac{e}{c}\mathcal{H}x\right)^2 + \frac{1}{2\mu}\hat{p}_z^2 - e\mathcal{E}x.$$

Introducing the notation

$$\frac{e\mathcal{H}}{c}x - \hat{p}_y - \frac{\mu c\mathcal{E}}{\mathcal{H}} = \hat{\pi},$$

we get for $\hat{H}$ the expression

$$\hat{H} = \frac{\hat{p}_x^2}{2\mu} + \frac{\hat{\pi}^2}{2\mu} - \frac{\hat{p}_y c\mathcal{E}}{\mathcal{H}} - \frac{\mu c^2\mathcal{E}^2}{2\mathcal{H}^2} + \frac{\hat{p}_z^2}{2\mu}$$

The commutation rule for $\hat{p}_x$ and $\hat{\pi}$ has the form

$$\hat{p}_x\hat{\pi} - \hat{\pi}\hat{p}_x = -i\frac{\hbar e\mathcal{H}}{c}.$$

From this we find that the eigenvalues of the operator $\hat{H}_1 = \frac{\hat{p}_x^2}{2\mu} + \frac{\hat{\pi}^2}{2\mu}$ coincide with the energy levels of an oscillator vibrating with a double Larmor frequency:

$$E_{1n} = \hbar\frac{e\mathcal{H}}{\mu c}\left(n + \frac{1}{2}\right).$$

The operators $\hat{p}_y$ and $\hat{p}_z$ appearing in the last components of the Hamiltonian commute with $\hat{H}_1$; hence the operator $\hat{H}_2 = \frac{1}{2\mu}\hat{p}_z^2 - \frac{\hat{p}_y c\mathcal{E}}{\mathcal{H}} - \frac{\mu c^2 \mathcal{E}^2}{2\mathcal{H}^2}$ can be diagonalized simultaneously with $\hat{H}_1$.

Therefore the energy spectrum of the particle is given by

$$E_{np_yp_z} = \hbar\frac{e\mathcal{H}}{\mu c}\left(n+\frac{1}{2}\right)+\frac{p_z^2}{2\mu}-\frac{p_y c\mathcal{E}}{\mathcal{H}}-\frac{\mu c^2\mathcal{E}^2}{2\mathcal{H}^2}.$$

Comparison with the result of the preceding problem shows that the electric field removes the degeneracy which occurs in the case in which only a magnetic field acts on the particle. In an electric field the energy levels depend on three quantum numbers.

5. $\psi_{np_yp_z}(x, y, z) = \exp\left[\frac{i}{\hbar}(p_y y + p_z z)\right]\times$

$$\times\exp\left[-\frac{e\mathcal{H}}{2\hbar c}\left(x-\frac{cp_y}{e\mathcal{H}}-\frac{\mu c^2\mathcal{E}}{e\mathcal{H}^2}\right)^2\right]\cdot H_n\left[\sqrt{\frac{e\mathcal{H}}{c\hbar}}\left(x-\frac{cp_y}{e\mathcal{H}}-\frac{\mu c\mathcal{E}}{e\mathcal{H}^2}\right)\right].$$

6.

$$E_{nmk} = \hbar\sqrt{\omega^2+\omega_0^2}(2n+|m|+1)+\hbar\omega m+\hbar\omega_0\left(k+\frac{1}{2}\right),$$

where $\omega = \frac{e\mathcal{H}}{2\mu c}$ $(n = 0, 1, 2, \ldots, m = 0, \pm 1, \pm 2, \ldots, k = 0, 1, 2 \ldots)$.

7.

$$\hat{x}(t) = \left(\frac{i\hbar}{\mu\omega}\frac{\partial}{\partial y}+\frac{x}{2}\right)\cos\omega t+\left(-\frac{i\hbar}{\mu\omega}\frac{\partial}{\partial x}+\frac{y}{2}\right)\sin\omega t+\left(-\frac{i\hbar}{\mu\omega}\frac{\partial}{\partial y}+\frac{x}{2}\right),$$

$$\hat{y}(t) = \left(-\frac{i\hbar}{\mu\omega}\frac{\partial}{\partial y}-\frac{x}{2}\right)\sin\omega t+\left(-\frac{i\hbar}{\mu\omega}\frac{\partial}{\partial x}+\frac{y}{2}\right)\cos\omega t+\left(+\frac{i\hbar}{\mu\omega}\frac{\partial}{\partial x}+\frac{y}{2}\right).$$

Here $\omega = \frac{e\mathcal{H}}{\mu c}$ (double the frequency of the Larmor precession).

8. The Schrödinger equation in cylindrical coordinates ϱ, φ, z has the form

$$-\frac{\hbar^2}{2\mu}\left\{\frac{\partial^2\psi}{\partial z^2}+\frac{\partial^2\psi}{\partial\varrho^2}+\frac{1}{\varrho}\frac{\partial\psi}{\partial\varrho}+\frac{1}{\varrho^2}\frac{\partial^2\psi}{\partial\varphi^2}\right\}-\frac{ie\hbar}{2\mu c}\mathcal{H}\frac{\partial\psi}{\partial\varphi}+\frac{e^2\mathcal{H}^2}{8\mu c^2}\varrho^2\psi = E\psi.$$

We shall seek a solution in the form

$$\psi(\varrho, \varphi, z) = \frac{1}{\sqrt{2\pi}} R(\varrho)\, e^{ik_z z}\, e^{im\varphi}.$$

We introduce the following notation:

$$\gamma = \frac{e\mathcal{H}}{2c\hbar}, \quad \beta = \frac{2\mu E}{\hbar^2} - k_z^2.$$

In the equation for the radial function $R(\varrho)$

$$R'' + \frac{1}{\varrho} R' + \left(\beta - \gamma^2\varrho^2 - 2\gamma m - \frac{m^2}{\varrho^2}\right) R = 0$$

we introduce a new independent variable $\xi = \gamma\varrho^2$. We then obtain

$$\xi R'' + R' + \left(-\frac{\xi}{4} + \lambda - \frac{m^2}{4\xi}\right) R = 0, \tag{1}$$

where $\lambda = \frac{\beta}{4\gamma} - \frac{m}{2}$.

The function we are seeking behaves like $e^{-\xi/2}$, as $\xi \to \infty$, and for small ξ is proportional to $\xi^{|m|/2}$. We seek the solution of the differential equation (1) in the form

$$R = e^{-\frac{1}{2}\xi}\, \xi^{\frac{1}{2}|m|}\, w(\xi).$$

We obtain $w(\xi)$ from the equation

$$\xi w'' + (1 + |m| - \xi) w' + \left(\lambda - \frac{|m|+1}{2}\right) w = 0,$$

which is satisfied by the confluent hypergeometric function

$$w = F\left\{-\left(\lambda - \frac{|m|+1}{2}\right),\ |m|+1,\ \xi\right\}.$$

For the wave function to be finite, the quantity $\lambda - \frac{|m|+1}{2}$ should be a non-negative integer n. Thus the energy levels are given by the expression

$$E = \hbar \frac{e\mathcal{H}}{\mu c}\left(n + \frac{|m|}{2} + \frac{m}{2} + \frac{1}{2}\right) + \frac{\hbar^2 k_z^2}{2\mu}.$$

9. In cylindrical coordinates

$$J_\varrho = 0,$$

$$J_\varphi = \left(\frac{e\hbar m}{\mu\varrho} - \frac{e^2\mathcal{H}}{2\mu c}\varrho\right)|\psi_{nmk_z}|^2,$$

$$J_z = \frac{e\hbar k_z}{\mu}|\psi_{nmk_z}|^2.$$

10. In the equation for the radial wave function R we introduce the change of variable $u = \sqrt{\varrho}\, R$ and obtain

$$u'' + \left[\frac{2\mu}{\hbar^2}E - k_z^2 - \frac{1}{\varrho^2}\left(m + \frac{e\mathcal{H}}{2\hbar c}\varrho^2\right)^2\right]u = 0,$$

where $m^2 - \frac{1}{4}$ becomes m^2.* The expression

$$U_{\text{eff}}(\varrho) = \frac{\hbar^2}{2\mu\varrho^2}\left(m + \frac{e\mathcal{H}}{2\hbar c}\varrho^2\right)^2$$

can be considered as an effective potential energy for one-dimensional motion.

From the quantization condition

$$\int_{\varrho_1}^{\varrho_2} \sqrt{\frac{2\mu}{\hbar^2}E - k_z^2 - \frac{1}{\varrho^2}\left(m + \frac{e\mathcal{H}}{2\hbar c}\varrho^2\right)^2}\, d\varrho = \pi\left(n + \frac{1}{2}\right)$$

we obtain the energy spectrum

$$E - \frac{\hbar^2 k_z^2}{2\mu} = \frac{e\hbar\mathcal{H}}{\mu c}\left(n + \frac{|m|}{2} + \frac{m}{2} + \frac{1}{2}\right).$$

The energy, measured from the minimum of $U_{\text{eff}}(\varrho)$

$$E' = \frac{e\hbar\mathcal{H}}{\mu c}\left(n + \frac{1}{2}\right)$$

represents the energy of the radial motion, and the energy

$$E'' = \frac{e\hbar\mathcal{H}}{2\mu c}(m + |m|)$$

represents the energy of the rotational motion. The transition to a classical circular orbit occurs when the condition $E' \ll E''$ or $n \ll \frac{1}{2}(m + |m|)$ is satisfied. This condition, of course, is satisfied only for positive m and therefore can be written in the form $n \ll m$.

* This change is similar to the change $l(l+1) \to (l+\frac{1}{2})^2$. See Prob. 2, §5 for the justification of this procedure.

11. Equating the expression under the radical to zero, we obtain for $m>0$

$$\varrho_{1,2}=\sqrt{\frac{2\hbar c}{e\mathcal{H}}}\left(\sqrt{n+m+\frac{1}{2}}\pm\sqrt{n+\frac{1}{2}}\right).$$

12.

$$\Delta\varrho\sim\sqrt{\frac{\hbar c}{e\mathcal{H}}}.$$

14. The Pauli equation has the form

$$i\hbar\frac{\partial}{\partial t}\begin{pmatrix}\psi_1\\ \psi_2\end{pmatrix}=\hat{H}_0\begin{pmatrix}\psi_1\\ \psi_2\end{pmatrix}-\mu_0(\hat{\boldsymbol{\sigma}}\boldsymbol{\mathcal{H}})\begin{pmatrix}\psi_1\\ \psi_2\end{pmatrix},$$

where $\hat{H}_0=\frac{1}{2\mu}\left(\mathbf{p}-\frac{e}{c}\mathbf{A}\right)^2+eU$, and μ_0 is the magnetic moment. We shall seek a wave function of the type

$$\begin{pmatrix}\psi_1\\ \psi_2\end{pmatrix}=\varphi(x,y,z,t)\begin{pmatrix}s_1(t)\\ s_2(t)\end{pmatrix}.$$

Here the function φ is a solution of the equation

$$i\hbar\frac{\partial\varphi}{\partial t}=\hat{H}_0\varphi.$$

Then for the spin function $\begin{pmatrix}s_1\\ s_2\end{pmatrix}$ we obtain the equation

$$i\hbar\frac{\partial}{\partial t}\begin{pmatrix}s_1\\ s_2\end{pmatrix}=-\mu_0(\hat{\boldsymbol{\sigma}}\boldsymbol{\mathcal{H}})\begin{pmatrix}s_1\\ s_2\end{pmatrix}.$$

15. Since $\mathcal{H}_x=\mathcal{H}_y=0$, $\mathcal{H}_z=\mathcal{H}(t)$, we have

$$i\hbar\frac{\partial s_1}{\partial t}=-\mu_0\mathcal{H}(t)s_1,$$

$$i\hbar\frac{\partial s_2}{\partial t}=\mu_0\mathcal{H}(t)s_2.$$

The solutions of these equations have the form

$$s_1=c_1\exp\left[\frac{i\mu_0}{\hbar}\int_0^t\mathcal{H}(t)\,dt\right],$$

$$s_2=c_2\exp\left[-\frac{i\mu_0}{\hbar}\int_0^t\mathcal{H}(t)\,dt\right].$$

The constants c_1 and c_2 are found from the initial conditions $c_1 = e^{-i\alpha} \cos \delta$, $c_2 = e^{i\alpha} \sin \delta$. From the form of the fuctions s_1 and s_2 it appears that the probability of a given orientation of the spin with respect to the z axis does not vary with time. The mean value of the spin component on the x axis is given by the expression

$$\bar{s}_x = \frac{\hbar}{2} \sin 2\delta \cos \left\{ \frac{2\mu_0}{\hbar} \int_0^t \mathcal{H}(t)\, dt - 2\alpha \right\},$$

and, similarly, on the y axis by

$$\bar{s}_y = -\frac{\hbar}{2} \sin 2\delta \sin \left\{ \frac{2\mu_0}{\hbar} \int_0^t \mathcal{H}(t)\, dt - 2\alpha \right\}.$$

The direction along which the spin component has the value $+\frac{1}{2}$ is described by the polar coordinates $\Theta = 2\delta$, $\Phi = 2\left(\alpha - \frac{\mu_0}{\hbar} \int_0^t \mathcal{H}(t)\, dt\right)$.

Therefore this direction generates in time a conical surface. With the intensity of the field constant, the straight line "along which the spin is directed" rotates uniformly around the direction of the magnetic field with a frequency of $\frac{2\mu_0 \mathcal{H}}{\hbar}$.

16. A state of arbitrary polarization of the incident beam can always be represented as the superposition of two states, one with the spin $\begin{pmatrix}1\\0\end{pmatrix}$ directed along the z axis, the other $\begin{pmatrix}0\\1\end{pmatrix}$ with the opposite spin orientation. We shall first consider the case in which the spins of the neutrons in the incident beam are oriented along the z axis. Then the incident, reflected, and refracted waves will be of the form

$$A\begin{pmatrix}1\\0\end{pmatrix} e^{i\mathbf{k}\mathbf{r}}, \quad B\begin{pmatrix}1\\0\end{pmatrix} e^{i\mathbf{k}_1\mathbf{r}}, \quad C\begin{pmatrix}1\\0\end{pmatrix} e^{i\mathbf{k}_2\mathbf{r}}.$$

The quantities $\mathbf{k}, \mathbf{k}_1\, \mathbf{k}_2$ are connected with the total energy E and the magnetic moment μ_0 of the neutron by means of the relations

$$\mathbf{k} = \frac{\mathbf{p}}{\hbar}, \quad \frac{\hbar^2 k^2}{2\mu} = E, \quad \frac{\hbar^2 k_1^2}{2\mu} = E, \quad \frac{\hbar^2 k_2^2}{2\mu} = E + \mu_0 \mathcal{H}.$$

From the continuity conditions for the wave function and its derivatives with respect to x at the boundary between the two regions ($x = 0$)

we obtain

$$k_y = k_{1y} = k_{2y}, \quad k_z = k_{1z} = k_{2z},$$

$$A + B = C,$$

$$k_x A + k_{1x} B = k_{2x} C.$$

From these relations it follows that $k_{1x} = -k_x$, i.e., the angle of incidence φ is equal to the angle of reflection φ_1. For the sake of simplicity, we set $k_y = 0$. Solving the equation, we obtain

$$\left(\frac{B}{A}\right) = \frac{k_x - k_{2x}}{k_x + k_{2x}}, \quad \left(\frac{C}{A}\right) = \frac{2k_x}{k_x + k_{2x}},$$

$$k_{2x} = k_x \sqrt{1 + \frac{2\mu}{\hbar^2 k_x^2} \mu_0 \mathcal{H}}.$$

Thus the reflection coefficient R is

$$R = \left(\frac{B}{A}\right)^2 = \left(\frac{k_x - k_{2x}}{k_x + k_{2x}}\right)^2.$$

If the direction of neutron spin is antiparallel to the z direction, we obtain the equation

$$k_{2x} = k_x \sqrt{1 - \frac{2\mu}{\hbar^2 k_x^2} \mu_0 \mathcal{H}},$$

the remaining results being similar. Since μ_0 is negative, then for the angle of refraction we have $\varphi_{2\uparrow} > \varphi > \varphi_{2\downarrow}$ (see Fig. 27).

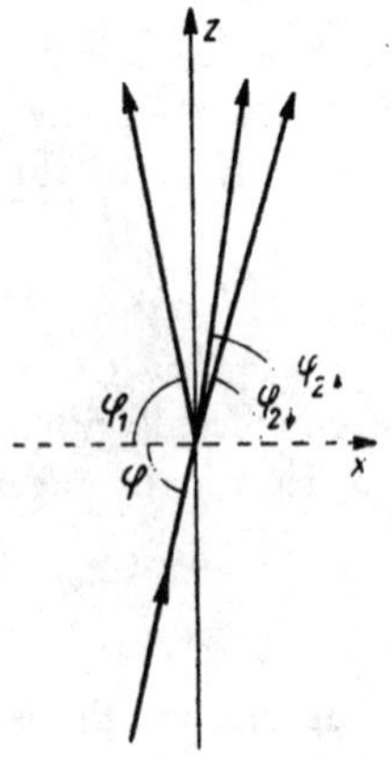

Fig. 27

In the case of an arbitrary orientation of the neutron spins the wave function in the region $x > 0$ will be of the form

$$C_{\uparrow}\begin{pmatrix}1\\0\end{pmatrix}e^{-i\mathbf{k}_{\uparrow}\mathbf{r}}+C_{\downarrow}\begin{pmatrix}0\\1\end{pmatrix}e^{i\mathbf{k}_{\downarrow}\mathbf{r}},$$

where $C_{\uparrow}$ and $C_{\downarrow}$ are the coefficients of the expansion of the initial spin state in terms of the states $\begin{pmatrix}1\\0\end{pmatrix}$ and $\begin{pmatrix}0\\1\end{pmatrix}$.

It may readily be estimated that even for $\mathcal{H} \sim 10^4$ there will be a significant reflection only for very slow (thermal) neutrons ($\lambda \sim 1$ Å) and for an angle of incidence φ differing from $\pi/2$ only by a fraction of a degree.

17. The Schrödinger equation for the spin function in the z-representation $\begin{pmatrix}s_1\\s_2\end{pmatrix}$ has the form

$$i\hbar\frac{\partial}{\partial t}\begin{pmatrix}s_1\\s_2\end{pmatrix}=-\mu\begin{pmatrix}\mathcal{H}_z & \mathcal{H}_x-i\mathcal{H}_y\\ \mathcal{H}_x+i\mathcal{H}_y & -\mathcal{H}_z\end{pmatrix}\begin{pmatrix}s_1\\s_2\end{pmatrix}$$

(here μ denotes the magnetic moment of the particle).

We introduce the notation

$$\frac{\mu}{\hbar}\mathcal{H}\cos\vartheta=a,\qquad \frac{\mu}{\hbar}\mathcal{H}\sin\vartheta=b.$$

After introduction of the new notation, the equations for determining the components s_1 and s_2 assume the form

$$\frac{ds_1}{dt}=ias_1+ibe^{-i\omega t}s_2,$$

$$\frac{ds_2}{dt}=ibe^{i\omega t}s_1-ias_2.$$

The solution of the set of equations is

$$s_1=Ae^{ip_1t}+Be^{ip_2t},$$

$$s_2=e^{i\omega t}\left[\frac{-a+p_1}{b}Ae^{ip_1t}+\frac{-a+p_2}{b}Be^{ip_2t}\right],$$

where

$$p_1=\sqrt{\frac{1}{4}\omega^2+a^2+b^2+\omega a}-\frac{\omega}{2},$$

$$p_2=-\sqrt{\frac{1}{4}\omega^2+a^2+b^2+\omega a}-\frac{\omega}{2}.$$

We determine the quantities A and B from the initial conditions and the normalization conditions $|s_1|^2 + |s_2|^2 = 1$.

After simple calculations we get the following values for the transition probability:

$$P\left(\frac{1}{2}, -\frac{1}{2}\right) = \frac{\sin^2\vartheta}{1+q^2-2q\cos\vartheta}\sin^2\left[\frac{t}{2}\omega(1-2q\cos\vartheta+q^2)^{1/2}\right],$$

where q denotes the ratio of the frequency of the Larmor precession to the frequency ω of the rotating magnetic field:

$$q = -\frac{2\mu\mathcal{H}}{\hbar\omega} = \frac{\omega_0}{\omega}.$$

The quantity q is positive if the magnetic field rotates in the direction of precession, and is negative in the opposite case.

If the angle ϑ is small, i.e., if $\sqrt{\mathcal{H}_x^2+\mathcal{H}_y^2}/\mathcal{H}_z \ll 1$, then the transition probability is approximately

$$P\left(\frac{1}{2}, -\frac{1}{2}\right) = \frac{\vartheta^2}{(1-q)^2+q\vartheta^2}\sin^2\left[\frac{t}{2}\omega[(1-q)^2+q\vartheta^2]^{1/2}\right].$$

From this formula it is seen that if the resonance relation $\omega = \omega_0$ is satisfied, i.e., if $q = +1$, the probability of a change in the orientation of the magnetic moment with respect to the magnetic field, which is equal to $P\left(\frac{1}{2}, -\frac{1}{2}\right) \approx \sin^2\frac{t\omega}{2}\vartheta$, may be close to unity for some value of t.

If, on the other hand, in the given case we change the direction of rotation of the magnetic field or if we change the sign of $\mathcal{H}_z$, then the transition probability is given by

$$P\left(\frac{1}{2}, -\frac{1}{2}\right) = \frac{\vartheta^2}{4}\sin^2\omega t,$$

which is considerably smaller than unity. The magnetic moment of the particle can be found from such a sharp qualitative difference.

18. (a) $\hat{x}(t) = \frac{\hat{p}_x}{m}t + x, \quad \hat{y}(t) = -\frac{\mu k}{2m}t^2\hat{\sigma}_y + \frac{\hat{p}_y}{m}t + y,$

$$\hat{z}(t) = \frac{\mu k}{2m}t^2\hat{\sigma}_z + \frac{\hat{p}_z}{m}t + z.$$

(b) $\overline{\hat{z}(t)} = (\bar{z})_0 + \frac{\mu k}{2m} t^2(\alpha\alpha^* - \beta\beta^*),$

$$\overline{\hat{y}(t)} = (\bar{y})_0 - \frac{\mu k}{2m} t^2(i\beta^*\alpha - i\beta\alpha^*),$$

$$\overline{\hat{x}(t)} = (\bar{x})_0 + \frac{p_0}{m} t,$$

$$\overline{(\Delta\hat{z})^2_t} = \overline{(\Delta z)^2_0} + \frac{\mu^2 k^2}{4m^2} t^4[1-(\alpha\alpha^*-\beta\beta^*)^2] + \frac{\hbar^2 t^2}{m^2}\int\left(\frac{\partial\varphi}{\partial z}\right)^2 d\tau,$$

$$\overline{(\Delta\hat{y})^2_t} = \overline{(\Delta y)^2_0} + \frac{\mu^2 k^2}{4m^2} t^4[1-(i\beta^*\alpha-i\beta\alpha^*)^2] + \frac{\hbar^2 t^2}{m^2}\int\left(\frac{\partial\varphi}{\partial y}\right)^2 d\tau,$$

$$\overline{(\Delta\hat{x})^2_t} = \overline{(\Delta x)^2_0} + \qquad\qquad + \frac{\hbar^2 t^2}{m^2}\int\left(\frac{\partial\varphi}{\partial x}\right)^2 d\tau.$$

Note: Consider, for example, a group of particles for which the spin component on the z axis for $t = 0$ has the value $+\frac{1}{2}$, i.e., $\alpha = 1$, $\beta = 0$. It is readily seen from the above results that such particles moving in a non-homogeneous magnetic field will produce two spots on a screen placed in their path. The z coordinates of these spots will be the same and the y coordinates will have opposite signs.

19. The direction of the magnetic field may be represented by the two polar angles ϑ, φ these angles being functions of time. The Hamiltonian for a neutral particle can be represented in the form

$$\hat{H} = -\mu\mathcal{H}(\hat{J}_x \sin\vartheta\cos\varphi + \hat{J}_y \sin\vartheta\sin\varphi + \hat{J}_z\cos\vartheta);$$

here $\mathcal{H}$ is the absolute value of the magnetic field intensity. We denote by $\hat{J}_\xi$ the operator of the angular momentum in the direction of the magnetic field:

$$\hat{J}_\xi = \hat{J}_x \sin\vartheta\cos\varphi + \hat{J}_y \sin\vartheta\sin\varphi + \hat{J}_z\cos\vartheta$$

and introduce the functions $\psi_m(t)$, which are eigenfunctions of the operator $\hat{J}_\xi$, i.e.,

$$\hat{J}_\xi\psi_m(t) = m\psi_m(t).$$

We shall seek a solution of the Schrödinger equation

$$i\hbar\frac{\partial\psi}{\partial t} = \hat{H}\psi$$

in the form

$$\psi = \sum a_m(t)\psi_m(t).$$

As is already known (see Prob. 20, §4),

$$\psi_m(t)=\exp(-i\hat{J}_z\varphi)\exp(-i\hat{J}_y\vartheta)\psi_m^{(0)},$$

where $\psi_m^{(0)}$ satisfies the relation

$$\hat{J}_z\psi_m^{(0)}=m\psi_m^{(0)}.$$

We shall first calculate $\dot{\psi}_m(t)$. Making use of the relations

$$\exp(i\hat{J}_y\vartheta)\hat{J}_z\exp(-i\hat{J}_y\vartheta)=\hat{J}_z\cos\vartheta-\hat{J}_x\sin\vartheta,$$

$$\hat{J}_x\psi_m^{(0)}=\frac{1}{2}\sqrt{(j+m)(j-m+1)}\,\psi_{m-1}^{(0)}+\frac{1}{2}\sqrt{(j+m+1)(j-m)}\,\psi_{m+1}^{(0)}.$$

$$\hat{J}_y\psi_m^{(0)}=\frac{i}{2}\sqrt{(j+m)(j-m+1)}\,\psi_{m-1}^{(0)}-\frac{i}{2}\sqrt{(j+m+1)(j-m)}\,\psi_{m+1}^{(0)},$$

we have

$$\dot{\psi}_m(t)=(-i\dot{\varphi}m\cos\vartheta)\psi_m(t)+\frac{1}{2}\sqrt{(j+m+1)(j-m)}\,(i\dot{\varphi}\sin\vartheta-\dot{\vartheta})\psi_{m+1}(t)+$$
$$+\frac{1}{2}\sqrt{(j+m)(j-m+1)}\,(i\dot{\varphi}\sin\vartheta+\dot{\vartheta})\psi_{m-1}(t).$$

Inserting the above expression for $\dot{\psi}_m(t)$ in the equation

$$i\hbar\sum_m\{\psi_m(t)\dot{a}_m(t)+a_m(t)\dot{\psi}_m(t)\}=-\mu\mathcal{H}\sum_m ma_m(t)\psi_m(t),$$

we obtain a system of equations which give us the time-dependence of the coefficients a_m:

$$i\hbar\frac{da_m}{dt}+m\mu\mathcal{H}a_m=-m\hbar\dot{\varphi}\cos\vartheta a_m+$$
$$+\frac{1}{2}\hbar(\dot{\varphi}\sin\vartheta+i\dot{\vartheta})\sqrt{(j+m)(j-m+1)}\,a_{m-1}+$$
$$+\frac{1}{2}\hbar(\dot{\varphi}\sin\vartheta-i\dot{\vartheta})\sqrt{(j-m)(j+m+1)}\,a_{m+1}.$$

If

$$\dot{\varphi}\ll\frac{\mu\mathcal{H}}{\hbar},\qquad\dot{\vartheta}\ll\frac{\mu\mathcal{H}}{\hbar},$$

i.e., if the angular velocity of the change in the magnetic field direction is much smaller than the frequency of precession, then we can neglect the right-hand side of each equation of the above system. We then obtain

$$a_m\sim\exp\left[i\frac{m\mu\mathcal{H}}{\hbar}t\right].$$

Therefore in this case the probability of finding different values of the component of the angular moment in the time-varying direction of the magnetic field does not change.

20.

$$E(m=-1)=2\mu\mathcal{H}-\mu^2a^{-3}(3\cos^2\theta-1),$$
$$E(m=0)=2\mu^2a^{-3}(3\cos^2\theta-1),$$
$$E(m=1)=-2\mu\mathcal{H}-\mu^2a^{-3}(3\cos^2\theta-1).$$

A similar problem arises when considering the nuclear magnetic resonance, for example, in gypsum $CaSO_4 \cdot 2H_2O$. Gypsum contains pairs of protons sufficiently far apart. In a first approximation, the crystal can be regarded as a number of independent nuclear magnetic pairs, where each nucleus interacts only with its partner.

22. Since the magnetic interaction between nuclei is considered to be small, the Hamilton operator has the following form

$$\hat{H}=g\beta\mathcal{H}\sum_{j=1}^{3}\hat{I}_{zj}+\hat{V},$$

where

$$\hat{V}=\sum A_{jk}(\hat{\mathbf{I}}_j\hat{\mathbf{I}}_k-3\hat{I}_{zj}\hat{I}_{zk})$$

(see the preceding problem).

The state of the system of three nuclei will be characterized by the following set of quantities:

$$\hat{\mathbf{J}}^2(\hat{\mathbf{J}}=\hat{\mathbf{I}}_1+\hat{\mathbf{I}}_2),\ \hat{\mathbf{I}}^2(\hat{\mathbf{I}}_1+\hat{\mathbf{I}}_2+\hat{\mathbf{I}}_3=\hat{\mathbf{I}}),$$
$$\hat{I}_z(\hat{I}_z=\hat{I}_{1z}+\hat{I}_{2z}+\hat{I}_{3z})$$

Calculating the matrix, we have

$$\begin{array}{lll|cccc|}
I_z=\frac{3}{2} & I=\frac{3}{2} & & -2a & 0 & 0 & 0\\
 & I=\frac{3}{2} & & 0 & 2a & \beta & \gamma\\
I_z=\frac{1}{2} & I=\frac{1}{2} & J=0 & 0 & \beta & 0 & 0\\
 & I=\frac{1}{2} & J=1 & 0 & \gamma & 0 & 0
\end{array};$$

here

$$\alpha = \frac{1}{4}(A_{12}+A_{23}+A_{31}), \quad \gamma = \frac{\sqrt{2}}{4}(2A_{12}-A_{13}-A_{21}),$$

$$\beta = \frac{\sqrt{6}}{4}(A_{13}-A_{23}), \qquad \delta^2 = \alpha^2+\beta^2+\gamma^2.$$

Matrix V for $I_z = -\frac{3}{2}, -\frac{1}{2}$ has the same form. The eigenvalues of the matrix for the dipole-dipole energy are equal to -2α, 0, $\alpha+\delta$, $\alpha-\delta$. We obtain the following values for the energy levels we are seeking

$$E_1 = \frac{3}{2}g\beta\mathcal{H} - 2\alpha, \qquad E_5 = -\frac{3}{2}g\beta\mathcal{H} - 2\alpha,$$

$$E_2 = \frac{1}{2}g\beta\mathcal{H}, \qquad E_6 = -\frac{1}{2}g\beta\mathcal{H},$$

$$E_3 = \frac{1}{2}g\beta\mathcal{H} + \alpha + \delta, \quad E_7 = -\frac{1}{2}g\beta\mathcal{H} + \alpha + \delta,$$

$$E_4 = \frac{1}{2}g\beta\mathcal{H} + \alpha - \delta, \quad E_8 = -\frac{1}{2}g\beta\mathcal{H} + \alpha + \delta.$$

23. We take the z axis in the direction of the magnetic field. The Hamilton operator has the form

$$\hat{H} = g\beta H\hat{I}_z + \hat{V}, \qquad \hat{V} = \frac{1}{2}A\sum \varphi_{ik}\hat{I}_i\hat{I}_k + \hat{I}_k\hat{I}_i).$$

Here

$$A = \frac{Q_0}{2I(2I-1)}, \qquad \varphi_{ik} = \frac{\partial^2\varphi}{\partial x_i\,\partial x_k}.$$

Calculating the matrix elements $\hat{V}$, we have

$$< m|\hat{V}|m > = \frac{AQ_{33}}{2}\{3m^2 - I(I+1)\},$$

$$< m|\hat{V}|m\pm 1) > = \frac{A(Q_{31}\pm iQ_{23})}{2}(2m\pm 1)\{(I\mp m)(I\pm m+1)\}^{\frac{3}{2}}$$

$$< m|\hat{V}|m\pm 2 > = \frac{A(\varphi_{11}-\varphi_{22}\pm 2i\varphi_{12})}{4}\{(I\mp m)(I\mp m-1)(I\pm m+1)\times$$

$$\times(I\pm m+2)\}^{\frac{1}{2}}.$$

Using the standard perturbation theory for the energy levels in the second approximation, we obtain the following values:

$$E_m = mg\beta\mathcal{H} + \frac{A}{2}\{3m^2 - I(I+1)\}\frac{\partial^2\varphi}{\partial z^2} + \frac{A}{2}m\frac{\left|\frac{\partial^2\varphi}{\partial x\partial z} - i\frac{\partial^2\varphi}{\partial y\partial z}\right|^2}{g\beta\mathcal{H}} \times$$

$$\times\{4I(I+1) - 8m^2 - 1\} + \frac{A^2}{8}m\frac{\left|\frac{\partial^2\varphi}{\partial x^2} - \frac{\partial^2\varphi}{\partial y^2} + 2i\frac{\partial^2\varphi}{\partial x\partial y}\right|^2}{g\beta\mathcal{H}} \times$$

$$\times\{2I(I+1) - 2m^2 - 1\}.$$

The derivative of the electrostatic potential φ is taken at the position of the nucleus.

§ 7. The atom

1. From the inequality we have

$$\int|\nabla\psi|^2\,d\tau + Z\int(\nabla|\psi|^2\nabla r)\,d\tau + Z^2\int(\nabla r)^2|\psi|^2\,d\tau \geqslant 0.$$

We integrate by parts the second term; since $(\nabla r)^2 = 1$ and $\Delta r = 2/r$, we obtain

$$\frac{1}{2}\int|\nabla\psi|^2\,d\tau - \int\frac{Z}{r}|\psi|^2\,d\tau \geqslant -\frac{Z^2}{2}\int|\psi|^2\,d\tau.$$

The left-hand side of the inequality represents the mean value of the Hamiltonian* $\hat{H} = -\frac{1}{2}\Delta - \frac{Z}{r}$ in the state ψ. The lowest value of the energy is $-Z^2/2$. The atom is then in the state with the wave function ψ_0 satisfying the first-order equation

$$\nabla\psi_0 + Z\psi_0\nabla r = 0,$$

from which it follows that

$$\psi_0 \sim e^{-Zr}.$$

5. First we calculate the wave functions in the momentum representation from the general formula

$$\varphi(\mathbf{p}) = \frac{1}{(2\pi\hbar)^{3/2}}\iiint e^{-i\frac{\mathbf{pr}}{\hbar}}\psi(\mathbf{r})\,d\tau.$$

---

* In units of $e = \hbar = \mu = 1$.

For state 1*s* we find

$$\varphi_{1s}(p)=\frac{1}{\pi}\left(\frac{2a}{\hbar}\right)^{3/2}\frac{1}{\left(\frac{p^2a^2}{\hbar^2}+1\right)^2}.$$

Analogously, for the 2*s* state we get

$$\varphi_{2s}(p)=\frac{1}{2\pi}\left(\frac{2a}{\hbar}\right)^{3/2}\frac{\frac{p^2a^2}{\hbar^2}-\frac{1}{4}}{\left(\frac{p^2a^2}{\hbar^2}+\frac{1}{4}\right)^3}.$$

For the 2*p* state we have three eigenfunctions ($m_z=-1,\ 0,\ +1$)

$$\varphi_{2p}^{(0)}(p)=-i\frac{1}{\pi}\left(\frac{a}{\hbar}\right)^{3/2}\frac{p_za}{\hbar\left(\frac{p^2a^2}{\hbar^2}+\frac{1}{4}\right)^3},$$

$$\varphi_{2p}^{(\pm1)}(p)=-\frac{i}{\sqrt{2}\pi}\left(\frac{a}{\hbar}\right)^{3/2}\frac{(p_x\pm ip_y)a}{\hbar\left(\frac{p^2a^2}{\hbar^2}+\frac{1}{4}\right)^3}.$$

With the help of these expressions we find the normalized distribution of the momenta

$$w(\mathbf{p})=|\varphi(\mathbf{p})|^2.$$

6. $$\sqrt{\overline{r^2}-\bar{r}^2}=\frac{\sqrt{n^2(n^2+2)-l^2(l+1)^2}}{2}.$$

For a given value of n this expression reaches a minimum for "circular orbits", i.e., for $l=n-1$

$$\sqrt{\overline{r^2}-\bar{r}^2}=\frac{1}{2}n\sqrt{2n+1},\quad \frac{\sqrt{\overline{r^2}-\bar{r}^2}}{\bar{r}}=\frac{1}{\sqrt{2n+1}}.$$

7. For $n_1=1,\ n_2=0,\ m=0$

$$\psi_{1,0,0}(\xi,\eta,\varphi)=-\frac{1}{\sqrt{2}}R_{20}(r)Y_{00}(\vartheta,\varphi)+\frac{1}{\sqrt{2}}R_{21}(r)Y_{10}(\vartheta,\varphi).$$

9. In non-quantum relativistic mechanics the Hamiltonian has the form

$$H=\sqrt{\mu^2c^4+p^2c^2}-\mu c^2+U(r)\approx\frac{p^2}{2\mu}+U(r)+H_1,$$

where

$$H_1 = -\frac{p^4}{8\mu^3 c^2}.$$

We now take $\mathbf{p}$ to be the operator $\mathbf{p} = -i\hbar\nabla$, and consider H_1 as a perturbation. In our approximation the Schrödinger equation has the form

$$\left[\frac{\hat{p}^2}{2\mu} + U(r)\right]\psi = E\psi,$$

and the correction we are seeking to the energy in the state n, l, m is

$$\Delta E_1 = -\frac{1}{8\mu^3 c^2}\int \psi^* \hat{p}^4 \psi \, d\tau = -\frac{1}{2\mu c^2}\int \psi^* (E - U(r))^2 \psi \, d\tau$$

$$= \frac{3E^2}{2\mu c^2} - \frac{\left(\frac{\mu e^4}{\hbar^2}\right)^2}{n^3 \mu c^2 (2l+1)} = \left[\frac{2}{8n^4} - \frac{1}{(2l+1)n^3}\right]\frac{\mu e^4}{\hbar^2}\left(\frac{e^2}{\hbar c}\right)^2.$$

10. Instead of starting from the unperturbed wave functions with a given l_z and s_z and then solving the secular equation, it is more convenient to choose as the unperturbed wave functions the eigenfunctions for given $\hat{\mathbf{l}}^2$ and $\hat{\mathbf{j}}^2$, where $\hat{\mathbf{j}} = \hat{\mathbf{l}} + \hat{\mathbf{s}}$ is the total angular momentum, which, as readily shown, commutes with $\hat{H}_2$. For these functions we have the relation

$$\hat{\mathbf{j}}^2 = j(j+1) = l(l+1) + s(s+1) + 2\hat{\mathbf{l}}\hat{\mathbf{s}}.$$

We then obtain

$$\Delta E_2 = \overline{H}_2 = \frac{j(j+1) - l(l+1) - s(s+1)}{4\mu^2 c^2}\,\hbar^2\left(\overline{\frac{1}{r}\frac{dU}{dr}}\right).$$

For the hydrogen atom $U = -e^2/r$, and since

$$\overline{\frac{1}{r^3}} = \frac{1}{n^3(l+1)\left(l+\frac{1}{2}\right)l}\left(\frac{\mu e^2}{\hbar^2}\right)^3,$$

we finally obtain

$$\Delta E_2 = \frac{\mu e^4}{\hbar^2}\left(\frac{e^2}{\hbar c}\right)^2 \frac{j(j+1) - l(l+1) - s(s+1)}{4n^3 l\left(l+\frac{1}{2}\right)(l+1)}.$$

This equation may be written more compactly, since for s we have here the value $\frac{1}{2}$ and for j the two cases $j = l-\frac{1}{2}$. $j = l+\frac{1}{2}$. It is readily shown that

$$2\mathbf{ls} = j(j+1)-l(l+1)-s(s+1) = \begin{cases} l \text{ for } j = l+\dfrac{1}{2}, \\ -(l+1) \text{ for } j = l-\dfrac{1}{2}, \end{cases}$$

so that for any j and l

$$\Delta E_2 = \frac{\mu e^4}{\hbar^2}\left(\frac{e^2}{\hbar c}\right)^2 \frac{1}{2n^3}\left(-\frac{1}{j+\dfrac{1}{2}}+\frac{1}{l+\dfrac{1}{2}}\right).$$

Adding ΔE_2 to the correction ΔE_1 resulting from the velocity dependence of the mass (see the preceding problem), we obtain

$$\Delta E = \frac{\mu e^4}{\hbar^2}\left(\frac{e^2}{\hbar c}\right)\frac{1}{n^3}\left(\frac{3}{8n}-\frac{1}{2j+1}\right).$$

This expression does not depend on l, i.e., the two levels with the same j and different l have the same energy (the levels are degenerate).

12. In β-decay the tritium nucleus goes over into the nucleus of the helium isotope He^3. The effect of the β-decay on the electron of the atom is that for a short time $t \ll \hbar^3/\mu e^4$ the potential energy of the electron in the atom is not $U = -e^2/r$, but $U = -2e^2/r$. We can estimate the time t by assuming that the β-electron travels through the atom in this time:

$$t \sim \frac{a_0}{v},$$

where $a_0 = \hbar^2/\mu c^2$, v is the velocity of the β-electron.

Since the energy of the β-electron is of the order of several keV, we therefore find that $t \sim 0 \cdot 1\, \hbar^3/\mu e^4$. At time t the wave function of the electron practically does not change, as can be seen from the Schrödinger equation:

$$\delta\psi \sim \frac{e^2}{r}\frac{t}{i\hbar}\psi \ll \psi.$$

We expand the wave function ψ of the electron into the eigenfunction of the electron in the field $Z = 2$:

$$\psi = \sum_n c_n\psi_n + \int c_{\mathbf{k}}\psi_{\mathbf{k}}\, d\mathbf{k}.$$

The expansion coefficients

$$c_n = \int \psi\psi_n \, d\tau,$$

$$c_{\mathbf{k}} = \int \psi\psi_{\mathbf{k}} \, d\tau,$$

define the probability of excitation

$$w_n = \sum |c_n|^2$$

and ionization

$$w_{\text{ion}} = \int |c_{\mathbf{k}}|^2 \, d\mathbf{k}.$$

Since ψ is a spherically symmetric function, then c_n and c_k differ from zero only in case n and k are s states ($l = 0$).
Since

$$R_{n0}^{(Z)} = 2\left(\frac{Z}{n}\right)^{3/2} e^{-\frac{Zr}{n}} F\left(-n+1, 2, \frac{2Zr}{n}\right),$$

we find that

$$c_n = \int_0^\infty R_{n0}^{(Z)} R_{10}^{(Z')} r^2 \, dr = \frac{8}{\left(Z' + \frac{Z}{n}\right)^3} \left(\frac{ZZ'}{n}\right)^{3/2} F\left(-n+1, 3, 2, \frac{2Z}{nZ'+Z}\right).$$

Setting $Z = 2$, $Z' = 1$, we obtain for $n = 1$

$$c_1 = \frac{16\sqrt{2}}{27},$$

i.e., the probability of the He^3 ion being in the ground state is equal to $w_1 = |c_1|^2 = \left(\frac{8}{9}\right)^3 = 0\cdot70$. According to this, the total probability of excitation and ionization will be equal to $1 - w_1 = 0\cdot30$. For $n = 2$, $c_2 = -\frac{1}{2}$ we have $w_2 = 0\cdot25$.
Using the formula

$$F(\alpha, \beta, \gamma, x) = (1-x)^{\gamma-\alpha-\beta} F(\gamma-\alpha, \gamma-\beta, \gamma, x),$$

we find

$$w_n = |c_n|^2 = \frac{2^9 n^5 (n-2)^{2n-4}}{(n+2)^{2n+4}}.$$

We give the value of the probability of excitation calculated with the help of this formula for the first few levels:

$$w_1 = \left(\frac{8}{9}\right)^3, \quad w_2 = \frac{1}{4}, \quad w_3 = \frac{2^9 3^5}{5^{10}} \approx 1\cdot 3\%,$$

$$w_4 = \frac{2^{23}}{2^{12}\, 3^{12}} \approx 0\cdot 39\%.$$

13. The Hamiltonian has the form (all calculations will be performed in the system of units in which $e = \hbar = \mu = 1$)

$$\hat{H} = -\frac{1}{2}\Delta_1 - \frac{1}{2}\Delta_2 - \frac{Z}{r_1} - \frac{Z}{r_2} + \frac{1}{r_{12}}.$$

Employing the variational method, we must evaluate the integral

$$E(Z') = \int \psi^*(\mathbf{r}_1, \mathbf{r}_2)\hat{H}\psi(\mathbf{r}_1, \mathbf{r}_2)\, d\tau_1\, d\tau_2,$$

and from the condition $dE/dZ' = 0$ find the value of Z'.

In our case

$$\psi(\mathbf{r}_1, \mathbf{r}_2) = c\, e^{-Z'(r_1+r_2)},$$

where for the normalization constant we have $c = Z'^3/\pi$.

The integrals of the first four terms are readily calculated:

$$\int \psi(\mathbf{r}_1, \mathbf{r}_2)\left\{-\frac{1}{2}\Delta_1 - \frac{1}{2}\Delta_2 - \frac{Z}{r_1} - \frac{Z}{r_2}\right\}\psi(\mathbf{r}_1, \mathbf{r}_2)\, d\tau_1\, d\tau_2 = Z'^2 - 2ZZ',$$

while the term with $1/r_{12}$ is more conveniently integrated in elliptical coordinates:

$$s = r_1 + r_2, \quad t = r_1 - r_2, \quad u = r_{12},$$

$$d\tau_1\, d\tau_2 = \pi^2(s^2 - t^2)u\, ds\, dt\, du,$$

$$-u \leqslant t \leqslant u, \quad 0 \leqslant u \leqslant s \leqslant \infty.$$

We then have

$$\int \psi^2(r_1, r_2)\frac{1}{r_{12}}\, d\tau_1\, d\tau_2 = \pi^2 c^2 \int_0^\infty ds \int_0^s du \int_{-u}^{+u} dt\, e^{-2Z's}\frac{s^2 - t^2}{u}u = \frac{5}{8}Z'.$$

Finally we obtain

$$E(Z') = Z'^2 - 2ZZ' + \frac{5}{8}Z'.$$

It is then found that $E(Z')$ is a minimum when

$$Z' = Z - \frac{5}{16}.$$

For this value of Z' the energy of the ground state is

$$E = -\left(Z - \frac{5}{16}\right)^2.$$

To obtain an idea of the accuracy of the above calculation we compute the ionization potential of helium ($Z = 2$) and compare the result with the observed value. The ionization potential of helium I_{He} is equal to the difference in the energy of a singly ionized helium atom and a neutral helium atom in the ground state. The calculated value of I_{He} is thus

$$I_{He} = 0\cdot8476 \text{ atomic units} = 1\cdot695 \text{ Ry}^* = 22\cdot9 \text{ eV}.$$

The observed value is $I_{He} = 1\cdot810$ *Ry*.

The ionization potentials of two-electron systems such as Li^+, Be^{++}, and others are also known from experiment. The observed and calculated values for several cases are compared in the table below.

Element	He ($Z = 2$)	Li^+ ($Z = 3$)	Be^{++} ($Z = 4$)	Be^{+++} ($Z=5$)	C^{++++} ($Z=6$)
Calculated I (*Ry*)	1·6952	5·445	11·195	18·945	28·695
Observed I (*Ry*)	1·810	5·560	11·307	19·061	28·816

The calculated ground-state energies are in satisfactory agreemen with the observed energies.

15. In a helium atom in the ground state the orbital angular momentum and the spin are equal to zero. As a result, helium exhibits diamagnetic properties. The diamagnetic susceptibility for one gram-atom is defined by the following formula:

$$\chi = -\frac{e^2 N_A}{6\mu c^2}(\overline{r_1^2} + \overline{r_2^2}),$$

where

$$\overline{r_1^2} + \overline{r_2^2} = \int (r_1^2 + r_2^2)\psi^2 \, d\tau_1 \, d\tau_2,$$

N_A is Avogadro's number. The approximate wave function of the ground state for the helium atom is

$$\psi(r_1, r_2) = \frac{Z'^3}{\pi a^3} e^{-Z'(r_1+r_2)/a}.$$

* 1 atomic unit of energy = 2 Ry = 27 eV.

Using this function to evaluate the mean value of $\overline{r_1^2}+\overline{r_2^2}$, we obtain the following result:

$$\overline{r_1^2}+\overline{r_2^2}=\frac{2a^2}{Z'^2}.$$

Substituting this value into the formula for diamagnetic susceptibility we find

$$\chi=-1.67\times 10^{-6}.$$

The observed value of diamagnetic susceptibility is

$$\chi=-(1.90\pm 0.02)\times 10^{-6}.$$

16. We introduce for convenience the notation

$$Z_1=\alpha,\quad Z_2=2\beta,\quad 2Z_1^{3/2}=a,\quad cZ_2^{3/2}=b.$$

From the conditions of orthonormality we obtain

$$\gamma Z_2=\frac{1}{2}(\alpha+\beta),\quad b^2=\frac{12\beta^5}{\alpha^2-\alpha\beta+\beta^2}.$$

In the new notation ψ_1 and ψ_2 have the form

$$\psi_1=\psi_{100}=ae^{-\alpha r}Y_{00},$$

$$\psi_2=\psi_{200}=b\left[1-\frac{1}{3}(\alpha+\beta)r\right]e^{-\beta r}Y_{00}.$$

The approximate wave function of the lithium atom in the ground state may be represented in the following way:

$$\Phi=\frac{1}{\sqrt{3!}}\begin{vmatrix}\psi_1(1)\eta_+(\sigma_1) & \psi_1(2)\eta_+(\sigma_2) & \psi_1(3)\eta_+(\sigma_3)\\ \psi_1(1)\eta_-(\sigma_1) & \psi_1(2)\eta_-(\sigma_2) & \psi_1(3)\eta_-(\sigma_3)\\ \psi_2(1)\eta_+(\sigma_1) & \psi_2(2)\eta_+(\sigma_2) & \psi_2(3)\eta_+(\sigma_3)\end{vmatrix},$$

where

$$\eta_+\left(\frac{1}{2}\right)=1,\ \eta_+\left(-\frac{1}{2}\right)=0,\ \eta_-\left(\frac{1}{2}\right)=0,\ \eta_-\left(-\frac{1}{2}\right)=1.$$

In this state $S=\frac{1}{2}$, $M=\frac{1}{2}$.

For this case the Hamiltonian* is

$$\hat{H}=\sum_{i=1}^{3}\left\{-\frac{1}{2}\Delta_i-\frac{Z}{r_i}\right\}+\frac{1}{r_{12}}+\frac{1}{r_{23}}+\frac{1}{r_{31}}.$$

* In Coulomb units $e=\hbar=\mu=1$.

We now calculate the energy of the state Φ. The kinetic energy of an electron in the 1s state is

$$T_1 = \int \left\{ -\psi_1 \frac{1}{2} \Delta \psi_1 \right\} d\tau = \int_0^\infty \left(\frac{d\psi_1}{dr} \right)^2 r^2 \, dr = \frac{1}{2} \alpha^2,$$

in the 2s state

$$T_2 = \frac{\beta^2}{6} + \frac{\beta^4}{\alpha^2 - \alpha\beta + \beta^2}.$$

The interaction energies of the inner and outer electrons with the nucleus are

$$U_1 = -\int \frac{Z\psi_1^2}{r} d\tau = Z\alpha^2 \int_0^\infty \frac{e^{-2\alpha r}}{r} r^2 \, dr = -Z\alpha$$

and

$$U_2 = -\frac{Z\beta}{2} + \frac{Z\beta^2}{2} \frac{\alpha - 2\beta}{\alpha^2 - \alpha\beta + \beta^2} \qquad \text{(for lithium } Z = 3\text{) respectively.}$$

The Coulomb interaction energy of the inner electrons is

$$K_{11} = \int\int \frac{1}{r_{12}} |\psi_1(r_1)|^2 \, |\psi_1(r_2)|^2 \, d\tau_1 \, d\tau_2 = \frac{5}{8} \alpha.$$

The interaction energy of the inner electrons with the outer one is

$$2K_{12} = 2 \int\int \frac{1}{r_{12}} |\psi_1(r_1)|^2 \, |\psi_2(r_2)|^2 \, d\tau_1 \, d\tau_2$$

$$= 2\alpha - \frac{2\alpha^3}{(\alpha+\beta)^2} - \frac{\alpha^4 \beta (3\alpha + \beta)}{(\alpha+\beta)^3 (\alpha^2 - \alpha\beta + \beta^2)}.$$

The exchange energy for two electrons with parallel spins is

$$A = 2a^2 b^2 \int_0^\infty e^{-(\alpha+\beta)r_2} \left[1 - \frac{1}{3} (\alpha+\beta) r_2 \right] r_2^2 \, dr_2 \times$$

$$\times \int_{r_2}^\infty e^{-(\alpha+\beta)r_1} \left[1 - \frac{1}{3} (\alpha+\beta) r_1 \right] r_1 \, dr_1$$

$$= \frac{4\alpha^3 \beta^5}{(\alpha+\beta)^5 (\alpha^2 - \alpha\beta + \beta^2)}.$$

Setting $\beta = \lambda\alpha$, we obtain

$$2T_1 + T_2 = T = \alpha^2 \varphi_1(\lambda),$$

$$2U_1 + U_2 + K_{11} + 2K_{12} - A = -\alpha\varphi_2(\lambda),$$

$$E = \alpha^2 \varphi_1(\lambda) - \alpha\varphi_2(\lambda).$$

The energy is a minimum for values of α and λ satisfying the conditions

$$\frac{\partial E}{\partial \alpha}=0, \qquad \frac{\partial E}{\partial \lambda}=0,$$

or

$$2\alpha\varphi_1(\lambda)-\varphi_2(\lambda)=0, \qquad \alpha\varphi_1'(\lambda)-\varphi_2'(\lambda)=0.$$

Eliminating α, we obtain

$$\frac{\varphi_1'(\lambda)}{\varphi_1(\lambda)}-\frac{2\varphi_2'(\lambda)}{\varphi_2(\lambda)}=0, \qquad \lambda=0\cdot 2846.$$

The corresponding values of α and β are

$$\alpha=2\cdot 694, \qquad \beta=0\cdot 767.$$

Inserting these values of the variational parameters in E, we obtain for the energy of the normal state of the lithium atom the value

$$E=-7\cdot 414 \text{ atomic units}=-200\cdot 8 \text{ eV}.$$

The observed value is $E_{obs}=-202\cdot 54$ eV. Using perturbation theory, i.e., setting $\alpha=3$, $\beta=\frac{3}{2}$, we obtain for the energy of the ground state a less accurate value $E=-7\cdot 05$, or in electron-volts

$$E=-190\cdot 84 \text{ eV}.$$

17. Let us investigate the operator of the kinetic energy of the nucleus $\hat{T}$. In the centre-of-mass system

$$\mathbf{P}+\sum \mathbf{p}_i=0,$$

where $\mathbf{P}$ is the momentum of the nucleus, $\mathbf{p}_i$ the momenta of the electrons, $\hat{T}$ has the form

$$\hat{T}=\frac{\hat{P}^2}{2M}=\frac{\left(\sum \hat{\mathbf{p}}_i\right)^2}{2M}=\sum \frac{\hat{p}_i^2}{2M}+\sum_{i>k}\frac{\hat{\mathbf{p}}_i\hat{\mathbf{p}}_k}{M}.$$

Since the ratio of the mass of the electron to the mass of the nucleus is $m/M \ll 1$, the displacements can be computed with the help of perturbation theory:

$$\Delta E=\int \psi^* \hat{T}\psi\, d\tau,$$

where ψ is the wave function of electrons in the field of an infinitely heavy stationary nucleus.

The first term in the formula for $\hat{T}$ differs only by the factor m/M from the kinetic energy of the electrons, which, on the basis of the virial theorem, is equal to the energy of the atom with the opposite sign. Therefore, ΔE can be represented as the sum of two expressions

$$\Delta E = \Delta E_1 + \Delta E_2,$$

$$\Delta E_1 = -\frac{m}{M} E, \quad \Delta E_2 = \frac{1}{M} \int \psi^* \sum_{i>k} \hat{\mathbf{p}}_i \hat{\mathbf{p}}_k \psi \, d\tau.$$

Let us investigate more closely the expression for ΔE_2. If we were to take as the function ψ the product of the wave functions of the individual electrons, then the term ΔE_2 would be equal to zero, since the average momentum of the electrons in a bound state is always zero. If, however, such a wave function is made symmetric then ΔE_2 will have a value other than zero. The symmetrized eigenfunction of the helium atom can be written in the form

$$\psi(r_1, r_2) = \frac{1}{\sqrt{2}} [\psi_1(r_1)\psi_2(r_2) \pm \psi_1(r_2)\psi_2(r_1)],$$

where the upper sign corresponds to parahelium (total spin $S = 0$) and the lower, to orthohelium ($S = 1$).

Substituting this expression into the formula for ΔE_2, we obtain

$$\Delta E_2 = \pm \frac{1}{M} \left| \int \psi_1^* \hat{\mathbf{p}} \psi_2 \, d\tau \right|^2.$$

The matrix element of the momentum is different from zero only when $\Delta l = 0, \pm 1$ and therefore ΔE_2 vanishes for the states $1snd$, $1snf$, etc.

Taking for the $1s$ electron and the np electron the hydrogen wave functions with effective charges Z_1 and Z_2

$$\psi_{1s}(r) = \sqrt{\frac{Z_1^3}{\pi}} e^{-Z_1 r},$$

$$\psi_{np} = Y_{10} \frac{2Z_2^2}{3} \frac{\sqrt{n^2-1}}{n^2} r e^{-Z_2 r/n} F\left(-n+2, 4; \frac{2Z_2 r}{n}\right),$$

we obtain

$$\Delta E_2 = \pm \frac{m}{M} \frac{64}{3} (Z_1 Z_2)^5 \frac{(Z_1 n - Z_2)^{2n-4}}{(Z_1 n + Z_2)^{2n+4}} n^3 (n^2 - 1),$$

where the upper sign refers to the paraterm $1snp\,^1P$, and the lower sign to the orthoterm $1snp\,^3P$.

18. Let $\psi(x, y, z)$ be the solution of the Schrödinger equation in the case of a discrete energy spectrum. Let us consider the one-parameter family of normalized functions $\lambda^{3/2}\psi(\lambda x, \lambda y, \lambda z)$.

The expression

$$I(\lambda) = \lambda^3 \int \left\{ \frac{\hbar^2}{2\mu} |\nabla\psi(\lambda x, \lambda y, \lambda z)|^2 + U(x, y, z)|\psi(\lambda x, \lambda y, \lambda z)|^2 \right\} dx\, dy\, dz,$$

as a function of λ, should reach an extreme value for $\lambda = 1$, i.e.,

$$\left(\frac{dI}{d\lambda}\right)_{\lambda=1} = 0.$$

Proceeding to new variables of integration λx, λy, λz, we obtain

$$I(\lambda) = \lambda^2 \overline{T} + \lambda^{-\nu} \overline{U},$$

from which we have

$$2\overline{T} - \nu\overline{U} = 0.$$

The virial theorem can be readily generalized to the case of a system of many particles.

19.

$$\text{(a) } Z^{-1/3}\frac{\hbar^2}{\mu e^2}, \quad \text{(b) } Z^{1/3}\frac{\mu e^4}{\hbar^2}, \quad \text{(c) } Z^{4/3}\frac{\mu e^4}{\hbar^2}, \quad \text{(d) } Z^{7/3}\frac{\mu e^4}{\hbar^2},$$

$$\text{(e) } Z^{2/3}\frac{e^2}{\hbar}, \quad \text{(f) } Z^{1/3}\hbar, \quad \text{(g) } Z^{1/3}.$$

20. The total energy consists of three parts: the kinetic energy of the electrons T, the interaction energy between the electrons and the nucleus U_{ne}, and the interaction energy between the electrons themselves U_{ee}. The latter two terms have the forms

$$U_{ne} = -\int \frac{Z}{r} \varrho \, d\tau,$$

$$U_{ee} = \frac{1}{2}\int \frac{\varrho(\mathbf{r})\varrho(\mathbf{r}')}{|\mathbf{r}-\mathbf{r}'|}\, d\tau\, d\tau'.$$

To calculate the kinetic energy we consider an infinitesimal volume element $d\tau$ in the atom. The number of electrons of momenta lying in the interval between $\mathbf{p}$ and $\mathbf{p} + d\mathbf{p}$ is proportional to the phase space available and equals

$$dn = \frac{8\pi p^2\, dp\, d\tau}{(2\pi)^3} = \frac{p^2\, dp\, d\tau}{\pi^2}.$$

We obtain the electron density by integrating over p from 0 to some maximum value $p = p_0$.

$$\varrho = \frac{p_0^3}{3\pi^2}.$$

The kinetic energy of the electrons in the volume $d\tau$ is

$$dT = \int_0^{p_0} \frac{p^2}{2}\, dn = \frac{p_0^5}{10\pi^2}\, d\tau.$$

Expressing p_0 in terms of ϱ and integrating over the volume of the atom, we obtain the kinetic energy of the electrons:

$$T = \frac{1}{10\pi^2}(3\pi^2)^{5/3} \int \varrho^{5/3}\, d\tau.$$

Finally, we obtain the total energy

$$E = T + U_{ne} + U_{ee}$$

$$= \frac{(3\pi^2)^{5/3}}{10\pi^2} \int \varrho^{5/3}\, d\tau - Z \int \frac{\varrho}{r}\, d\tau + \frac{1}{2} \int \int \frac{\varrho(\mathbf{r})\varrho(\mathbf{r}')}{|\mathbf{r}-\mathbf{r}'|}\, d\tau\, d\tau'.$$

22. The element of volume expressed in terms of x has the form

$$d\tau = 4\pi r^2\, dr = 8\pi\lambda^3 x^5\, dx.$$

Let us compute the kinetic energy

$$T = \frac{12(3\pi^2)^{5/3}}{25\pi} \lambda^3 A^{5/3}$$

and the interaction energy between the electrons and the nucleus

$$U_{ne} = -8\pi A\lambda^2 Z.$$

In order to calculate $U_{ee} = \frac{1}{2} \int \int \frac{\varrho(\mathbf{r})\,\varrho(\mathbf{r}')}{|\mathbf{r}-\mathbf{r}'|}\, d\tau\, d\tau'$, we first find the potential φ_e produced by the electrons. Solving Poisson's equation

$$\Delta\varphi_e = 4\pi\varrho,$$

we get

$$\varphi_e = -\frac{16\pi A\lambda^2}{x^2}\,[1 - e^{-x}(x+1)].$$

Calculating U_{ee} by using Green's theorem, we find

$$U_{ee} = -\frac{1}{2} \int \varphi_e \varrho\, d\tau = 16\pi^2 A^2 \lambda^5.$$

We determine A from the normalization condition:

$$\int \varrho\, d\tau = 16\pi A\lambda^3 = N.$$

Substituting $A = \dfrac{N}{16\pi\lambda^3}$ into the expressions for T, U_{ne}, U_{ee}, we obtain

$$T = \frac{12}{25\pi}\left(\frac{3\pi N}{16}\right)^{5/3}\frac{1}{\lambda^2},$$

$$U_{ne} = -\frac{ZN}{2\lambda},$$

$$U_{ee} = \frac{N^2}{16\lambda}.$$

$E = T + U_{ne} + U_{ee}$ attains a minimum for

$$\lambda = \frac{9}{25}\left(\frac{3\pi N}{16}\right)^{2/3}\frac{1}{Z-\dfrac{N}{8}},$$

for which it has the value

$$E = \frac{25}{36}\left(\frac{16}{3\pi}\right)^{2/3} N^{1/3}\left(Z-\frac{N}{8}\right)^2 \text{ atomic units.}$$

For a neutral atom

$$E = \frac{25\times 49}{36\times 64}\left(\frac{16}{3\pi}\right)^{2/3} Z^{7/3} = 0{\cdot}758\,Z^{7/3} \text{ atomic units.}$$

23. Let $\varrho(\mathbf{r})$ be the expression for the electron density in the Thomas-Fermi model. For a function of this form the energy of the atom

$$E = \frac{(3\pi^2)^{5/3}}{10\pi^2}\int \varrho^{5/3}\,d\tau - Z\int\frac{\varrho}{r}\,d\tau + \frac{1}{2}\int\int\frac{\varrho(\mathbf{r})\,\varrho(\mathbf{r}')}{|\mathbf{r}-\mathbf{r}'|}\,d\tau\,d\tau'.$$

is a minimum.

If, into the above expression, instead of ϱ, we put the function $\lambda^3\varrho(\lambda\mathbf{r})$ satisfying the same normalization condition as $\varrho(\mathbf{r})$, we then obtain $E(\lambda) = \lambda^2 T + \lambda U$, where T is the kinetic energy and U the potential energy of the electrons in the atom. Since $E(\lambda)$ should attain a minimum for $\lambda = 1$, then we have the following relation

$$2T + U = 0,$$

which expresses the virial theorem.

24. The interaction energy between the electrons can be written in the form

$$U_{ee} = -\frac{1}{2}\int\varphi_e\varrho\,d\tau = \frac{Z}{2}\int\frac{\varrho}{r}\,d\tau - \frac{1}{2}\int\varphi\varrho\,d\tau, \tag{1}$$

where φ_e is the potential energy resulting from the interaction between the electrons and φ is the potential of a self-consistent field containing the nuclear field:

$$\varphi = \varphi_e + \frac{Z}{r}.$$

For the Thomas-Fermi model we have the relations

$$\frac{p_0^2}{2} = \varphi - \varphi_0, \quad \varrho = \frac{p_0^3}{3\pi^2},$$

where p_0 is the maximum momentum, φ_0 is the potential at the boundary of the atom. Eliminating p_0 and expressing φ in Eq. (1) in terms of ϱ, we obtain

$$U_{ee} = \frac{Z}{2}\int \frac{\varrho}{r}\,d\tau - \frac{(3\pi^2)^{2/3}}{4}\int \varrho^{5/3}\,d\tau - \frac{\varphi_0 N}{2}.$$

The first two terms differ only by multiplicative factors from the interaction energy between the electrons and the nucleus

$$U_{ne} = -Z\int \frac{\varrho}{r}\,d\tau$$

and from the kinetic energy

$$T = \frac{3(3\pi^2)^{2/3}}{10}\int \varrho^{5/3}\,d\tau.$$

Therefore

$$U_{ee} = -\frac{1}{2}U_{ne} - \frac{5}{6}T - \frac{\varphi_0 N}{2}.$$

Inserting here the value of T from the virial theorem, namely

$$2T = -U_{ne} - U_{ee},$$

we finally obtain

$$U_{ee} = -\frac{1}{7}U_{ne} - \frac{6}{7}\varphi_0 N.$$

For a neutral atom ($Z = N$) we have $\varphi_0 = 0$ and

$$U_{ee} = -\frac{1}{7}U_{ne}.$$

25. The total ionization energy is equal to the total energy of the electrons, but is of opposite sign. Using the virial theorem, we find (see preceding problem)

$$E_{\text{ion}} = -\frac{3}{7}U_{ne} + \frac{3}{7}\varphi_0 N.$$

We rearrange the expression

$$U_{ne} = -Z\int \frac{\varrho}{r}\, d\tau$$

in the following manner. We introduce the potential φ_e produced by the electrons

$$\Delta\varphi_e = 4\pi\varrho$$

and make use of Green's theorem to obtain the equation

$$U_{ne} = -\frac{Z}{4\pi}\int \frac{\Delta\varphi_e}{r}\, d\tau = Z\varphi_e(0).$$

[The surface integral over the boundary of an atom is equal to zero and $\Delta \frac{1}{r} = -4\pi\delta(\mathbf{r})$].

Consequently,

$$E_{\text{ion}} = -\frac{3}{7} Z\varphi_e(0) + \frac{3}{7} N\varphi_0.$$

Going over to Thomas-Fermi units

$$r = xbZ^{-1/3}, \quad b = \frac{1}{2}\left(\frac{3\pi}{4}\right)^{2/3}, \quad \varphi - \varphi_0 = \frac{Z^{4/3}}{b}\,\frac{\chi(x)}{x},$$

we find

$$\varphi_e(r) = \varphi - \frac{Z}{r} = \varphi_0 - \frac{Z^{4/3}}{b}\,[1-\chi(x)],$$

and since for small values of x

$$\chi(x) = 1 - ax + \frac{4}{3}x^{3/2},$$

where $a = a_0 = 1{\cdot}58$ for a neutral atom (for a positive ion $a > a_0$), then

$$E_{\text{ion}} = \frac{3}{7}\frac{Z^{7/3}}{b}\,a - \frac{3}{7}\,\frac{(Z-N)^2 Z^{1/3}}{bx_0},$$

x_0 is the radius of a $(Z-N)$-fold ionized atom.

26. The potential due to a point charge and the potential due to a charge distributed over the surface of a sphere is the same outside this sphere if the total charge in both cases is the same. Inside the sphere the difference between two potentials is

$$\Delta\varphi = -Ze\left(\frac{1}{r} - \frac{1}{a}\right).$$

The change in the potential energy of electrons of the atom is

$$\Delta U = Ze^2 \sum_{i=1}^{N} \left(\frac{1}{r_i} - \frac{1}{a}\right) \varepsilon(r_i),$$

where we have introduced the auxiliary function

$$\varepsilon(r) = \begin{cases} 1 & r < a, \\ 0 & r > a. \end{cases}$$

The displacement of the energy levels in first-order perturbation theory is

$$\Delta E = Ze^2 \int |\psi|^2 \sum_{i=1}^{N} \left(\frac{1}{r_i} - \frac{1}{a}\right) \varepsilon(r_i)\, d\tau_1 \dots d\tau_N.$$

Integration over all the variables, but one, gives

$$\int |\psi(r_1, r_2, \dots, r_N)|^2 \, d\tau_2 \dots d\tau_N = \frac{1}{N} \varrho(r),$$

where $\varrho(r)$ is the electron density.

Hence

$$\Delta E = Ze^2 \int \varrho(r) \left(\frac{1}{r} - \frac{1}{a}\right) \varepsilon(r)\, d\tau.$$

We note that $\varrho(r)$ changes slowly in the region $r<a$. We can, therefore, take r outside the integral at the point $r = 0$, and we obtain

$$\Delta E = Ze^2 \varrho(0) \frac{2\pi}{3} a^2.$$

27. The wave function of the s electron is

$$\psi_n(r) = \frac{1}{\sqrt{4\pi}} \frac{\chi_n(r)}{r},$$

where χ_n also satisfies the equation

$$\chi_n'' + \frac{2\mu}{\hbar^2} [E_n - U(r)]\, \chi_n = 0$$

and the normalization condition

$$\int_0^\infty \chi_n^2 \, dr = 1.$$

The approximate quasi-classical solution of this equation has the form

$$\chi_n = \frac{A_n}{\sqrt{p_n}} \cos\left(\frac{1}{\hbar} \int_0^r p_n \, dr + \varphi\right), \tag{1}$$

where

$$p_n = \sqrt{2\mu [E_n - U(r)]}.$$

This solution, however, is inconvenient in the region of small values of r. Indeed, if r is small $\left(r \ll \frac{\hbar^2}{Z^{1/3}\mu e^2}\right)$ then, firstly, we can neglect the screening of the field of the nucleus and set $U(r) = -Ze^2/r$; secondly, we can also neglect E_n in comparison to $U(r)$. Substituting $p = \sqrt{\frac{2\mu Ze^2}{r}}$ in the condition determining the range of applicability of the quasi-classical approximation

$$\frac{d\left(\frac{\hbar}{p}\right)}{dr} \ll 1,$$

we have

$$Z \gg \frac{\hbar^2}{Z\mu e^2}.$$

In order to obtain for χ_n a formula applicable in the region where r is small, we return to the original equation and replace in it $U(r)$ by $\frac{-Ze^2}{r}$ and neglect E_n:

$$\chi_n'' + \frac{2\mu}{\hbar^2}\frac{Ze^2}{r}\chi_n = 0.$$

The solution of this equation has the form

$$\chi_n = C_n \sqrt{r}\, J_1\left(2\sqrt{\frac{2\mu Ze^2 r}{\hbar^2}}\right). \tag{2}$$

To find the relation between the constants C_n and A_n we note that the region of applicability of the quasi-classical solution (1)

$$r \gg \frac{\hbar^2}{Z\mu e^2}$$

and the region of applicability of the solution (2) in which we neglect the screening of the field of the nucleus

$$r \ll \frac{\hbar^2}{Z^{1/3}\mu e^2}$$

coincide for large values of Z, and therefore the solutions (1) and (2) should be identical in the region $\hbar^2/Z\mu e^2 \ll r \ll \hbar^2/Z^{1/3}\mu e^2$.

Let us show that the solutions (1) and (2) have the same form in the entire region of applicability. For this purpose we set $p = \sqrt{\frac{2\mu Ze^2}{r}}$ in (1). We then obtain

$$\chi_n = \frac{A_n\sqrt[4]{r}}{\sqrt[4]{2\mu Ze^2}} \cos\left(\frac{2\sqrt{2\mu Ze^2 r}}{\hbar} + \varphi\right), \qquad \frac{\hbar^2}{Z\mu e^2} < r < \frac{\hbar^2}{Z^{1/3}\mu e^2}. \quad (3)$$

The condition $r \gg \frac{\hbar^2}{Z\mu e^2}$ signifies that the argument of the Bessel function in (2) is large. But for large values of arguments ($x \gg 1$)

$$J_1(x) \approx \sqrt{\frac{2}{\pi x}} \cos\left(x - \frac{3\pi}{4}\right).$$

Consequently, the solution (2) takes the form

$$\chi_n = C_n \sqrt{r} \sqrt{\frac{\hbar}{\pi\sqrt{2\mu Ze^2 r}}} \cos\left(\frac{2\sqrt{2\mu Ze^2 r}}{\hbar^2} - \frac{3\pi}{4}\right),$$

$$\frac{\hbar^2}{Z\mu e^2} \ll r \ll \frac{\hbar^2}{Z^{1/3}\mu e^2}.$$

Comparing this formula with (3), we find

$$\varphi = -\frac{3\pi}{4}, \qquad A_n = C_n\sqrt{\frac{\hbar}{\pi}}.$$

We can now calculate $\psi^2(0)$. For $x \ll 1$, we have $J_1(x) = x/2$ and with the formula (2) we find that

$$\left.\frac{\chi_n}{r}\right|_{r\to 0} = C_n\sqrt{\frac{2\mu Ze^2}{\hbar^2}} = \sqrt{\frac{2\pi\mu Ze^2}{\hbar^3}}\, A_n.$$

Therefore

$$\psi_{n}^2(0) = \frac{\mu Zc^2}{2\hbar^3} A_{n}^2.$$

The constant A_n is found from the normalization condition

$$\int_0^\infty \chi_n^2\, dr = A_n^2 \int_0^\infty \frac{\cos^2\left(\frac{1}{\hbar}\int_0^r p_n\, dr - \frac{3\pi}{4}\right)}{\sqrt{2\mu[E_n - U(r)]}}\, dr \approx \frac{A_n^2}{2} \int \frac{dr}{\sqrt{2\mu[E_n - U(r)]}} = 1.$$

Differentiating with respect to n the quantum condition

$$\int \sqrt{2\mu[E_n - U(r)]}\, dr = \pi(n + \gamma)\hbar,$$

which defines E_n as a function of n, we find that

$$\mu \frac{dE_n}{dn} \int \frac{dr}{\sqrt{2\mu[E_n - U(r)]}} = \pi\hbar .$$

Comparing the last expression with the normalization condition, we obtain

$$A_n^2 = \frac{2\mu}{\pi\hbar} \frac{dE_n}{dn},$$

and, finally,

$$\psi_n^2(0) = \frac{Ze^2\mu^2}{\pi\hbar^4} \frac{dE_n}{dn}. \tag{4}$$

For an unscreened Coulomb field $E_n = -\frac{Z^2}{2n^2}\frac{\mu e^4}{\hbar^2}$ and $\psi_n^2(0) = \frac{Z^3}{\pi n^3}\left(\frac{\mu e^2}{\hbar^2}\right)^3$ found from formula (4) are identical with the results which we would have obtained from an exact computation. In atomic spectroscopy we often use the formula

$$E_n = -\frac{\mu e^4}{2\hbar^2} \frac{1}{(n-\sigma)^2},$$

for the energy levels of excitation states of a valence electron, where σ, called the *Rydberg correction*, depends weakly on n. Taking this formula for E_n, we obtain

$$\psi^2(0) = \frac{Z}{\pi} \frac{\mu^3 e^6}{\hbar^6} \frac{\left(1 - \frac{d\sigma}{dn}\right)}{(n-\sigma)^3}.$$

28. Let the electron be in a stationary state with the magnetic quantum number m. The wave function for such a state is

$$\psi_{nlm}(r, \vartheta, \varphi) = R_{nl}(r) P_l^{(m)}(\cos\vartheta) e^{im\varphi}.$$

The density of the electric current in the state ψ_{nlm} in the spherical coordinate system has the form

$$J_r = J_\vartheta = 0, \quad J_\varphi = -\frac{|e|\hbar m}{\mu r \sin\vartheta} |\psi_{nlm}|^2.$$

The magnetic field intensity vector is obviously directed along the z axis.

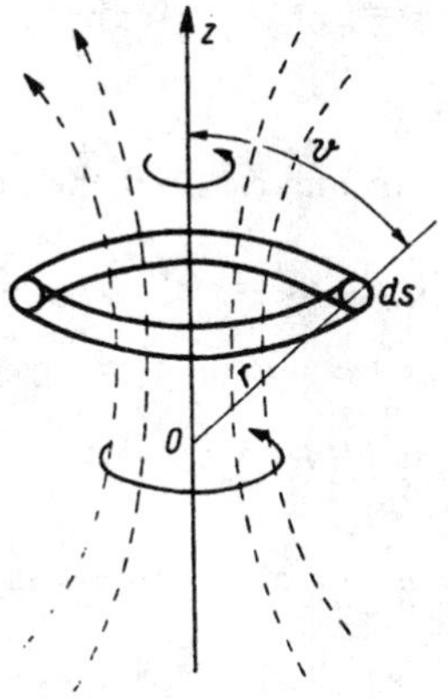

Fig. 28

The circular current dJ produces at point 0 (see Fig. 28) a magnetic field of intensity $d\mathcal{H}_z = \frac{dJ}{rc} 2\pi \sin^2\vartheta$; since $dJ = J_\varphi r \, d\vartheta \, dr$, then

$$\mathcal{H}_z = -\frac{2\pi |e| \hbar m}{\mu c} \int_0^\infty \frac{R^2}{r} dr \int_0^\pi |P_l^{(m)}|^2 \sin\vartheta \, d\vartheta = -\frac{m |e| \hbar}{\mu c} \int_0^\infty \frac{R^2}{r} dr.$$

The same result can be obtained in a different way. We know from electrodynamics that the magnetic field intensity produced by the motion of a charge (neglecting retardation effects) is

$$\mathcal{H} = \frac{e}{\mu c} \frac{1}{r^3} [\mathbf{rp}] = \frac{e}{\mu c} \frac{1}{r^3} \mathbf{l},$$

where $\mathbf{r}$ is the radius vector to the point of observation and $\mathbf{l}$ is the angular momentum.

To obtain the mean value of the magnetic field intensity in quantum theory one must evaluate the integral

$$\overline{\mathcal{H}}_z = -\frac{|e| \hbar}{\mu c} \int \psi^* \frac{\hat{l}_z}{r^3} \psi \, d\tau.$$

Since $\hat{l}_z \psi = m\psi$, then

$$\overline{\mathcal{H}}_z = -m \frac{|e|\hbar}{\mu c} \overline{\frac{1}{r^3}} = -m \frac{|e|\hbar}{\mu c} \left(\frac{\mu e^2}{\hbar^2}\right)^3 \frac{1}{n^3 l \left(l+\frac{1}{2}\right)(l+1)}.$$

The mean value $\mathcal{H}_x$, $\mathcal{H}_y$ in our case will be zero owing to the fact that

$$\int \psi^*_{nlm} \hat{l}_y \psi_{nlm}\, d\tau = \int \psi^*_{nlm} \hat{l}_x \psi_{nlm}\, d\tau = 0.$$

For the $2p$ state ($m = 1$) we obtain

$$\overline{\mathcal{H}}_z = -\frac{1}{12} \frac{|e|\hbar}{2\mu c} \left(\frac{\mu e^2}{\hbar^2}\right)^3, \quad \text{i.e.} \quad \overline{\mathcal{H}}_z \sim 10^4 \text{ gauss}.$$

29. The magnetic moment of a particle is

$$\mathfrak{M} = \frac{e}{2\mu c} \int \psi^* \hat{\mathbf{l}} \psi\, d\tau.$$

In the case of two particles we introduce new variables: the coordinates of the centre of mass (X, Y, Z) and the coordinates (x, y, z) representing the distance between the particles.

In the new variables the mean value of the magnetic moment is equal to

$$\overline{\mathfrak{M}}_z = \frac{e}{2c} \int \psi^* \left\{ \left(\frac{1}{m} + \frac{1}{M}\right) \left(X \frac{\partial}{\partial y} - Y \frac{\partial}{dx}\right) + \right.$$

$$\left. + \frac{1}{m+M} \left(x \frac{\partial}{\partial Y} - y \frac{\partial}{\partial X}\right) + \frac{m-M}{mM} \left(x \frac{\partial}{\partial y} - y \frac{\partial}{\partial x}\right) \right\} \psi\, d\tau.$$

Similar expressions are obtained for $\overline{\mathfrak{M}}_x$ and $\overline{\mathfrak{M}}_y$.

In the stationary state the mean values of the coordinates x, y, z and the momenta $-i\hbar \frac{\partial}{\partial x}$, $-i\hbar \frac{\partial}{\partial y}$, $-i\hbar \frac{\partial}{\partial z}$ are equal to zero. As a result, the last expression can be simplified to

$$\overline{\mathfrak{M}}_z = -\frac{e}{2mc} \left(1 - \frac{m}{M}\right) \int \psi^* \left\{ -i\hbar \left(x \frac{\partial}{\partial y} - y \frac{\partial}{\partial x}\right) \right\} \psi\, d\tau.$$

In this problem m denotes the mass of the electron and M the mass of the nucleus.

30. $\Delta E = 0{\cdot}00844 \times 2{\cdot}79\ Z^3/n^3\ \text{cm}^{-1}$.

For the ground state of the hydrogen atom ($Z = n = 1$)

$$\Delta E = 0{\cdot}0235\ \text{cm}^{-1}.$$

31. To determine the energy it is necessary to find the intensity of the magnetic field produced by the electron. Owing to the orbital motion of the electron, a magnetic field is produced at the nucleus. The intensity of this field is given by the Biot-Savart law

$$\mathcal{H}_l = \frac{1}{c}\frac{[\mathbf{rj}]}{r^3},$$

where $\mathbf{r}$ is the radius vector from the nucleus to the electron, and $\mathbf{j} = -e\mathbf{v}$ ($-e$ is the electronic charge). We introduce the orbital angular momentum operator $\hat{\mathbf{l}}$. Then for $\mathcal{H}_l$ we obtain

$$\hat{\mathcal{H}}_l = -\frac{e\hbar}{\mu c}\frac{1}{r^3}\hat{\mathbf{l}}.$$

Since the electron also has a spin magnetic moment, the total magnetic field intensity at the nucleus will be given by

$$\hat{\mathcal{H}} = \hat{\mathcal{H}}_l + \hat{\mathcal{H}}_s = \frac{e\hbar}{\mu c r^3}\left\{\hat{\mathbf{s}} - \frac{3(\mathbf{r}\hat{\mathbf{s}})\mathbf{r}}{r^2} - \hat{\mathbf{l}}\right\}.$$

Thus the operator of the hyperfine-structure energy can be represented by

$$\hat{w} = -\beta\hat{\mathbf{i}}\,\hat{\mathcal{H}}.$$

Here, $\hat{\mathbf{i}}$ is the nuclear spin operator and $\beta\hat{\mathbf{i}}$ is the magnetic moment. We shall consider $\hat{w}$ as a small perturbation. The unperturbed state is characterized by the quantum numbers n, j ($j = l + \frac{1}{2}$, $j = l - \frac{1}{2}$), l (we assume LS coupling).

To determine the hyperfine-structure energy we should take the average of the operator $\hat{w}$ over the state with the quantum numbers

$$f, j, l \quad (\hat{\mathbf{f}} = \hat{\mathbf{j}} + \hat{\mathbf{i}}).$$

From the basic formula

$$(\hat{\mathbf{A}})_{fm}^{fm'} = \frac{(\hat{\mathbf{f}}\hat{\mathbf{A}})_f^f}{f(f+1)}(\hat{\mathbf{f}})_{fm}^{fm'}$$

we have

$$(\hat{\mathcal{H}})_{nljm_j}^{nljm'_j} = -\frac{e\hbar}{\mu c}\left(\frac{1}{r^3}\right)_{nljm_j}^{nljm_j}\frac{l(l+1)}{j(j+1)}(\hat{\mathbf{j}})_{jm_j}^{jm'_j}.$$

Using this relation, it may readily be shown that the operator $\hat{w}$ can be represented in the form

$$\hat{w} = \frac{e\hbar}{\mu c}\beta\frac{1}{r^3}\frac{l(l+1)}{j(j+1)}(\hat{\mathbf{i}}\,\hat{\mathbf{j}}).$$

It thus follows that the hyperfine-structure energy E is given by the expression

$$E=\frac{e\hbar}{2\mu c}\beta\overline{\left(\frac{1}{r^3}\right)}\frac{l(l+1)}{j(j+1)}\{f(f+1)-j(j+1)-i(i+1)\}.$$

Hence, if we take into account the hyperfine-structure, every term associated with a given set of quantum numbers n, l, j splits into $2i+1$ components (if $j>i$). The hyperfine multiplet structure intervals are then determined by the various values of f.

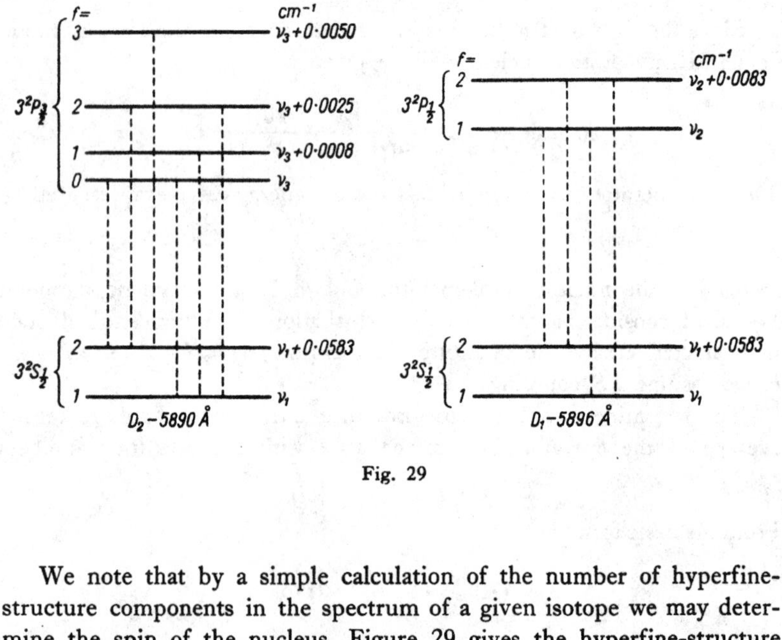

Fig. 29

We note that by a simple calculation of the number of hyperfine-structure components in the spectrum of a given isotope we may determine the spin of the nucleus. Figure 29 gives the hyperfine-structure multiplets for the D lines of sodium.

The fine structure, i.e., the existence of the doublet (the lines $D_1 = 5896$ Å corresponding to the transition $3^2P_{3/2}\rightarrow 3^2S_{1/2}$ and $D_2 = 5890$ Å corresponding to the transition $3^2P_{1/2}\rightarrow 3^2S_{1/2}$) is due to the spin-orbit interaction (see Prob. 10, §7).

32. In diamagnetic atoms not only is the total angular momentum equal to zero but the resultant orbital angular momentum and the resultant spin angular momentum of the electrons also vanish. Owing to precession, the electron gains the additional velocity

$$\mathbf{v}' = \frac{e}{2\mu c}[\mathcal{H}\mathbf{r}].$$

If by **A** we denote the vector potential of the external magnetic field $\mathbf{A} = \frac{1}{2}(\mathcal{H}\mathbf{r})$ then the last relation can be rewritten in the form

$$\mathbf{v}' = \frac{e}{\mu c}\mathbf{A}.$$

The density of the current due to the precession of the electrons is

$$\mathbf{J} = \frac{e}{\mu c}\mathbf{A}\varrho(\mathbf{r}),$$

where $\varrho(\mathbf{r})$ is the density of the charge at the point $\mathbf{r}$ (the charge of the electron is $-e$).

Let us first find the vector potential $\mathbf{A}'$ of the induced magnetic field

$$\mathbf{A}'(r) = \frac{e}{2\mu c^2}\int [\mathcal{H}\mathbf{r}']\frac{\varrho(r')}{|\mathbf{r}-\mathbf{r}'|}d\tau'.$$

Using the formula

$$\frac{1}{r_{12}} = \frac{1}{\sqrt{R^2+r^2-2Rr\{\cos\theta\cos\vartheta+\sin\theta\sin\vartheta\cos(\Phi-\varphi)}}$$

$$= \sum_{l,m}\sqrt{\frac{4\pi}{2l+1}}\,\frac{Y^*_{lm}(\theta,\Phi)\,Y_{lm}(\vartheta,\varphi)}{Y_{l0}(0)}\begin{cases}\dfrac{r^l}{R^{l+1}}, & r<R,\\[2mm] \dfrac{R^l}{r^{l+1}}, & R<r,\end{cases}$$

we obtain the following expression for $\mathbf{A}'$:

$$\mathbf{A}'(\mathbf{r}) = \frac{e}{6\mu c^2}[\mathcal{H}\mathbf{r}]\left\{\frac{1}{r^3}\int_{r'<r} r'^2\varrho(r')d\tau' + \int_{r'>r}\frac{\varrho(r')}{r'}d\tau'\right\}.$$

With this formula we can readily calculate the intensity of the magnetic field in the direction of the z axis.

This intensity is equal to

$$\mathcal{H}'_z = \left(\frac{e\mathcal{H}}{2\mu c^2 r^5}\right)\left(z^2-\frac{1}{3}r^2\right)\int_{r'<r} r'^2\varrho(r')\,d\tau' + \frac{e\mathcal{H}}{3\mu c^2}\int_{r'>r}\frac{\varrho(r')}{r'}\,d\tau'.$$

The induced field at the centre of the atom, i.e., the field acting on the nucleus

$$\mathcal{H}_z'(0) = \frac{e\mathcal{H}}{3\mu c^2}\int \frac{\varrho(r')}{r'}\,d\tau' = \frac{e\mathcal{H}}{3\mu c^2}\varphi(0)$$

depends on the electrostatic potential $\varphi(0)$ produced by the electrons.
In the Thomas-Fermi model

$$\varphi(0) = -1.588\frac{Ze}{b}, \quad \text{where} \quad b = 0.858\frac{a}{Z^{1/3}}, \quad a = \frac{\hbar^2}{\mu e^2}$$

and therefore

$$\mathcal{H}_z'(0) = -0.319\times 10^{-4}\,Z^{4/3}\,\mathcal{H}.$$

33. $$\mathcal{H}_z'(0) = -\frac{27}{24}\frac{e^2\mathcal{H}}{\mu c^2 a} = -0.599\times 10^{-4}\mathcal{H}.$$

34. ${}^1S_0$, ${}^3S_1$, ${}^3P_{0,1,2}$, ${}^2D_{3/2,5/2}$, ${}^4D_{1/2,3/2,5/2,7/2}$.

35. (a) ${}^1S_0{}^3S_1$, (b) ${}^1P_1{}^3P_{012}$, (c) ${}^1D_2{}^3D_{123}$,
(d) ${}^1S_0{}^3S_1{}^1P_1$, ${}^3P_{012}{}^1D_2{}^3D_{123}$.

36. (a) ${}^4S{}^2P{}^2D$, (b) ${}^1S{}^3P{}^1D{}^3F{}^1G$, (c) ${}^2S{}^2P{}^4P{}^2D$.

37. $O\,{}^3P_2$, $Cl\,{}^2P_{3/2}$, $Fe\,{}^5D_4$, $Co\,{}^4F_{9/2}$, $As\,{}^4S_{3/2}$, $La\,{}^2D_{3/2}$.

38. K, Zn, C, O — even; B, N, Cl —odd.

39. If all three quantum numbers are different, the number of states is equal to the number of combinations of N numbers taken $N/2 + M_s$ at a time, i.e.,

$$g(M_s) = C_N^{\frac{1}{2}N+M_s}.$$

If there are N' pairs of identical sets of the three quantum numbers n, l, m_l, then

$$g(M_s) = C_{N-2N'}^{\frac{1}{2}(N-2N')+M_s}.$$

40. The number of states is equal to

$$\frac{N_l(N_l-1)\dots(N_l-x+1)}{x!}, \quad \text{where} \quad N_l = 2(2l+1).$$

42. The antisymmetric wave function of the form

$$\Phi = \frac{1}{N!}\begin{vmatrix} \psi_{n^1l^1m_l^1m_s^1}(\xi_1) & \psi_{n^1l^1m_l^1m_s^1}(\xi_2)\dots\psi_{n^1l^1m_l^1m_s^1}(\xi_N) \\ \psi_{n^2l^2m_l^2m_s^2}(\xi_1) & \psi_{n^2l^2m_l^2m_s^2}(\xi_2)\dots\psi_{n^2l^2m_l^2m_s^2}(\xi_N) \\ \dots\dots & \dots\dots\dots\dots \\ \psi_{n^Nl^Nm_l^Nm_s^N}(\xi_1) & \psi_{n^Nl^Nm_l^Nm_s^N}(\xi_2)\dots\psi_{n^Nl^Nm_l^Nm_s^N}(\xi_N) \end{vmatrix},$$

constructed from the wave functions of the one-electron problem will be denoted for the sake of brevity as

$$\Phi(n^1l^1m_l^1m_s^1,\ n^2l^2m_l^2m_s^2, \dots, n^Nl^Nm_l^Nm_s^N).$$

We shall investigate the action of the symmetric operator

$$(\hat{L}_x - i\hat{L}_y) = \sum_{i=1}^{N} (\hat{l}_x^i - i\hat{l}_y^i)$$

on the antisymmetric function

$$\Phi(n^1l^1m_l^1m_s^1,\ n^2l^2m_l^2m_s^2 \dots, n^Nl^Nm_l^Nm_s^N).$$

It is readily seen that

$$\begin{aligned}(\hat{L}_x - i\hat{L}_y)&\Phi(n^1l^1m_l^1m_s^1,\ n^2l^2m_l^2m_s^2, \dots)\\ &= \sqrt{(l^1+m_l^1)(l^1-m_l^1+1)}\,\Phi(n^1l^1m_l^1-1m_s^1,\ n^2l^2m_l^2m_s^2, \dots)+\\ &+\sqrt{(l^2+m_l^2)(l^2-m_l^2+1)}\,\Phi(n^1l^1m_l^1m_s^1\ n^2l^2m_l^2-1m_s^2, \dots)+\\ &\dots+\sqrt{(l^N+m_l^N)(l^N-m_l^N+1)}\,\Phi(n^1l^1m_l^1m_s^1, \dots, n^Nl^Nm_l^N-1m_s^N).\end{aligned}$$

The operator $\hat{S}_x - i\hat{S}_y$ has analogous properties, except that m_s, instead of m_l, is reduced by 1.

If the wave function is an eigenfunction of the four commuting operators $\hat{S}^2$, $\hat{L}^2$, $\hat{S}_z$, $\hat{L}_z$, the action of the operators $(\hat{L}_x - i\hat{L}_y)$ and $(\hat{S}_x - i\hat{S}_y)$ reduces to

$$\begin{aligned}(\hat{L}_x-i\hat{L}_y)\,\Phi(SLM_SM_L) &= \sqrt{(L+M_L)(L-M_L+1)}\,\Phi(SLM_SM_L-1),\\ (\hat{S}_x-i\hat{S}_y)\,\Phi(SLM_SM_L) &= \sqrt{(S+M_S)(S-M_S+1)}\,\Phi(SLM_S-1M_L).\end{aligned} \tag{1}$$

After these preliminary remarks, we now proceed to the solution of the problem.

In the case under consideration we have to do with the same type of electrons, and therefore we can everywhere omit the quantum numbers nl. The value of the spin component will sometimes be denoted by the superscript $(\pm)$ added to m_l.

For the configuration p^3 we introduce a classification of the states with respect to M_S and M_L; we restrict ourselves to non-negative numbers only and obtain in each case the terms:

M_S	M_L			
$\frac{3}{2}$	0	$\Phi(1^+, -1^+, 0^+)$	$[^4S]$	Φ_1

$$\frac{1}{2} \quad 0 \quad \begin{matrix} \Phi(1^+, -1^+, 0^-) \\ \Phi(1^-, -1^+, 0^+) \\ \Phi(1^+, -1^-, 0^+) \end{matrix} \begin{bmatrix} {}^4S \\ {}^2P \\ {}^2D \end{bmatrix} \begin{matrix} \Phi_2 \\ \Phi_3 \\ \Phi_4 \end{matrix}$$

$$\frac{1}{2} \quad 1 \quad \begin{matrix} \Phi(1^+, 1^-, -1^+) \\ \Phi(1^+, 0^+, 0^-) \end{matrix} \begin{bmatrix} {}^2P \\ {}^2D \end{bmatrix} \begin{matrix} \Phi_5 \\ \Phi_6 \end{matrix}$$

$$\frac{1}{2} \quad 2 \quad \Phi(1^+, 1^-, 0^+) [{}^2D] \Phi_7$$

We now investigate the action of the operators $(\hat{L}_x - i\hat{L}_y)$, $(\hat{S}_x - i\hat{S}_y)$ on some of the states given above:

$$(\hat{L}_x - i\hat{L}_y) \Phi_5 = \sqrt{2}(\Phi_3 - \Phi_2),$$
$$(\hat{L}_x - i\hat{L}_y) \Phi_6 = \sqrt{2}(\Phi_2 - \Phi_4),$$
$$(\hat{L}_x - i\hat{L}_y) \Phi_7 = \sqrt{2}(\Phi_5 - \Phi_6),$$
$$(\hat{S}_x - i\hat{S}_y) \Phi_1 = \Phi_2 + \Phi_3 + \Phi_4;$$

Φ_1 is the wave function of the state ${}^4S$ with $M_S = \frac{3}{2}$, $M_L = 0$. Operating on this function with $(\hat{S}_x - i\hat{S}_y)$, we obtain, by (1),

$$(\hat{S}_x - i\hat{S}_y) \Phi \left({}^4S, \frac{3}{2}, 0\right) = \sqrt{3} \, \Phi \left({}^4S, \frac{1}{2}, 0\right).$$

Since

$$(\hat{S}_x - i\hat{S}_y) \Phi_1 = \Phi_2 + \Phi_3 + \Phi_4,$$

then

$$\Phi \left({}^4S, \frac{1}{2}, 0\right) = \frac{1}{\sqrt{3}} (\Phi_2 + \Phi_3 + \Phi_4).$$

Similarly, for the term ${}^2D$ we obtain the following wave functions:

$$\Phi \left({}^2D, \frac{1}{2}, 2\right) = \Phi_7,$$

$$\Phi \left({}^2D, \frac{1}{2}, 1\right) = \frac{1}{\sqrt{2}} (\Phi_5 - \Phi_6),$$

$$\Phi \left({}^2D, \frac{1}{2}, 0\right) = \frac{1}{\sqrt{6}} (\Phi_3 - 2\Phi_2 + \Phi_4).$$

The state $^2P, \frac{1}{2}, 1$ ($M_S = \frac{1}{2}$, $M_L = 1$) represents the linear combination of the states Φ_5 and Φ_6 orthogonal to $^2D, \frac{1}{2}, 1$. We thus find

$$\Phi\left(^2P, \frac{1}{2}, 1\right) = \frac{1}{\sqrt{2}}(\Phi_5 + \Phi_6)$$

and further

$$\Phi\left(^2P, \frac{1}{2}, 0\right) = \frac{1}{\sqrt{2}}(\Phi_3 - \Phi_4).$$

We can obtain in the same way the wave functions corresponding to the negative values of the components.

43. As may be seen from the table below, in order to determine the eigenfunction of the two terms ^2D we must first calculate the eigenfunctions of the terms ^2H, ^2G, ^4F, ^2F.

M_S	M_L			
$\frac{1}{2}$	5	$\Phi(2^+, 2^-, 1^+)$	Φ_1	[^2H]
$\frac{1}{2}$	4	$\Phi(2^+, 2^-, 0^+)$	Φ_2	^2H
		$\Phi(2^+, 1^+, 1^-)$	Φ_3	^2G
$\frac{1}{2}$	3	$\Phi(2^+, 2^-, -1^+)$	Φ_4	^2H
		$\Phi(2^+, 1^+, 0^-)$	Φ_5	^2G
		$\Phi(2^+, 1^-, 0^+)$	Φ_6	^4F
		$\Phi(2^-, 1^+, 0^+)$	Φ_7	^2F
$\frac{1}{2}$	2	$\Phi(2^+, 2^-, -2^+)$	Φ_8	^2H
		$\Phi(2^+, 1^+, -1^-)$	Φ_9	^2G
		$\Phi(2^+, 1^-, -1^+)$	Φ_{10}	^4F
		$\Phi(2^-, 1^+, -1^+)$	Φ_{11}	^2F
		$\Phi(2^+, 0^+, 0^-)$	Φ_{12}	^2D
		$\Phi(1^+, 1^-, 0^+)$	Φ_{13}	^2D
$\frac{3}{2}$	3	$\Phi(2^+, 1^+, 0^+)$	Φ_{14}	[^4F]
$\frac{3}{2}$	2	$\Phi(2^+, 1^+, -1^+)$	Φ_{15}	[^4F]

We first find the result of the action of the operators $(\hat{L}_x - i\hat{L}_y)$, $(\hat{S}_x - i\hat{S}_y)$ on some of the above states:

$$\begin{aligned}
(\hat{L}_x - i\hat{L}_y)\Phi_1 &= -2\Phi_3 + \sqrt{6}\,\Phi_2,\\
(\hat{L}_x - i\hat{L}_y)\Phi_2 &= -2\Phi_7 + 2\Phi_6 + \sqrt{6}\,\Phi_4,\\
(\hat{L}_x - i\hat{L}_y)\Phi_3 &= -\sqrt{6}\,\Phi_6 + \sqrt{6}\,\Phi_5,\\
(\hat{S}_x - i\hat{S}_y)\Phi_{14} &= \Phi_7 + \Phi_6 + \Phi_5,\\
(\hat{L}_x - i\hat{L}_y)\Phi_4 &= -2\Phi_{11} + 2\Phi_{10} + 2\Phi_8,\\
(\hat{L}_x - i\hat{L}_y)\Phi_5 &= \sqrt{6}\,\Phi_{12} + \sqrt{6}\,\Phi_9,\\
(\hat{S}_x - i\hat{S}_y)\Phi_{15} &= \Phi_{11} + \Phi_{10} + \Phi_9,\\
(\hat{L}_x - i\hat{L}_y)\Phi_6 &= 2\Phi_{13} - \sqrt{6}\,\Phi_{12} + \sqrt{6}\,\Phi_{10},\\
(\hat{L}_x - i\hat{L}_y)\Phi_7 &= -2\Phi_{13} + \sqrt{6}\,\Phi_{11}.
\end{aligned}$$

The wave function of the state ^2H, $\frac{1}{2}$, $5(M_S = \frac{1}{2},\ M_L = 5)$ is equal to Φ_1, i.e., $\Phi(^2H, \frac{1}{2}, 5) = \Phi_1$. Applying the operator $(\hat{L}_x - i\hat{L}_y)$ we have

$$(\hat{L}_x - i\hat{L}_y)\Phi\left(^2H, \frac{1}{2}, 5\right) = \sqrt{10}\,\Phi\left(^2H, \frac{1}{2}, 4\right).$$

Since $(\hat{L}_x - i\hat{L}_y)\Phi_1 = -2\Phi_3 + \sqrt{6}\,\Phi_2$, it follows that

$$\Phi\left(^2H, \frac{1}{2}, 4\right) = \frac{1}{\sqrt{10}}\left\{\sqrt{6}\,\Phi_2 - 2\Phi_3\right\}.$$

The other states of the term ^2H which we need to solve the problem are found to be

$$\Phi\left(^2H, \frac{1}{2}, 3\right) = \frac{1}{\sqrt{30}}\left\{\sqrt{6}\,\Phi_4 - 2\Phi_5 + 4\Phi_6 - 2\Phi_7\right\},$$

$$\Phi\left(^2H, \frac{1}{2}, 2\right) = \frac{1}{\sqrt{30}}\left\{\Phi_8 - \Phi_9 + 3\Phi_{10} - 2\Phi_{11} - 3\Phi_{12} + \sqrt{6}\,\Phi_{13}\right\}.$$

The state $\Phi(^2G, \frac{1}{2}, 4)$ represents a linear combination of the states Φ_2 and Φ_3, orthogonal to the state $\Phi(^2H, \frac{1}{2}, 4)$. From these conditions we find the wave function $\Phi(^2G, \frac{1}{2}, 4)$:

$$\Phi\left(^2G, \frac{1}{2}, 4\right) = \frac{e^{i\alpha}}{\sqrt{10}}\left\{2\Phi_2 + \sqrt{6}\,\Phi_3\right\}.$$

Since the states of different terms are not related by phase relations, then we can set $\alpha = 0$.

The other states of the term ^2G are obtained by successive application of the operator $(\hat{L}_x - i\hat{L}_y)$. We then obtain

$$\Phi\left({}^2G, \frac{1}{2}, 4\right) = \frac{1}{\sqrt{10}}\{2\Phi_2 + \sqrt{6}\Phi_3\},$$

$$\Phi\left({}^2G, \frac{1}{2}, 3\right) = \frac{1}{\sqrt{20}}\{\sqrt{6}\Phi_4 + 3\Phi_5 - \Phi_6 - 2\Phi_7\},$$

$$\Phi\left({}^2G, \frac{1}{2}, 2\right) = \sqrt{\frac{3}{140}}\left\{2\Phi_8 + 3\Phi_9 + \Phi_{10} - 4\Phi_{11} + 4\Phi_{12} + \sqrt{\frac{2}{3}}\Phi_{13}\right\}.$$

In order to solve the problem we must still determine the functions

$$\Phi\left({}^4F, \frac{1}{2}, 2\right) \quad \text{and} \quad \Phi\left({}^2F, \frac{1}{2}, 2\right).$$

Since $\Phi({}^4F, \frac{3}{2}, 3) = \Phi_{14}$, then by applying the operator $(\hat{S}_x - i\hat{S}_y)$ we obtain the state $\Phi({}^4F, \frac{1}{2}, 3)$

$$(\hat{S}_x - i\hat{S}_y)\,\Phi\left({}^4F, \frac{3}{2}, 3\right) = \sqrt{3}\,\Phi\left({}^4F, \frac{1}{2}, 3\right) = (\hat{S}_x - i\hat{S}_y)\,\Phi_{14},$$

$$\Phi\left({}^4F, \frac{1}{2}, 3\right) = \frac{1}{\sqrt{3}}\{\Phi_5 + \Phi_6 + \Phi_7\}.$$

From the state $\Phi({}^4F, \frac{1}{2}, 3)$, with the help of the operator $(\hat{L}_x - i\hat{L}_y)$, we find the state $\Phi({}^4F, \frac{1}{2}, 2)$

$$\Phi\left({}^4F, \frac{1}{2}, 2\right) = \frac{1}{\sqrt{3}}\{\Phi_9 + \Phi_{10} + \Phi_{11}\}.$$

The wave function $\Phi({}^2F, \frac{1}{2}, 3)$ is found from the fact that it is orthogonal to the three functions

$$\Phi\left({}^2H, \frac{1}{2}, 3\right), \quad \Phi\left({}^2G, \frac{1}{2}, 3\right), \quad \Phi\left({}^4F, \frac{1}{2}, 3\right).$$

The normalized function $\Phi({}^2F, \frac{1}{2}, 3)$ found from these conditions is equal to

$$\Phi\left({}^2F, \frac{1}{2}, 3\right) = \frac{1}{\sqrt{12}}\{-\sqrt{6}\Phi_4 + \Phi_5 + \Phi_6 - 2\Phi_7\}.$$

Finally,

$$\Phi\left({}^2F, \frac{1}{2}, 2\right) = \frac{1}{\sqrt{12}}\{-2\Phi_8 + \Phi_9 - \Phi_{10} + \sqrt{6}\Phi_{13}\}.$$

We now have four states with $M_S = \frac{1}{2}$ and $M_L = 2$:

$$\Phi\left({}^2H, \frac{1}{2}, 2\right) = \frac{1}{\sqrt{30}}\{\Phi_8 - \Phi_9 + 3\Phi_{10} - 2\Phi_{11} - 3\Phi_{12} + \sqrt{6}\,\Phi_{13}\},$$

$$\Phi\left({}^2G, \frac{1}{2}, 2\right) = \sqrt{\frac{3}{140}}\left\{2\Phi_8 + 3\Phi_9 + \Phi_{10} - 4\Phi_{11} + 4\Phi_{12} + \sqrt{\frac{2}{3}}\,\Phi_{13}\right\},$$

$$\Phi\left({}^4F, \frac{1}{2}, 2\right) = \frac{1}{\sqrt{3}}\{\Phi_9 + \Phi_{10} + \Phi_{11}\},$$

$$\Phi\left({}^2F, \frac{1}{2}, 2\right) = \frac{1}{\sqrt{12}}\{-2\Phi_8 + \Phi_9 - \Phi_{10} + \sqrt{6}\,\Phi_{13}\}.$$

Two more ${}^2D$ states enter this group. These two orthogonal states are also orthogonal to the four states written above.

From the conditions of orthogonality and normalization we get the following orthogonal functions:

$$\Phi\left(a^2D, \frac{1}{2}, 2\right) = \frac{1}{2}\left\{-\Phi_8 - \Phi_9 + \Phi_{10} + \Phi_{12}\right\},$$

$$\Phi\left(b^2D, \frac{1}{2}, 2\right) = \frac{1}{\sqrt{84}}\left\{-5\Phi_8 + 3\Phi_9 + \Phi_{10} - 4\Phi_{11} - 3\Phi_{12} - 2\sqrt{6}\,\Phi_{13}\right\}.$$

The remaining wave functions of the ${}^2D$ states corresponding to other values of the angular momentum components can readily be obtained by successive application of the operators $(\hat{L}_x - i\hat{L}_y)$ and $(\hat{S}_x - i\hat{S}_y)$.

44. We first list all states belonging to the configuration $npn'p$. We restrict ourselves to non-negative values of M_S and M_L.

pp		M_S	
		1	0
M_L	2	$\Phi_1(1^+,1^+)$	$\Phi_2(1^+,1^-)$ $\Phi_3(1^-,1^+)$
	1	$\Phi_4(1^+,0^+)$ $\Phi_5(0^+,1^+)$	$\Phi_6(1^+,0^-)$ $\Phi_7(0^+,1^-)$ $\Phi_8(1^-,0^+)$ $\Phi_9(0^-,1^+)$
	0	$\Phi_{10}(1^+,-1^+)$ $\Phi_{11}(0^+,0^+)$ $\Phi_{12}(-1^+,1^+)$	$\Phi_{13}(1^+,-1^-)$ $\Phi_{14}(0^+,0^-)$ $\Phi_{15}(-1^+,1^-)$ $\Phi_{16}(1^-,-1^+)$ $\Phi_{17}(0^-,0^+)$ $\Phi_{18}(-1^-,1^+)$

It is seen from the table that the given configuration has the terms ^1S, ^3S, ^1P, ^3P, ^1D, and ^3D.

In the solution of the problem we take as the functions for the zero-order approximation the functions given in the table.

Since the energy does not depend on the value of the components M_S and M_L, the perturbation matrix can be built up from submatrices in the following way:

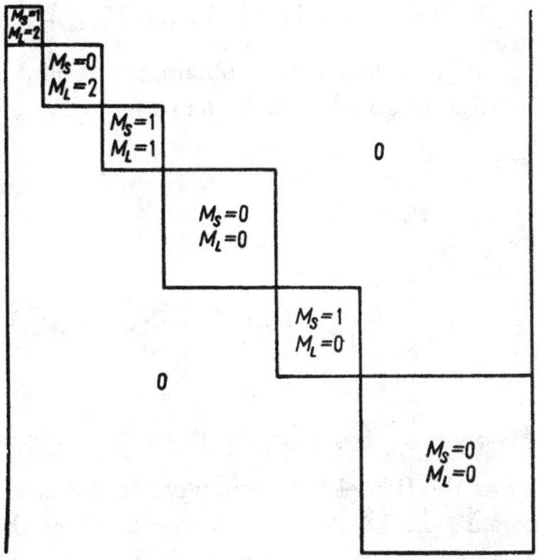

We denote the perturbation operator by $\hat{V}$. In the first submatrix there appears only the term ^3D. Consequently,

$$E^{(1)}(^3D) = V_{11}.$$

In the second submatrix ($M_L = 2$, $M_S = 0$) there appears the combination of the two terms ^3D and ^1D:

$$E^{(1)}(^3D) + E^{(1)}(^1D) = V_{22} + V_{33}.$$

In the third submatrix ($M_L = 1$, $M_S = 1$) we have the combination of the two terms ^3D and ^3P:

$$E^{(1)}(^3D) + E^{(1)}(^3P) = V_{44} + V_{55}.$$

In the fourth submatrix ($M_L = 0$, $M_S = 0$) we have the four terms ^3D, ^1D, ^3P, ^1P:

$$E^{(1)}(^3D) + E^{(1)}(^1D) + E^{(1)}(^3P) + E^{(1)}(^1P) = V_{66} + V_{77} + V_{88} + V_{99}.$$

In the fifth submatrix ($M_L = 0$, $M_S = 1$) we have the three terms ^3D, ^3P, ^3S:

$$E^{(1)}(^3D)+E^{(1)}(^3P)+E^{(1)}(^3S)=V_{10,10}+V_{11,11}+V_{12,12}.$$

And finally, in the sixth submatrix ($M_L = 0$, $M_S = 0$) we have all the terms: ^3D, ^3P, ^3S, ^1D, ^1P, ^1S:

$$E^{(1)}(^3D)+E^{(1)}(^1D)+E^{(1)}(^3P)+E^{(1)}(^1P)+E^{(1)}(^3S)+E^{(1)}(^1S)$$
$$=V_{13,13}+V_{14,14}+V_{15,15}+V_{16,16}+V_{17,17}+V_{18,18}.$$

From these equations one can readily obtain expressions for the terms with the help of the diagonal matrix elements

$$E^{(1)}(^3D)=V_{11},$$

$$E^{(1)}(^1D)=V_{22}+V_{33}-V_{11},$$

$$E^{(1)}(^3P)=V_{44}+V_{55}-V_{11},$$

$$E^{(1)}(^1P)=V_{66}+V_{77}+V_{88}+V_{99}+V_{11}-V_{22}-V_{33}-V_{44}-V_{55},$$

etc.

46. The expression $\frac{e\hbar}{2\mu c}(\hat{L}_z+2\hat{S}_z)\mathcal{H}$ is a small perturbation. Let us consider coupling of the Russel-Saunders type. In this case $\hat{H}_0$ commutes with the operators $\hat{\mathbf{J}}^2$, $\hat{J}_z$, $\hat{\mathbf{L}}^2$, $\hat{\mathbf{S}}^2$. The energy levels of the unperturbed state are characterized by the numbers J, L, S. Each of these levels is degenerate with respect to the direction of the vector $\hat{\mathbf{J}}$, the multiplicity of the degeneracy being $2J + 1$.

Since the matrix elements of the operator $(\hat{L}_z + 2\hat{S}_z)$ that are not diagonal with respect to $\hat{J}_z$ vanish, the correction to the energy is simply equal to the mean value of the operator $(\hat{L}_z + 2\hat{S}_z)$ in the state characterized by the quantum numbers J, J_z, L, S. To calculate this mean value we use the formula obtained in Prob. 35, §4; setting $g_1 = 1$, $g_2 = 2$, $\hat{\mathbf{J}}_1 = \hat{\mathbf{L}}$, $\hat{\mathbf{J}}_2 = \hat{\mathbf{S}}$, we obtain

$$\overline{\hat{L}_z+2\hat{S}_z}=J_z\left\{\frac{3}{2}+\frac{S(S+1)-L(L+1)}{2J(J+1)}\right\}=J_z g.$$

According to the discussion above, the splitting of the levels will be given by

$$E^{(1)}_{JLS}=\frac{e\hbar\mathcal{H}}{2\mu c}gJ_z.$$

51. The energy of an atom in a magnetic field is of the same order of magnitude as the spin-orbit interaction energy. Therefore we combine the spin-orbit interaction operator $\varphi(r)\,\hat{\mathbf{l}}\,\hat{\mathbf{s}}$ (see Prob. 10, §7) with the operator $\frac{e\hbar}{2\mu c}(\hat{l}_z+2\hat{s}_z)$, and their sum

$$V=\varphi(r)\hat{\mathbf{l}}\hat{\mathbf{s}}+\frac{e\hbar}{2\mu c}\mathcal{H}(\hat{l}_z+2\hat{s}_z)$$

will be regarded as a small perturbation. In the unperturbed state the constants of motion will be: the square of the orbital angular momentum, the component of the orbital angular momentum, the square and component of the spin. In this case it is more convenient to employ other conserved quantities. We shall describe the unperturbed stationary state by the quantum numbers n, l, j, m_j ($\hat{\mathbf{l}}^2$, $\hat{\mathbf{j}}^2$, $\hat{j}_z$ commute with $\hat{H}_0$). The degree of degeneracy in the case of a Coulomb field is equal to $2n^2$, while in the case of an arbitrary centrally symmetric field it is $2(2\,l+1)$. We do not need to solve the secular equation of such a high degree. We note that in the perturbed state the square of the orbital angular momentum and the component of the total angular momentum are constants of motion. As a result, the wave function of the perturbed case should be a combination of the functions

$$\psi^{(0)}_{nljm_j},$$

corresponding to the same values of n, l, m_j, i.e.,

$$\psi=c_1\psi^{(0)}\left(n,\,l,\,j=l-\frac{1}{2},\,m_j\right)+c_2\psi^{(0)}\left(n,\,l,\,j=l+\frac{1}{2},\,m_j\right),$$

or in another form

$$\psi=c_1\frac{R^{(0)}_{nl}}{\sqrt{2l+1}}\begin{pmatrix}\sqrt{l+m_j+\frac{1}{2}}\,Y_{l,m_j-1/2}\\ \sqrt{l-m_j+\frac{1}{2}}\,Y_{l,m_j+1/2}\end{pmatrix}+$$

$$+c_2\frac{R^{(0)}_{nl}}{\sqrt{2l+1}}\begin{pmatrix}\sqrt{l-m_j+\frac{1}{2}}\,Y_{l,m_j-1/2}\\ -\sqrt{l+m_j+\frac{1}{2}}\,Y_{l,m_j+1/2}\end{pmatrix}.$$

The matrix elements of the operator $\hat{\mathbf{V}}$ are equal to

$$(\hat{V})^{nl(j=l+\frac{1}{2})m_j}_{nl(j=l+\frac{1}{2})m_j}=A\frac{l}{2}+\mathcal{H}\mu_0 m_j\left(1+\frac{1}{2l+1}\right),$$

$$(\hat{V})^{nl(j=l-\frac{1}{2})m_j}_{nl(j=l-\frac{1}{2})m_j}=-A\frac{l+1}{2}+\mathcal{H}\mu_0 m_j\left(1-\frac{1}{2l+1}\right),$$

$$(\hat{V})^{nl(j=l-\frac{1}{2})m_j}_{nl(j=l+\frac{1}{2})m_j}=(\hat{V})^{nl(j=l+\frac{1}{2})m_j}_{nl(j=l-\frac{1}{2})m_j}=\frac{\mathcal{H}\mu_0}{2l+1}\sqrt{\left(l+\frac{1}{2}\right)^2-m_j^2},$$

where

$$A=\int_0^\infty R_{nl}(r)\varphi(r)R_{nl}(r)r^2\,dr \quad\text{and}\quad \mu_0=\frac{e\hbar}{2\mu c}.$$

Solving the secular equation

$$\begin{vmatrix} E^0_{nl}+A\frac{l}{2}+\mathcal{H}\mu_0 m_j\left(1+\frac{1}{2l+1}\right)-E, & \frac{\mathcal{H}\mu_0}{2l+1}\sqrt{\left(l+\frac{1}{2}\right)^2-m_j^2} \\ \frac{\mathcal{H}\mu_0}{2l+1}\sqrt{\left(l+\frac{1}{2}\right)^2-m_j^2}, & E^{(0)}_{nl}-A\frac{l+1}{2}+\mathcal{H}\mu_0 m_j\left(1-\frac{1}{2l+1}\right)-E, \end{vmatrix}=0,$$

we find the value of the energy E. We denote by E_+ and E_- the energy of a one-electron atom with spin-orbit coupling; E_+ refers to the state with $j=l+\frac{1}{2}$, while E_- refers to the state with $j=l-\frac{1}{2}$.

From the solution of Prob. 10, § 7 it follows that

$$E_+=E^{(0)}_{nl}+A\frac{l}{2},\qquad E_-=E^{(0)}_{nl}-A\frac{l+1}{2}.$$

Solving the secular equation, we find the value of E:

$$E=\frac{1}{2}(E_++E_-)+\mathcal{H}\mu_0 m_j\pm$$

$$\pm\sqrt{\frac{1}{4}(E_+-E_-)^2+\mathcal{H}\mu_0\frac{m_j}{2l+1}(E_+-E_-)+\frac{1}{4}\mathcal{H}^2\mu_0^2}.$$

Let us consider the limiting cases:

(a) In the case of a weak field, i.e., for $\mu_0\mathcal{H}\ll E_+-E_-$, we obtain the following expressions:

$$E=E_++\mathcal{H}\mu_0 m_j\frac{2l+2}{2l+1},$$

$$E=E_--\mathcal{H}\mu_0 m_j\frac{2l}{2l+1}.$$

The first value of the energy corresponds to the energy of the nth level of the state with $j = l+\frac{1}{2}$, the other refers to the state with $j = l-\frac{1}{2}$ (see Prob. 46, §7).

(b) In the case of a strong field, i.e., for $\mu_0\mathcal{H} \gg E_+ - E_-$,

$$E = \frac{1}{2}(E_+ + E_-) + \mathcal{H}\mu_0 m_j \pm \frac{1}{2}\mathcal{H}\mu_0 \pm \frac{m_j}{2l+1}(E_+ - E_-).$$

We denote by E_c the energy of the centre of gravity of the levels in the absence of a magnetic field, i.e.,

$$E_c = \frac{E_+(l+1) + E_- l}{2l+1}$$

$\left(\text{the ratio of statistical weights of the } E_+ \text{ and } E_- \text{ states is equal to } \frac{2l+1}{2l}\right)$, while ΔE denotes the difference $E_+ - E_-$. In the new notation E will be given by

$$E = E_c + \mathcal{H}\mu_0\left(m_j \pm \frac{1}{2}\right) \pm \frac{\Delta E}{2l+1}\left(m_j \mp \frac{1}{2}\right).$$

As may readily be seen, this expression is identical to that in Prob. 53, §7.

The upper sign indicates the state with $m_l = m_j - \frac{1}{2}$, $m_s = \frac{1}{2}$ and the lower sign, the state with $m_l = m_j + \frac{1}{2}$, $m_s = -\frac{1}{2}$.

52.

$$c_1 = \sqrt{\frac{1}{2}(1+\gamma)}, \quad c_2 = \sqrt{\frac{1}{2}(1-\gamma)} \quad \text{for the upper level,}$$

$$c_1 = \sqrt{\frac{1}{2}(1-\gamma)}, \quad c_2 = -\sqrt{\frac{1}{2}(1+\gamma)} \quad \text{for the lower level,}$$

where

$$\gamma = \frac{\frac{1}{2}\Delta E + \frac{m_j}{2l+1}\mathcal{H}\mu_0}{\sqrt{\frac{1}{4}(\Delta E)^2 + \frac{m_j}{2l+1}\Delta E\mathcal{H}\mu_0 + \frac{1}{4}\mathcal{H}^2\mu_0^2}}.$$

We shall investigate the limiting cases.

(a) Vanishingly small magnetic field, i.e., $\Delta E \gg \mathcal{H}_{\mu_0}$. Then $\gamma \approx 1$. For the upper level we have $c_1 = 1$, $c_2 = 0$, for the lower level $c_1 = 0$, $c_2 = 1$.

(b) A strong magnetic field, i.e., $\Delta E \ll \mathcal{H}\mu_0$. In this case $\gamma = \dfrac{m_j}{l+\frac{1}{2}}$ and

$$c_1 = \sqrt{\frac{l+m_j+\frac{1}{2}}{2l+1}}, \quad c_2 = \sqrt{\frac{l-m_j+\frac{1}{2}}{2l+1}} \quad \text{for the upper level,}$$

$$c_1 = \sqrt{\frac{l-m_j+\frac{1}{2}}{2l+1}}, \quad c_2 = -\sqrt{\frac{l+m_j+\frac{1}{2}}{2l+1}} \quad \text{for the lower lewel.}$$

Inserting the values of c_1 and c_2 in the expression for the wave functions (see Prob. 51, §7), we have

$$\psi = R_{nl}^{(0)} \begin{pmatrix} Y_{l,\, m-\frac{1}{2}}(\vartheta, \varphi) \\ 0 \end{pmatrix} \quad \text{for the upper level,}$$

$$\psi = R_{nl}^{(0)} \begin{pmatrix} 0 \\ Y_{l,\, m-\frac{1}{2}}(\vartheta, \varphi) \end{pmatrix} \quad \text{for the lower level.}$$

53. Since the energy in the magnetic field is much greater than the spin-orbit interaction energy, we can neglect the spin-orbit interaction in the first-order approximation.

In this case $\hat{l}_z$ and $\hat{s}_z$ are constants of motion, while the energy of the splitting is determined by the formula

$$E^{(1)} = \frac{e\hbar}{2\mu c}\mathcal{H}(m_l + 2m_s).$$

In the second-order approximation we take into account the spin-orbit interaction. The multiplet splitting which is superimposed on the splitting in the magnetic field is determined by the mean value of the operator $\dfrac{e^2}{2\mu^2 c^2}\dfrac{1}{r^3}(\hat{\mathbf{l}}\hat{\mathbf{s}})$ (see Prob. 10, §7) for the state with given values of m_l and m_s. When one of the components of the angular momentum has a definite value, the mean value of the two remaining components is zero, and therefore

$$\overline{\hat{\mathbf{l}}\hat{\mathbf{s}}} = m_l m_s.$$

Hence, if the spin-orbit interaction is taken into account, the displacement of the levels is

$$E^{(1)} = \frac{e\hbar}{2\mu c}\mathcal{H}(m_l + 2m_s) + \frac{e^2\hbar^2}{2\mu^2 c^2}\overline{\frac{1}{r^3}}\, m_l m_s.$$

We substitute into the formula the value $\overline{\frac{1}{r^3}}$ expressed in terms of the splitting of the fine structure in the case of no magnetic field. As may readily be shown, (see Prob. 10, § 7)

$$\frac{e^2\hbar^2}{2\mu^2c^2}\overline{\frac{1}{r^3}} = \frac{E_{n,l,j=l+\frac{1}{2}} - E_{n,l,j=l-\frac{1}{2}}}{l+\frac{1}{2}} = \frac{\Delta E}{l+\frac{1}{2}}.$$

Finally, for $E^{(1)}$ we obtain

$$E^{(1)} = \frac{e\hbar}{2\mu c}\mathcal{H}(m_l + 2m_s) + \frac{\Delta E}{l+\frac{1}{2}} m_l m_s.$$

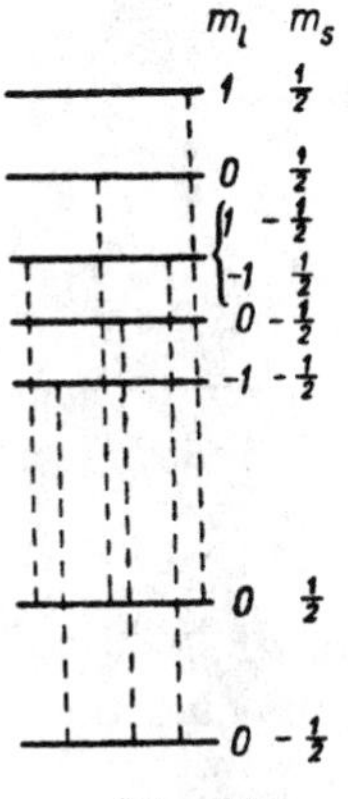

Fig. 30

Fig. 30 gives the scheme for the splitting of the $1s$ and $2p$ terms for an alkali metal atom in a strong magnetic field.

54. The perturbation energy in this case is

$$\hat{V} = \frac{e\hbar}{\mu c}\beta\frac{1}{r^3}\hat{\mathbf{i}}\left\{\frac{3(\mathbf{r}\hat{\mathbf{s}})\mathbf{r}}{r^2} - \hat{\mathbf{s}}\right\} + \frac{e\hbar}{\mu c}\mathcal{H}\hat{s}_z + \beta\mathcal{H}\hat{i}_z.$$

In this expression we may neglect the last term, since the magnetic moment of the nucleus is small in comparison to the magnetic moment of the electron $\left(\beta < \frac{e\hbar}{2\mu c}\right)$.

Proceeding as in the solution of Prob. 51, § 7, we find the secular equation

$$\begin{vmatrix} E_+ + \dfrac{\mu_0 \mathcal{H} m_f}{i+\frac{1}{2}} - E & \dfrac{\mathcal{H}\mu_0}{i+\frac{1}{2}}\sqrt{\left(i+\frac{1}{2}\right)^2 - m_f^2} \\ \dfrac{\mathcal{H}\mu_0}{i+\frac{1}{2}}\sqrt{\left(i+\frac{1}{2}\right)^2 - m_f^2} & E_- - \dfrac{\mathcal{H}\mu_0}{i+\frac{1}{2}} m_f - E \end{vmatrix} = 0.$$

Here E_+ and E_- are the energies of terms in which the hyperfine structure is taken into account, where E_+ refers to a state with $f = i + \frac{1}{2}$, and E_- to one with $f = i - \frac{1}{2}$; m_f is the component of the total angular momentum ($m_f = f, f-1, \ldots, -f$) and $\mu_0 = \dfrac{e\hbar}{2\mu c}$. Solving the secular equation, we find

$$E = \frac{E_+ + E_-}{2} \pm \frac{\Delta E}{2}\sqrt{1 + \frac{2\xi}{i+\frac{1}{2}} m_f + \xi^2},$$

where $\Delta E = E_+ - E_-$, and $\xi = \dfrac{2\mu_0 \mathcal{H}}{\Delta E}$.

We shall determine the order of magnitude of the magnetic field intensity for which the observed splitting will be described by the obtained formula. From the conditions stated in the problem we have $\mathcal{H} \sim \dfrac{|\Delta E_{ff'}|}{\mu_0}$. In the case of the sodium atom $\Delta E_{ff'} = 0\cdot0583\ \text{cm}^{-1} = 1\cdot962 \times 10^{-16}$ erg; since $\mu_0 = 0\cdot922 \times 10^{-20}$ gauss-cm³, the intensity $\mathcal{H}$ should be of the order of 600 gauss.

We now consider the limiting cases.

(a) In the case of a weak field, i.e., for $\mu_0 \mathcal{H} \ll \Delta E$, we obtain the following expressions for the energy:

$$E = E_+ + \frac{\mathcal{H}\mu_0}{i+\frac{1}{2}} m_f,$$

$$E = E_- - \frac{\mathcal{H}\mu_0}{i+\frac{1}{2}} m_f.$$

(b) In the case of a strong field, i.e., for ($\mathcal{H}\mu_0 \gg \Delta E$)

$$E = \frac{1}{2}(E_+ + E_-) \pm \mathcal{H}\mu_0.$$

57. The normalized eigenfunctions of the hydrogen atom in the unperturbed state have the form (see Prob. 21, §4)

$$\left.\begin{aligned} \psi_{nl\left(j=l+\frac{1}{2}\right)m_j} &= \frac{R_{n,\,j-\frac{1}{2}}}{\sqrt{2j}}\begin{pmatrix}\sqrt{j+m_j}\,Y_{j-\frac{1}{2},\,m_j-\frac{1}{2}}\\ \sqrt{j-m_j}\,Y_{j-\frac{1}{2},\,m_j+\frac{1}{2}}\end{pmatrix} = u_- \\ \psi_{nl\left(j=l-\frac{1}{2}\right)m_j} &= \frac{R_{n,\,j+\frac{1}{2}}}{\sqrt{2j+2}}\begin{pmatrix}\sqrt{j+1-m_j}\,Y_{j+\frac{1}{2},\,m_j-\frac{1}{2}}\\ -\sqrt{j+1+m_j}\,Y_{j+\frac{1}{2},\,m_j+\frac{1}{2}}\end{pmatrix} = u_+ \end{aligned}\right\} \tag{1}$$

The energy is determined by two quantum numbers n, j. When the atom is in a homogeneous electric field ($\mathcal{E}_x=\mathcal{E}_y=0$, $\mathcal{E}_z=\mathcal{E}$) then, as before, $\hat{j}_z$ remains a constant of motion, while the orbital angular momentum ceases to be a constant of motion. For transitions between states with different values of m_j the matrix elements of the perturbation operator $\hat{V}=e\mathcal{E}z$ are equal to zero. The diagonal elements of the operator $\hat{V}$ also vanish, i.e.,

$$\sum_\sigma \int u_+^* z u_+\,d\tau = \sum_\sigma \int u_-^* z u_-\,d\tau. \tag{2}$$

Therefore in order to determine the splitting we must calculate the matrix elements of $\hat{V}$ corresponding to the transition from the state n, j, m_j, $l=j+\frac{1}{2}$ to the state n, j, m_j, $l=j-\frac{1}{2}$.

The matrix element we are seeking is

$$\begin{aligned} V_{21} = V_{12} &= e\mathcal{E}\sum_\sigma \int u_-^* z u_+\,d\tau \\ &= \int_0^\infty r^3 R_{n,\,j-\frac{1}{2}}(r) R_{n,\,j+\frac{1}{2}}(r)\,dr\,\frac{1}{2\sqrt{j(j+1)}}\times\Big\{\sqrt{(j+m_j)(j-m_j+1)}\times \\ &\times\int Y^*_{j-\frac{1}{2},\,m_j-\frac{1}{2}}\,Y_{j+\frac{1}{2},\,m_j-\frac{1}{2}}\cos\vartheta\,d\Omega - \sqrt{(j-m_j)(j+m_j+1)}\times \\ &\times\int Y^*_{j-\frac{1}{2},\,m_j+\frac{1}{2}}\,Y_{j+\frac{1}{2},\,m_j+\frac{1}{2}}\cos\vartheta\,d\Omega\Big\}. \end{aligned} \tag{3}$$

We first integrate over the angles. From the formula

$$\cos\vartheta\, Y_{lm}(\vartheta,\varphi) = \sqrt{\frac{(l+m+1)(l-m+1)}{(2l+1)(2l+3)}}\,Y_{l+1,\,m}(\vartheta,\varphi) + \sqrt{\frac{(l+m)(l-m)}{(2l+1)(2l-1)}}\,Y_{l-1,\,m}(\vartheta,\varphi)$$

we find that the bracketed expression in formula (3) is

$$\frac{1}{2\sqrt{j(j+1)}}\{(j+m_j)(j-m_j+1)-(j+m_j+1)(j-m_j)\}=\frac{m_j}{\sqrt{j(j+1)}}.$$

Next, integrating over r, we have

$$-\frac{3}{2}n\sqrt{n^2-\left(j+\frac{1}{2}\right)^2}.$$

Hence, for the matrix element of perturbation (3) we finally obtain

$$V_{21}=V_{12}=-\frac{3}{4}n\frac{\sqrt{n^2-\left(j+\frac{1}{2}\right)^2}}{j(j+1)}m_je\mathcal{E}.$$

The correction to the energy is found by solving the secular equation

$$\begin{vmatrix}-\varepsilon & V_{12}\\ V_{21} & -\varepsilon\end{vmatrix}=0,\quad \varepsilon=\pm V_{12};$$

$$\varepsilon=\pm\frac{3}{4}\sqrt{n^2-\left(j+\frac{1}{2}\right)^2}\frac{nm_j}{j(j+1)}e\mathcal{E}.$$

For a given n the term with $j=n-\frac{1}{2}$ does not split in the electric field, since it is not degenerate with respect to the quantum number l (l has a definite value $l=j-\frac{1}{2}=n-1$). All of the remaining terms of the fine structure split up into $2j+1$ equidistant levels ($m_j=-j,\ldots\ldots\ldots,+j$).

58.

$$\mu_0=\frac{e\hbar}{2\mu c}\frac{\left(j+\frac{1}{2}\right)^2}{j(j+1)}m_j.$$

59. In this case the spin-orbit interaction V_1, the relativistic correction for the change of mass V_2, and the energy of the electron in an external uniform electric field $V_3=-Fz$ are all quantities of the same order. We shall therefore consider the sum of these effects as a small perturbation of the system. In our calculations we take as the unperturbed function the state characterized by the orbital angular momentum L, its component m_l and the spin component m_s in the direction of the electric field (z axis).

Calculating the matrix elements of the quantities V_1, V_2, and V_3, we have (in atomic units)

$$(V_1)_{lm_l}^{l'm_l'} = \begin{cases} \frac{1}{2}\alpha^2 \dfrac{m_l(m-m_l)}{n^3 l\left(l+\frac{1}{2}\right)(l+1)}\delta_{ll'} & \text{for} \quad m_l' = m_l, \\ \frac{1}{4}\alpha^2 \dfrac{\sqrt{\left(l+\frac{1}{2}\right)^2 - m^2}}{n^3 l\left(l+\frac{1}{2}\right)(l+1)}\delta_{ll'} & \text{for} \quad m_l' = m + \frac{1}{2} \text{ and} \\ & m_l = m - \frac{1}{2} \text{ or } \textit{vice versa}, \\ 0 \text{ for all other states,} & \end{cases}$$

$$(V_2)_{lm_l}^{l'm_l'} = -\frac{1}{2}\frac{\alpha^2}{n^3}\left(\frac{1}{l+\frac{1}{2}} - \frac{3}{4n}\right)\delta_{ll'}\delta_{m_l m_l'},$$

$$(V_3)_{lm_l}^{l'm_l'} = -\frac{3n}{2}\sqrt{\frac{(n^2-l^2)(l^2-m_l^2)}{4l^2-1}}F\delta_{l',l-1}\delta_{m_l' m_l}.$$

In the case $n = 2$ the energy of the states with the quantum numbers $l = 1$, $m = m_l + m_s = \pm\frac{3}{2}$ $(j = \frac{3}{2})$ in the electric field does not change. The displacement of this level, owing to the inclusion of V_1 and V_2, is $\alpha^2/128$ atomic units (see Prob. 10, §7).

The splitting of the level whose quantum numbers are $n = 2$, $m = \pm\frac{1}{2}$ is found from the solution of the secular equation

$$\begin{vmatrix} -\frac{11}{4}\delta - E^{(1)} & \delta\sqrt{2} & 0 \\ \delta\sqrt{2} & -\frac{7}{4}\delta - E^{(1)} & -3F \\ 0 & -3F & -\frac{15}{4}\delta - E^{(1)} \end{vmatrix} = 0,$$

where $3\delta = \alpha^2/32$ atomic units is the splitting of the fine structure level with $n = 2$ in the absence of an external field. We introduce in this equation the quantity ε related to $E^{(1)}$ by $E^{(1)} = \varepsilon - \frac{11}{4}\delta$ $\left(-\frac{11}{4}\delta\right.$ is the energy of the centre of gravity of the three energy levels $\dfrac{E_1^{(1)}+E_2^{(1)}+E_3^{(1)}}{3}$ $=-\dfrac{11}{4}\delta\Big)$ and obtain

$$\begin{vmatrix} -\varepsilon & \delta\sqrt{2} & 0 \\ \delta\sqrt{2} & \delta-\varepsilon & -3F \\ 0 & -3F & -\delta-\varepsilon \end{vmatrix}=0,$$

or

$$\varepsilon^3-\varepsilon(3\delta^2+9F^2)-2\delta^3=0.$$

We solve the latter equation for the weak field case ($F \ll \delta$) and the strong field case ($F \gg \delta$). In the first case we obtain

$$\varepsilon_1=-\delta-\sqrt{3}F-\frac{F^2}{\delta},$$

$$\varepsilon_2=-\delta+\sqrt{3}F-\frac{F^2}{\delta},$$

$$\varepsilon_3=2\delta+2\frac{F^2}{\delta}.$$

In the second case

$$\varepsilon_1=-3F-\frac{1}{2}\frac{\delta^2}{F}+\frac{1}{9}\frac{\delta^3}{F^2},$$

$$\varepsilon_2=\qquad -\frac{2}{9}\frac{\delta^3}{F^2},$$

$$\varepsilon_3=\quad 3F+\frac{1}{2}\frac{\delta^2}{F}+\frac{1}{9}\frac{\delta^3}{F^2}.$$

60. The mean value of the total energy is

$$\overline{\hat{H}}=\frac{\int\psi_0^*(1+\lambda u)\hat{H}\psi_0(1+\lambda u)\,d\tau}{\int\psi_0^*\psi_0(1+\lambda u)^2\,d\tau}. \tag{1}$$

Integrating by parts, we bring the numerator of this expression to a more convenient form.

The kinetic energy operator of the electrons has the form

$$\hat{T}=-\frac{1}{2}\sum_{i=1}^{n}\Delta_i,$$

where n is the number of electrons, while Δ_i is the Laplacian acting on the coordinates of the ith electron. (We use atomic units.) We write the expression for the mean value of the kinetic energy in a form that is symmetric with respect to ψ_0^* and ψ_0, that is,

$$\bar{T} = -\frac{1}{2}\sum_{i=1}^{n}\frac{1}{2}\int \{\psi_0^*(1+\lambda u)\Delta_i(1+\lambda u)\psi_0 + \psi_0(1+\lambda u)\Delta_i(1+\lambda u)\psi_0^*\}\,d\tau.$$

Differentiating under the integral sign, we get

$$\bar{T} = -\frac{1}{2}\sum_{i=1}^{n}\frac{1}{2}\int\{\psi_0^*(1+\lambda u)^2\Delta_i\psi_0 + \psi_0(1+\lambda u)^2\Delta_i\psi_0^* + 2\lambda\psi_0\psi_0^*(1+\lambda u)\Delta_i u + 2\lambda(1+\lambda u)\nabla_i(\psi_0^*\psi_0)\nabla_i u\}\,d\tau. \quad (2)$$

Let us simplify the last two expressions. For this purpose we consider the identity

$$\nabla_i\{\psi_0^*\psi(1+\lambda u)\nabla_i u\} = \psi_0^*\psi_0(1+\lambda u)\Delta_i u + (1+\lambda u)\nabla_i(\psi_0^*\psi_0)\nabla_i u + \lambda\psi_0^*\psi(\nabla_i u)^2. \quad (3)$$

Integrating the identity (3) over the entire configuration space, we have

$$\int\{\psi_0^*\psi_0(1+\lambda u)\Delta_i u + (1+\lambda u)\nabla_i(\psi_0^*\psi_0)\nabla_i u\}\,d\tau = -\lambda\int\psi_0^*\psi(\nabla_i u)^2\,d\tau. \quad (4)$$

Inserting (4) in (2), we obtain

$$\bar{T} = -\frac{1}{2}\sum_{i=1}^{n}\frac{1}{2}\int\{\psi_0^*(1+\lambda u)^2\Delta_i\psi_0 + \psi_0(1+\lambda u)^2\Delta_i\psi_0^*\}\,d\tau + \frac{\lambda^2}{2}\sum_{i=1}^{n}\int\psi_0^*\psi_0(\nabla_i u)^2\,d\tau.$$

The Hamiltonian is $\hat{H} = \hat{H}_0 + u = \hat{T} + \hat{V} + u$. Taking into account the fact that $\hat{V}$ commutes with $(1+\lambda u)$, we can represent expression (1) in the form

$$\bar{H} = E_0 + \frac{\frac{1}{2}\int(1+\lambda u)^2(\psi_0^*\hat{H}\psi_0 + \psi_0\hat{H}\psi_0^*)\,d\tau + \frac{\lambda^2}{2}\sum_{i=1}^{n}\int\psi_0^*\psi_0(\nabla_i u)^2\,d\tau}{\int(1+\lambda u)^2\psi_0^*\psi_0\,d\tau}.$$

Since $\hat{H}_0\psi_0 = E_0\psi_0$, then

$$\bar{\hat{H}} = E_0 + \frac{\int \psi_0^* u(1+\lambda u)^2 \psi_0\, d\tau + \frac{\lambda^2}{2}\sum_{i=1}^{n}\int \psi_0^*\psi_0 (\nabla_i u)^2\, d\tau}{\int (1+\lambda u)^2 \psi_0^*\psi_0\, d\tau},$$

or

$$\bar{\hat{H}} = E_0 + \frac{(u)_{00} + 2\lambda(u^2)_{00} + \lambda^2(u^3)_{00} + \frac{\lambda^2}{2}\sum_{i=1}^{n}\{(\nabla_i u)^2\}_{00}}{1 + 2\lambda(u)_{00} + \lambda^2(u^2)_{00}}, \tag{5}$$

where $(u)_{00} = \int \psi_0^* u\psi_0\, d\tau$, $(u^2)_{00} = \int \psi_0^* u^2\psi_0\, d\tau$, etc. Let us expand the second term in formula (5) into a series and neglect the terms with $(u^3)_{00}$, $(u)_{00}^3$, etc.

For the energy correction ΔE we get an approximate expression

$$\Delta E \approx (u)_{00} + 2\lambda(u^2)_{00} - 2\lambda(u)_{00}^2 + \frac{\lambda^2}{2}\sum_{i=1}^{n}\{(\nabla_i u)^2\}_{00}. \tag{6}$$

The value of the variational parameter λ is found from the condition

$$\frac{d\Delta E}{d\lambda} = 2(u^2)_{00} - 2(u)_{00}^2 + \lambda\sum_{i=1}^{n}\{(\nabla_i u)^2\}_{00} = 0,$$

whence for λ we obtain the value

$$\lambda = 2\,\frac{(u)_{00}^2 - (u^2)_{00}}{\sum_{i=1}^{n}\{(\nabla_i u)^2\}_{00}}.$$

Inserting λ in (6), we obtain the relation

$$\Delta E = (u)_{00} - 2\,\frac{\{(u)_{00}^2 - (u^2)_{00}\}^2}{\sum_{i=1}^{n}\{(\nabla_i u)^2\}_{00}}. \tag{7}$$

61. If the atom is in a uniform electric field of intensity $\mathcal{E}$ in the direction of the z axis, the perturbation operator is

$$u = -\mathcal{E}\sum_{i=1}^{n} z_i = -\mathcal{E}z.$$

The matrix element $(u)_{00}$ is zero.

By Eq. (7) of the preceding problem, we have

$$\Delta E \approx -2\mathcal{E}^2 \frac{\{(z^2)_{00}\}}{n},$$

where n is the number of electrons.

It thus follows that the coefficient of polarization is

$$\alpha = \frac{4\{(z^2)_{00}\}^2}{n}.$$

It should be noted that this formula, obtained by introducing only a single variational parameter λ, is a satisfactory approximation only in the case of hydrogen and helium atoms. In the case of atoms with several electron shells the deformation of the shells will not be the same. Consequently, to obtain a more satisfactory result with the variational method we should introduce a separate variational parameter λ for each electron shell.

For the hydrogen atom

$$(z^2)_{00} = \frac{1}{3}(r^2)_{00} = \frac{1}{3}\frac{4\pi}{\pi}\int_0^\infty e^{-2r} r^4 \, dr = 1,$$

and therefore

$$\alpha = 4 \text{ atomic units.}$$

In the CGSE system

$$\alpha = 4\left(\frac{\hbar^2}{\mu e^2}\right) \text{cm}^3.$$

For the helium atom we take the wave function of the ground state in the form

$$\psi_0 = \frac{Z_{\text{eff}}^2}{\pi} e^{-Z_{\text{eff}}(r_1+r_2)}, \quad \text{where } Z_{\text{eff}} = \tfrac{27}{16} \quad \text{(see Prob. 14, §7).}$$

The calculation gives

$$\alpha = 0.98 \text{ atomic units or } \alpha = 0.98\left(\frac{\hbar^2}{\mu e^2}\right)^3 \text{cm}^3.$$

For the value found for α the dielectric constant of helium in standard conditions should be

$$\varepsilon = 1.00049.$$

The observed value is 1.00074.

The comparatively large difference between the calculated and observed values is explained mainly by the fact that we employed a rough approximation for the unperturbed function.

62. (a) $$E^{(1)} = \frac{m_j}{8j(j+1)}\frac{e\hbar}{\mu c}\{(2j+1)^2\mathcal{H} \pm$$
$$\pm\sqrt{(2j+1)^2\mathcal{H}^2 + \frac{36\,n^2}{\alpha^2}\left\{n^2 - \left(j+\frac{1}{2}\right)^2\right\}\mathcal{E}^2}$$
$$\left(\alpha = \frac{e^2}{\hbar c}\right)$$

(see Prob. 57, 59, §7).

(b) $$E^{(1)}\left(n = 2,\, m_j = \pm\frac{3}{2}\right) = -\frac{\alpha^2}{128}\left(\frac{\mu e^4}{\hbar^2}\right) \pm \frac{e\hbar}{\mu c}\mathcal{H},$$
$$E^{(1)}\left(n = 2,\, m_j = \pm\frac{1}{2}\right) = \varepsilon - \frac{11}{384}\alpha^2\left(\frac{\mu e^4}{\hbar^2}\right),$$

where ε is found by solving the following equation of the third degree:
$$\varepsilon^3 \mp 2\beta\varepsilon^2 - \varepsilon(3\delta^2 + 9F^2 - \beta^2) \pm 2\delta^2\beta - 2\delta^3 = 0,$$
Here,
$$\beta = \frac{e\hbar}{2\mu c}\mathcal{H}, \quad F = e\mathcal{E}\frac{\hbar^2}{\mu e^2}, \quad \delta = \frac{\alpha^2}{96}\frac{\mu e^4}{\hbar^2}.$$

In the case of strong fields ($\delta \ll F$, $\delta \ll \beta$)
$$\varepsilon_1 = \pm\beta - 3F - \frac{1}{2}\frac{9F \mp \beta}{9F^2 \mp 3F\beta}\delta^2,$$
$$\varepsilon_2 = \pm\frac{2\beta}{9F^2 - \beta^2}\delta^2,$$
$$\varepsilon_3 = \pm\beta + 3F + \frac{1}{2}\frac{9F \pm \beta}{9F \pm 3F\beta}\delta^2$$
(see Prob. 59, §7).

63. We take the z axis in the direction of the magnetic field and the x axis in the direction of the electric field. The potential energy operator for an electron in these fields has the following form:
$$w = \frac{e\hbar}{2\mu c}(\hat{l}_z + 2\hat{s}_z)\mathcal{H} + e\mathcal{E}x.$$

We shall regard this operator as a small perturbation of the unperturbed stationary state characterized by the quantum numbers n, l, m, σ (m

and σ are the projections of the orbital and spin angular momenta on the z axis). The non-vanishing matrix elements of x have the form

$$(x)_{l,\,m-1}^{l-1,\,m} = (x)_{l-1,\,m}^{l,\,m-1} = \frac{3}{4} n \sqrt{\frac{(n^2-l^2)(l-m+1)(l-m)}{(2l+1)(2l-1)}}\, a,$$

$$(x)_{l-1,\,m-1}^{l,\,m} = (x)_{l,\,m}^{l-1,\,m-1} = -\frac{3}{4} n \sqrt{\frac{(n^2-l^2)(l+m-1)(l+m)}{(2l+1)(2l-1)}}\, a,$$

$$\left(a = \frac{\hbar^2}{\mu e^2}\right).$$

We shall consider the case of $n = 2$. Let us take the z component of the spin to be $+\frac{1}{2}$. We introduce the following notation: $\beta = \frac{e\hbar}{2\mu c}\mathcal{H}$, $\gamma = \frac{3}{\sqrt{2}} e\mathcal{E}a$. In this notation the matrix of the perturbation operator has the form

$$\begin{pmatrix} 2\beta & 0 & 0 & -\gamma \\ 0 & \beta & 0 & 0 \\ 0 & 0 & 0 & \gamma \\ -\gamma & 0 & \gamma & \beta \end{pmatrix}.$$

The states are numbered by decreasing values of l and m. The state with quantum numbers $l = 1$, $m = 0$ does not combine with the remaining states. The energy of this state is $E_1^{(1)} = \beta$. The three remaining eigenvalues of the perturbation matrix are found by solving the secular equation

$$E^3 - 3\beta E^2 + 2(\beta^2 - \gamma^2)E + 2\gamma^2\beta = 0.$$

Solving it, we have

$$E_{2,4}^{(1)} = \beta, \qquad E_{3,4}^{(1)} = \beta \pm \sqrt{\beta^2 + 2\gamma^2},$$

or, inserting the values of β and γ, we obtain

$$E_2^{(1)} = \frac{e\hbar}{2\mu c}\mathcal{H}, \qquad E_{3,4}^{(1)} = \frac{e\hbar}{2\mu c}\left\{\mathcal{H} \pm \sqrt{\mathcal{H}^2 + \frac{36}{a^2}\mathcal{E}^2}\right\}.$$

§8. The molecule

1. If we neglect the difference between the centre of gravity of the molecule and the centre of gravity of the nuclei and assume that the centre of gravity lies at the origin of the coordinate system, then the

Schrödinger equation for a diatomic molecule can be written

$$\left\{-\frac{\hbar^2}{2m}\sum_i\left(\frac{\partial^2}{\partial x_i^2}+\frac{\partial^2}{\partial y_i^2}+\frac{\partial^2}{\partial z_i^2}\right)-\right.$$

$$-\frac{\hbar^2}{2M\varrho^2}\left[\frac{\partial}{\partial\varrho}\left(\varrho^2\frac{\partial}{\partial\varrho}\right)+\frac{1}{\sin\theta}\frac{\partial}{\partial\theta}\left(\sin\theta\frac{\partial}{\partial\theta}\right)+\frac{1}{\sin^2\theta}\frac{\partial^2}{\partial\varphi^2}\right]+$$

$$\left.+V(x_i, y_i, z_i; \varrho, \theta, \varphi)\right\}\psi(\ldots x_i, y_i, z_i, \ldots; \varrho, \theta, \varphi)=E\psi.$$

Here x_i, y_i, z_i are the coordinates of the ith electron in the stationary coordinate system, the angles θ, φ describe the position in space of the line joining the nuclei; ϱ is the distance between the nuclei, and M is the reduced mass of the two nuclei. This equation is inconvenient, since the angles θ and φ occur in the expression for the potential energy V of the electrostatic interaction. To put the Schrödinger equation in more convenient form we introduce a new system of coordinates ξ, η, ζ rotating together with the nuclei. The axis ζ will be directed along the line joining the nuclei, the axis ξ lying in the xy plane. We choose the positive direction of the axis ξ so that axes z, ζ, ξ form a right-hand coordinate system. The relations between the old and new coordinate systems are now as follows:

$$\begin{aligned}\xi_i &= -x_i\sin\varphi+y_i\cos\varphi,\\ \eta_i &= -x_i\cos\theta\cos\varphi-y_i\cos\theta\sin\varphi+z_i\sin\theta,\\ \zeta_i &= x_i\sin\theta\cos\varphi+y_i\sin\theta\sin\varphi+z_i\cos\theta.\end{aligned}$$

Differentiation for constant x_i, y_i, z_i will be distinguished from differentiation for constant ξ_i, η_i, ζ_i by a prime on the ∂. We then have

$$\frac{\partial'}{\partial\theta}=\frac{\partial}{\partial\theta}+\sum_i\left(\zeta_i\frac{\partial}{\partial\eta_i}-\eta_i\frac{\partial}{\partial\zeta_i}\right),$$

$$\frac{\partial'}{\partial\varphi}=\frac{\partial}{\partial\varphi}-\sum_i\left\{\sin\theta\left(\zeta_i\frac{\partial}{\partial\xi_i}-\xi_i\frac{\partial}{\partial\zeta_i}\right)+\cos\theta\left(\xi_i\frac{\partial}{\partial\eta_i}-\eta_i\frac{\partial}{\partial\xi_i}\right)\right\}.$$

It is readily seen that

$$\sum_i\left(\frac{\partial^2}{\partial x_i^2}+\frac{\partial^2}{\partial y_i^2}+\frac{\partial^2}{\partial z_i^2}\right)=\sum_i\left(\frac{\partial^2}{\partial\xi_i^2}+\frac{\partial^2}{\partial\eta_i^2}+\frac{\partial^2}{\partial\zeta_i^2}\right).$$

The potential energy in the new coordinate system will have the form

$$V=\frac{Z_1Z_2e^2}{\varrho}+\sum_{i>k}\frac{e^2}{r_{ik}^*}-\sum_{k=1}^{n}\frac{Z_1e^2}{r_{1k}^*}-\sum_{k=1}^{n}\frac{Z_2e^2}{r_{2k}^*},$$

where

$$r^*_{ik}\sqrt{(\xi_i-\xi_k)^2+(\eta_i-\eta_k)^2+(\zeta_i-\zeta_k)^2}$$

is the distance between the ith and kth electrons in the new coordinates;

$$r^*_{1k}=\sqrt{\xi_k^2+\eta_k^2+\left(\zeta_k+\varrho\frac{M_2}{M_1+M_2}\right)^2}$$

is the distance between the kth electron to the first nucleus, and

$$r^*_{2k}=\sqrt{\xi_k^2+\eta_k^2+\left(\zeta_k-\frac{\varrho M_1}{M_1+M_2}\right)^2}$$

is the distance between the kth electron and the second nucleus. Thus, in the new coordinates the potential energy does not depend on the angles θ and φ.

The Schrödinger equation now takes the form

$$\left\{-\frac{\hbar^2}{2m}\sum_i\left(\frac{\partial^2}{\partial\xi_i^2}+\frac{\partial^2}{\partial\eta_i^2}+\frac{\partial^2}{\partial\zeta_i^2}\right)-\right.$$

$$-\frac{\hbar^2}{2M}\frac{1}{r^2}\left[\frac{\partial}{\partial r}\left(r^2\frac{\partial}{\partial r}\right)+\cot\theta\left(\frac{\partial}{\partial\theta}-i\hat{L}_\xi\right)+\right.$$

$$\left.+\left(\frac{\partial}{\partial\theta}-i\hat{L}_\xi\right)^2+\frac{1}{\sin^2\theta}\left(\frac{\partial}{\partial\varphi}-i\sin\theta\hat{L}_\eta-i\cos\theta\hat{L}_\zeta\right)^2\right]+$$

$$\left.+V(\xi_i,\eta_i,\zeta_j;\varrho)-E\right\}\psi(\xi_i,\eta_i,\zeta_i;\varrho,\theta,\varphi)=0.$$

Here $\hat{L}_\xi$, $\hat{L}_\eta$ and $\hat{L}_\zeta$ are (in units of $\hbar$) the operators of the components of the orbital angular momentum of the electrons in the system of coordinates ξ, η, ζ.

2. We denote the spin variable of the ith electron with respect to the rest system by s_i' and with respect to the moving coordinate system by s_i The functions $\psi(\ldots s_i\ldots)$ with spins referring to the coordinate system ξ, η, ζ are related to the functions $\psi(\ldots s_i'\ldots)$ with spins referring to the coordinate system x, y, z by a linear transformation

$$\psi(\ldots s_i\ldots)=\sum_{s_1',\ldots s_i'\ldots}S(s_1,\ldots,s_i\ldots,s_1',\ldots,s_i',\ldots)\psi(\ldots s_i'\ldots)=\hat{S}\psi(\ldots s_i'\ldots),$$

where

$$S(s_1,s_2,\ldots,s_i,\ldots,s_1',\ldots,s_i',\ldots)=S(s_1;s_1')S(s_2;s_2')\ldots S(s_i;s_i')\ldots,$$

while

$$S\left(\frac{1}{2};\frac{1}{2}\right)=\cos\frac{\theta}{2}e^{\frac{i}{2}\left(\varphi+\frac{\pi}{2}\right)},$$

$$S\left(\frac{1}{2};-\frac{1}{2}\right)=i\sin\frac{\theta}{2}e^{-\frac{i}{2}\left(\varphi+\frac{\pi}{2}\right)},$$

$$S\left(-\frac{1}{2};\frac{1}{2}\right)=i\sin\frac{\theta}{2}e^{\frac{i}{2}\left(\varphi+\frac{\pi}{2}\right)},$$

$$S\left(-\frac{1}{2};-\frac{1}{2}\right)=\cos\frac{\theta}{2}e^{-\frac{i}{2}\left(\varphi+\frac{\pi}{2}\right)}.$$

The Schrödinger equation we are seeking will have the form

$$[\hat{S}\hat{H}\hat{S}^{-1}-E]\psi(\ldots\xi_i,\eta_i,\zeta_i,s_i,\ldots;\varrho,\theta,\varphi)=0,$$

where $\hat{H}$ is the Hamiltonian which we found in the preceding problem.

After simple calculations the Schrödinger equation takes the form

$$\Bigg\{-\frac{\hbar^2}{2m}\sum_i\left(\frac{\partial^2}{\partial\xi_i^2}+\frac{\partial^2}{\partial\eta_i^2}+\frac{\partial^2}{\partial\zeta_i^2}\right)-$$

$$-\frac{\hbar^2}{2M}\frac{1}{r^2}\left[\frac{\partial}{\partial r}\left(r^2\frac{\partial}{\partial r}\right)+\cot\theta\left(\frac{\partial}{\partial\theta}-i\hat{M}_\xi\right)+\left(\frac{\partial}{\partial\theta}-iM_\xi\right)^2+\right.$$

$$\left.+\frac{1}{\sin^2\theta}\left(\frac{\partial}{\partial\varphi}-i\sin\theta\hat{M}_\eta-i\cos\theta\hat{M}_\zeta\right)^2\right]+$$

$$+V-E\Bigg\}\psi(\ldots\xi_i,\eta_i,\zeta_i,s_i,\ldots;\varrho,\theta,\varphi)=0.$$

Here, unlike the preceding problem, $\hat{M}_\xi$, $\hat{M}_\eta$, $\hat{M}_\zeta$ denote the operators of the components of the total (orbital and spin) angular momentum of the electrons.

3. We assume that we know the solution of the problem with fixed centres, i.e., we know the electronic terms $E^{el}(\varrho)$ and the wave function Φ_{el}. Let us consider case *a*; we assume that in the state described by the wave function Φ_{el} the component of the total (orbital and spin) angular momentum along the axis of the molecule is Ω. We multiply the Schrödinger equation

$$\hat{H}\Phi_{el}(\xi_i,\eta_i,\zeta_i;\sigma_i;\varrho)f(\varrho)\Theta(\theta,\varphi)=E\Phi_{el}(\xi_i,\eta_i,\zeta_i;\sigma_i;\varrho)f(\varrho)\Theta(\theta,\varphi)$$

on the left by Φ_{el}^* and integrate over the coordinates in the problem with fixed centres, and we sum over σ_i. Noting that

$$\int \Phi_{el}^* \hat{M}_\xi \Phi_{el}\, d\tau = \int \Phi_{el}^* \hat{M}_\eta \Phi_{el}\, d\tau = 0,$$

we obtain

$$\left[B \frac{\partial}{\partial \varrho}\left(\varrho^2 \frac{\partial}{\partial \varrho}\right) - E^{el}(\varrho) - U(\varrho) - E^{rot} + E\right] f(\varrho) = 0,$$

$$B\left[\frac{1}{\sin\theta}\frac{\partial}{\partial\theta}\left(\sin\theta\frac{\partial}{\partial\theta}\right) + \frac{1}{\sin^2\theta}\left(\frac{\partial}{\partial\varphi} - i\Omega\cos\theta\right)^2\right]\Theta(\theta,\varphi) + $$
$$+E^{rot}\,\Theta(\theta,\varphi) = 0.$$

In the last two equations we have introduced the following notation:

$$U(\varrho) = \frac{1}{M}\int \Phi_{el}^* \left\{\frac{1}{\varrho^2}(\hat{M}_\xi^2 + \hat{M}_\eta^2) - \frac{\partial^2}{\partial\varrho^2}\right\}\Phi_{el}\, d\tau,$$

$$B = \frac{\hbar^2}{2M}\frac{1}{\varrho^2}.$$

The quantity B is called the *rotational constant*.

4. N_2 molecule (atoms in the ^4S state):

$$^1\Sigma_g^+,\quad ^3\Sigma_u^+,\quad ^5\Sigma_g^+,\quad ^7\Sigma_u^+.$$

Br_2 molecule (atoms in the ^2P state):

$$2\,^1\Sigma_g^+,\quad ^1\Sigma_u^-,\quad ^1\Pi_g,\quad ^1\Pi_u,\quad ^1\Delta_g;$$
$$2\,^3\Sigma_u^+,\quad ^3\Sigma_g^-,\quad ^3\Pi_g,\quad ^3\Pi_u,\quad ^1\Delta_u.$$

LiH molecule (Li atom in ^2S_g state, H atom in ^2S_g state):

$$^1\Sigma^+,\quad ^3\Sigma^+.$$

HBr molecule (Br atom in the ^2P_u state):

$$^1\Sigma^+,\quad ^3\Sigma^+,\quad \Pi^1,\quad ^3\Pi.$$

CN molecule (C atom in the ^3P_g state, N atom in the ^4S_u state):

$$^2\Sigma^+,\quad ^4\Sigma^+,\quad ^6\Sigma^+,\quad ^2\Pi,\quad ^4\Pi,\quad ^6\Pi.$$

(The number in front of the term symbol indicates the number of terms.)

5. A helium atom in the ground state is characterized by the fact that both of its electrons are in the lowest energy level (parahelium). The complete eigenfunction of the ground state of an helium atom can be represented approximately in the form

$$\frac{1}{\sqrt{2}}\psi_a(1)\psi_a(2)\{\eta_+(\sigma_1)\eta_-(\sigma_2)-\eta_+(\sigma_2)\eta_-(\sigma_1)\},$$

where ψ_a is the hydrogen function.

The eigenfunction of the hydrogen atom has the form

$$\psi_b(3)\eta_+(\sigma_3) \quad \text{or} \quad \psi_b(3)\eta_-(\sigma_3).$$

If the distance between the two atoms is large, the wave function o the system can be written in the form of the product

$$\frac{1}{\sqrt{2}}\psi_a(1)\psi_a(2)\psi_b(3)\{\eta_+(\sigma_1)\eta_-(\sigma_2)-\eta_+(\sigma_2)\eta_-(\sigma_1)\}\eta_+(\sigma_3).$$

If we take into account the interchange of the electrons, the eigenfunction of the system should be anti-symmetrized with respect to the interchange of all the electrons. Only one anti-symmetric function is an eigenfunction of the system in the zero-order approximation:

$$\Psi=\frac{1}{\sqrt{6(1-S)}}\{\psi_a(1)\psi_a(2)\psi_b(3)\,[\eta_+(1)\eta_-(2)-\eta_+(2)\eta_-(1)]\,\eta_+(3)+$$
$$+\psi_a(3)\psi_a(1)\psi_b(2)\,[\eta_+(3)\eta_-(1)-\eta_+(1)\eta_-(3)]\,\eta_+(2)+$$
$$+\psi_a(2)\psi_a(3)\psi_b(1)\,[\eta_+(2)\eta_-(3)-\eta_+(3)\eta_-(2)]\,\eta_+(1)\}.$$

In the quantity (1) $1/\sqrt{6(1-S)}$ is a normalizing factor and

$$S=\int\psi_a(1)\psi_a(2)\psi_b(3)\,\psi_a(2)\psi_a(3)\psi_b(1)\,d\tau_1\,d\tau_2\,d\tau_3$$
$$=\int\psi_a(1)\psi_a(2)\psi_b(3)\psi_a(1)\psi_a(3)\psi_b(2)\,d\tau_1\,d\tau_2\,d\tau_3$$
$$=\int\psi_a(1)\psi_a(3)\psi_b(2)\psi_a(2)\psi_a(3)\psi_b(1)\,d\tau_1\,d\tau_2\,d\tau_3.$$

Making use of perturbation theory, we have

$$\varepsilon=\sum_\sigma\int\Psi\hat{H}\Psi\,d\tau,$$

where Ψ is the eigenfunction in the zero-order approximation and $\hat{H}$ is the energy of the perturbation. We sum over the spin variables. It should be borne in mind that $\hat{H}$ has a different form for the different parts of the function Ψ, namely, for $\psi_a(1)\psi_a(2)\psi_b(3)$ the perturbation energy is

$$\hat{H}=e^2\left(\frac{2}{R}-\frac{1}{r_{a3}}-\frac{1}{r_{b1}}-\frac{1}{r_{b2}}+\frac{1}{r_{13}}+\frac{1}{r_{23}}\right),$$

and for $\psi_a(1)\psi_a(3)\psi_b(2)$

$$\hat{H} = e^2\left(\frac{2}{R} - \frac{1}{r_{a2}} - \frac{1}{r_{b1}} - \frac{1}{r_{b3}} + \frac{1}{r_{12}} + \frac{1}{r_{32}}\right).$$

Since the integrals differ only in the labelling of electrons and are therefore identical, we have

$$\varepsilon = \frac{K-A}{1-S}, \tag{2}$$

where

$$K = e^2 \int \left(\frac{2}{R} + \frac{1}{r_{13}} + \frac{1}{r_{23}} - \frac{1}{r_{a3}} - \frac{1}{r_{b1}} - \frac{1}{r_{b2}}\right) \times$$
$$\times \psi_a^2(1)\psi_a^2(2)\psi_b^2(3)\, d\tau_1\, d\tau_2\, d\tau_3,$$

$$A = e^2 \int \left(\frac{2}{R} + \frac{1}{r_{13}} + \frac{1}{r_{23}} - \frac{1}{r_{a3}} - \frac{1}{r_{b1}} - \frac{1}{r_{b2}}\right) \times$$
$$\times \psi_a(1)\psi_a(2)\psi_b(3)\psi_a(1)\psi_a(3)\psi_b(2)\, d\tau_1\, d\tau_2\, d\tau_3.$$

The integrals K, A, S have, in general, the same character as the corresponding integrals for the hydrogen molecule problem. Calculation of the integrals indicates that formula (2) corresponds to a repulsion curve. This holds not only for the helium atom, but for all inert gases.

6. After separating the motion of the centre of mass from the relative motion of the nuclei, we obtain for the wave function of the latter the equation

$$\Delta\psi + \frac{2\mu}{\hbar^2}\left[E + 2D\left(\frac{1}{\varrho} - \frac{1}{2\varrho^2}\right)\right]\psi = 0.$$

Using spherical coordinates and setting

$$\psi = \frac{\chi(\varrho)}{\varrho} Y_{KM}(\theta, \varphi),$$

we obtain for χ the differential equation

$$\frac{d^2\chi}{d\varrho^2} + \left[-\lambda^2 + \frac{2\gamma^2}{\varrho} - \frac{\gamma^2 + K(K+1)}{\varrho^2}\right]\chi = 0,$$

$$\lambda = \sqrt{-\frac{2\mu a^2 E}{\hbar^2}}, \qquad \gamma^2 = \frac{2\mu a^2}{\hbar^2} D.$$

If we now introduce the change of variables $\chi(\varrho)=\varrho^s e^{-\lambda\varrho}u(\varrho)$, where $s=\frac{1}{2}+\sqrt{\gamma^2+\left(K+\frac{1}{2}\right)^2}$, we obtain the hypergeometric equation

$$\varrho u''+(2s-2\lambda\varrho)u'+(-2s\lambda+2\gamma^2)u=0.$$

This equation has the following solution finite at $\varrho=0$

$$u=cF\left(s-\frac{\gamma^2}{\lambda},\ 2s,\ 2\lambda\varrho\right).$$

In states with a discrete set of eigenfunctions the wave function should tend to zero as $r\to\infty$. This means that the expression for u should reduce to the polynomial

$$s-\frac{\gamma^2}{\lambda}=-v,$$

where v is a non-negative integer. From this condition we obtain the energy levels:

$$E_{vK}=-\frac{\hbar}{2\mu a^2}\frac{\gamma^4}{\left[v+\frac{1}{2}+\sqrt{\gamma^2+\left(K+\frac{1}{2}\right)^2}\right]^2}.$$

The dimensionless parameter γ^2 is proportional to the reduced mass of the nuclei μ, so that $\gamma^2\gg 1$. If v and K are not very large

$$v\ll\gamma,\qquad K\ll\gamma,$$

the expression E_{vK} takes the form

$$E_{vK}=-D+\hbar\omega_0\left(v+\frac{1}{2}\right)+\frac{\hbar^2}{2\mu a^2}\left(K+\frac{1}{2}\right)^2-\frac{3\hbar^2}{2\mu a^2}\left(v+\frac{1}{2}\right)^2-\frac{3\hbar^3\left(K+\frac{1}{2}\right)\left(v+\frac{1}{2}\right)}{2\mu^2a^4\omega_0},$$

where

$$\omega_0=\sqrt{\frac{2D}{\mu a^2}}.$$

The dissociation energy is approximately equal to

$$E_0=D-\frac{\hbar\omega_0}{2}.$$

The second and third terms in the expression for E_{vK} give the energy of the vibrational and rotational motion. The fourth term takes into account anharmonic oscillations, and finally the fifth term gives the correction to the energy for the interaction between the rotational and vibrational motions of the nuclei.

Since D is a quantity of the order of unity in atomic units ($e = m = \hbar = 1$) it follows from the results above that

$$D : \hbar\omega_0 : \frac{\hbar^2}{2\mu a^2} \sim 1 : \sqrt{\frac{m}{\mu}} : \frac{m}{\mu},$$

where m is the mass of the electron.

It is thus clear that the difference in energy of the electrons in two different quantum states (a quantity of the order D) is large in comparison to the difference in energy between two vibrational states, which, in turn, is large in comparison to the distance between rotational levels.

7. Let us find the minimum effective potential of

$$W = -2D\left(\frac{1}{\varrho} - \frac{1}{2\varrho^2}\right) + \frac{A^2}{\varrho^2}, \quad \text{where} \quad A^2 = \frac{\hbar^2 K(K+1)}{2\mu a^2},$$

from the condition that the derivative of W is zero:

$$W' = -2D\left(-\frac{1}{\varrho^2} + \frac{1}{\varrho_0^3}\right) - \frac{2A^2}{\varrho_0^3} = 0.$$

From this we get

$$\varrho_0 = 1 + \frac{A^2}{D}.$$

In the vicinity of this position of equilibrium we expand the effective potential:

$$W(\varrho) \approx -2D\left(\frac{1}{\varrho_0} - \frac{1}{\varrho_0^2}\right) + \frac{A^2}{\varrho_0^2} + \frac{D + A^2}{\varrho_0^4}(\varrho - \varrho_0)^2.$$

Retaining terms of the order A^2, we obtain

$$W(\varrho) \approx -D + A^2 + (D - 3A^2)(\varrho - \varrho_0)^2.$$

Let us now substitute this expression into the equation for χ:

$$\frac{d^2\chi}{d\varrho^2} + \frac{2\mu a^2}{\hbar^2}[E + D - A^2 - (D - 3A^2)(\varrho - \varrho_0)^2]\chi = 0.$$

From this we find the energy levels

$$E_{vK} = -D + A^2 + \hbar\omega\left(v + \frac{1}{2}\right).$$

In the same approximation

$$\omega = \omega_0\left(1 - \frac{3}{2}\frac{A^2}{D}\right), \qquad \omega_0 = \sqrt{\frac{2D}{\mu a^2}}.$$

Finally, for the energy E_{vK} we get

$$E_{vK} = -D + \hbar\omega_0\left(v + \frac{1}{2}\right) + \frac{\hbar^2 K(K+1)}{2\mu a^2} - \frac{3}{2}\frac{\hbar K(K+1)\left(v + \frac{1}{2}\right)}{\mu^2 a^4 \omega_0}.$$

Since our calculations have not taken the anharmonic effect into account, this formula lacks the term $-\frac{3\hbar^2}{2\mu a^2}\left(v + \frac{1}{2}\right)^2$, obtained in the solution of the preceding problem.

8. In the infrared band we have to do with transitions in which the vibrational and rotational quantum numbers change, the electrons remaining in the ground state. For the transition between two states $v', J' \to v'', J''$ we have

$$\omega = \omega_0(v' - v'') + \frac{\hbar}{2\mu a^2}[J'^2 + J' - J''^2 - J''].$$

From the selection rule for J we have $J'' = J' \pm 1$. We thus obtain the following frequencies

$$\omega = \omega_0(v' - v'') - \frac{\hbar}{2\mu a^2}2(J'+1) \qquad \begin{pmatrix} J'' = J'+1, \\ J' = 0, 1, 2, \ldots \end{pmatrix}$$

and

$$\omega = \omega_0(v' - v'') + \frac{\hbar}{2\mu a^2}2J' \qquad \begin{pmatrix} J'' = J'-1, \\ J' = 1, 2, \ldots \end{pmatrix}.$$

We note that these two series of frequencies in molecular spectroscopy are called the P and R branches.

From the obtained expression it is seen that the difference in the frequencies of the two neighbouring lines for fixed v' and v'' is equal to (in cm^{-1})

$$\Delta\nu = \frac{\Delta\omega}{2\pi c} = \frac{\hbar}{2\pi c \mu a^2}.$$

The moment of inertia of the HCl^{35} molecule is

$$I = \mu a^2 = \frac{\hbar}{2\pi c \Delta\nu} = 2 \cdot 65 \times 10^{-40}\, g \cdot cm^2.$$

Usin e value of the reduced mass

$$\mu = M_H \frac{1\times 35}{1+35} = 0\cdot 972\, M_H = 1\cdot 61\times 10^{-24}\ \text{g},$$

we find the distance between the nuclei in HCl

$$a = \sqrt{\frac{I}{\mu}} = 1\cdot 29\times 10^{-8}\ \text{cm}.$$

The distance a corresponding to the equilibrium of the molecules DCl and HCl is the same, since the shape of the potential curves is determined by the electron states. It thus follows that

$$\frac{\Delta\nu_{DCl}}{\Delta\nu_{HCl}} = \frac{\mu_{HCl}}{\mu_{DCl}}; \qquad \Delta\nu_{DCl} = 10\cdot 7\ \text{cm}^{-1}.$$

9. The distance between the two first rotational levels is

$$\Delta\nu_{rot} = \frac{\hbar}{2\pi c I} = 41\cdot 5\ \text{cm}^{-1}.$$

We thus find that

$$\frac{\Delta\nu_{rot}}{\Delta\nu_{vib}} = 0\cdot 0104.$$

10. The energy of dissociation of the D_2 molecule is $4\cdot 54$ eV.

11. After making a change of variables $\xi = \frac{r-a}{a}$, we write the equation for the radial function χ/r:

$$\frac{d^2\chi}{d\xi^2} + \frac{2\mu a^2}{\hbar^2}(E-V)\chi = 0.$$

Setting $z = \alpha e^{-2\beta\xi}$, we obtain:

$$\chi'' + \frac{1}{z}\chi' + \left(-\frac{s^2}{z^2} - \frac{1}{4} + \frac{v+s+\frac{1}{2}}{z}\right)\chi = 0,$$

where

$$s = \sqrt{\frac{\mu a^2(D-E)}{\beta^2\hbar^2}}, \qquad v+s+\frac{1}{2} = \frac{\mu a^2 D}{\alpha\beta^2\hbar^2}, \qquad \alpha^2 = \frac{2\mu a^2 D}{\beta^2\hbar^2}.$$

If we make the substitution $\chi = e^{-\frac{z}{2}} z^s u(z)$, the equation above reduces to the hypergeometric equation $zu'' + (2s+1-z)u' + nu = 0$, whose solution is the confluent hypergeometric function $u = F(-v, 2s+1, z)$. This function satisfies the condition that χ vanishes as $r \to +\infty$ for positive values of s (discrete spectrum). For $r \to -\infty$ the wave function should vanish so that F reduces to a polynomial, i.e., so that v is a non-negative integer. This condition gives the energy spectrum:

$$E_n = \hbar\omega\left(v+\frac{1}{2}\right) - \frac{\hbar^2\omega^2}{4D}\left(v+\frac{1}{2}\right)^2, \quad \text{where} \quad \omega = 4\beta\sqrt{\frac{D}{2\mu a^2}}.$$

Thus the interval between the vibrational levels decreases with an increase in the quantum number v.

The dissociation energy is

$$E_0 = D - \frac{\hbar\omega}{2} + \frac{\hbar^2\omega^2}{16D}.$$

13. For the parameters characterizing the rotational motion let us take the Euler angles (θ, ψ, φ); in this case the coordinates of a point x, y, z in a stationary system are related to the coordinates ξ, η, ζ in a moving system as follows:

$$\left.\begin{aligned} x &= \xi(\cos\psi\cos\varphi - \sin\psi\sin\varphi\cos\theta) - \\ &\qquad -\eta(\cos\psi\sin\varphi + \sin\psi\cos\varphi\cos\theta) + \zeta\sin\psi\sin\theta, \\ y &= \xi(\sin\psi\cos\varphi + \cos\psi\sin\varphi\cos\theta) + \\ &\qquad +\eta(-\sin\psi\sin\varphi + \cos\psi\cos\varphi\cos\theta) - \zeta\cos\psi\sin\theta, \\ z &= \xi\sin\varphi\sin\theta + \eta\cos\varphi\sin\theta + \zeta\cos\theta. \end{aligned}\right\} \quad (1)$$

In order to find the form of the operators $\hat{J}_\xi$, $\hat{J}_\eta$, $\hat{J}_\zeta$, let us make use of the fact that the operator $\hat{J}_\xi$ is equal to $-i\frac{\partial}{\partial a}$ where a denotes the angle taken in a plane perpendicular to the ξ axis. Since a rotation of the system of coordinates ξ, η, ζ, e.g., with respect to the ξ axis by an infinitesimal angle da changes the angles, then the operator $\hat{J}_\xi$ in units of $\hbar$ can be written

$$\hat{J}_\xi = -i\left(\frac{\partial\theta}{\partial a}\frac{\partial}{\partial\theta} + \frac{\partial\varphi}{\partial a}\frac{\partial}{\partial\varphi} + \frac{\partial\psi}{\partial a}\frac{\partial}{\partial\psi}\right).$$

For an infinitesimal rotation about the ξ axis by the angle da we have

$$\left.\begin{aligned} \xi &= \xi', \\ \eta &= \eta' - \zeta'\,da, \\ \zeta &= \eta'\,da + \zeta' \end{aligned}\right\} \quad (2)$$

and

$$z = \xi'\sin(\varphi + d\varphi)\sin(\theta + d\theta) + \\ + \eta'\cos(\varphi + d\varphi)\sin(\theta + d\theta) + \zeta'\cos(\theta + d\theta). \quad (3)$$

On the other hand, inserting (2) in (1), we get

$$z = \xi' \sin\varphi \sin\theta + \eta'(\cos\varphi \sin\theta + \cos\theta\, d\alpha) + \\ + \zeta'(\cos\theta - \cos\varphi \sin\theta\, d\alpha). \quad (4)$$

After comparing (3) and (4), we find that

$$\frac{d\theta}{d\alpha} = \cos\varphi, \qquad \frac{d\varphi}{d\alpha} = -\sin\varphi \cot\theta.$$

Similarly, we get $\frac{d\psi}{d\alpha} = \frac{\sin\varphi}{\sin\theta}$.

Finally for $\hat{J}_\xi$ we have the formula

$$\hat{J}_\xi = -i\left(\cos\varphi \frac{\partial}{\partial\theta} - \sin\varphi \cot\theta \frac{\partial}{\partial\varphi} + \frac{\sin\varphi}{\sin\theta}\frac{\partial}{\partial\psi}\right).$$

Proceeding in similar fashion, we find the form of the other two operators

$$\hat{J}_\eta = -i\left(-\sin\varphi \frac{\partial}{\partial\theta} - \cos\varphi \cot\theta \frac{\partial}{\partial\varphi} + \frac{\cos\varphi}{\sin\theta}\frac{\partial}{\partial\psi}\right),$$

$$\hat{J}_\zeta = -i\frac{\partial}{\partial\varphi}.$$

16.

$$E_J = \frac{\hbar^2}{2A} J(J+1).$$

Each level is $(2J+1)$-fold degenerate with respect to the directions of the angular momentum in the stationary system and with respect to the angular momentum relative to the body itself.

17. Since $\hat{H} = \frac{1}{2A}\hat{\mathbf{J}}^2 + \frac{1}{2}\left(\frac{1}{C} - \frac{1}{A}\right)\hat{J}_\zeta^2$, then $E = \frac{1}{2A} J(J+1) + \frac{1}{2}\left(\frac{1}{C} - \frac{1}{A}\right)k^2$; $J_\zeta = k$ $|k| \leqslant J$.

In this case, the total multiplicity of the degeneracy is equal to $2(2J+1)$. The multiplicity of the degeneracy with respect to the directions of the angular momentum in the stationary system is, as above, $(2J+1)$.

18.

$$-\frac{\hbar^2}{2A}\left\{\frac{1}{\sin\theta}\frac{\partial}{\partial\theta}\left(\sin\theta\frac{\partial u}{\partial\theta}\right) + \frac{1}{\sin^2\theta}\left(\frac{\partial^2 u}{\partial\varphi^2} + \frac{\partial^2 u}{\partial\psi^2}\right) - \\ - 2\frac{\cos\theta}{\sin^2\theta}\frac{\partial^2 u}{\partial\psi\,\partial\varphi}\right\} - \frac{\hbar^2}{2}\left(\frac{1}{C} - \frac{1}{A}\right)\frac{\partial^2 u}{\partial\varphi^2} = Eu.$$

19. Since $\hat{J}_\zeta = -i\dfrac{\partial}{\partial\varphi}$ and $\hat{J}_z = -i\dfrac{\partial}{\partial\psi}$ commute with $\hat{J}^2$ we shall seek an eigenfunction in the form

$$\Phi_{kJM_J} = \Theta_{kJM_J}(\theta)e^{iM_J\psi}e^{ik\varphi},$$

where M_J, k are the components of the angular momenta on the stationary z axis and on the rotating ξ axis respectively.

Since ψ and φ appear in equation (14) in a symmetric way and $|k| \leqslant J$, then $|M_J| \leqslant J$.

We consider the operators

$$\hat{J}_\xi + i\hat{J}_\eta = -ie^{-i\varphi}\left(\frac{\partial}{\partial\theta} - i\cot\theta\frac{\partial}{\partial\varphi} + \frac{i}{\sin\theta}\frac{\partial}{\partial\psi}\right), \tag{1}$$

$$\hat{J}_\xi - i\hat{J}_\eta = -ie^{i\varphi}\left(\frac{\partial}{\partial\theta} + i\cot\theta\frac{\partial}{\partial\varphi} - \frac{i}{\sin\theta}\frac{\partial}{\partial\psi}\right). \tag{2}$$

It is readily shown that

$$\hat{J}_\zeta(\hat{J}_\xi - i\hat{J}_\eta)\Phi_{kJM_J} = (k+1)(\hat{J}_\xi - i\hat{J}_\eta)\Phi_{kJM_J},$$

i.e., the expression $(\hat{J}_\xi - i\hat{J}_\eta)\Phi_{kJM_J}$ is an eigenfunction corresponding to the eigenvalue $(k+1)$ of the operator $\hat{J}_\zeta$. We set $k = J$; we then have

$$(\hat{J}_\xi - i\hat{J}_\eta)\Phi_{JJM_J} \equiv 0.$$

The last relation can be written in the form

$$\left(\frac{\partial}{\partial\theta} + i\cot\theta\frac{\partial}{\partial\varphi} - \frac{i}{\sin\theta}\frac{\partial}{\partial\psi}\right)\Theta_{JJM_J}(\theta)e^{iM_J\psi}e^{iJ\varphi} \equiv 0.$$

We thus obtain an ordinary differential equation of the first order for Θ_{JJM_J}:

$$\frac{d\Theta_{JJM_J}}{d\theta} + \frac{M_J - J\cos\theta}{\sin\theta}\Theta_{JJM_J} = 0.$$

The general solution of this equation is

$$\left.\begin{aligned} \Theta_{JJM_J} &= c\frac{(\sin\theta)^J}{\left(\tan\dfrac{\theta}{2}\right)^{M_J}}, \\ &\text{or} \\ \Theta(\theta) &= c(1-\cos\theta)^{\frac{J-M_J}{2}}(1+\cos\theta)^{\frac{J+M_J}{2}}. \end{aligned}\right\} \tag{3}$$

Since the function Θ should be finite, $|M^J| \leqslant J$.

To determine the function Θ_{kJM_J} let us consider the action of the $(\hat{J}_\xi + i\hat{J}_\eta)$. operator on it. Since

$$\hat{J}_\zeta(\hat{J}_\xi + i\hat{J}_\eta)\,\Theta_{kJM_J} = (k-1)(\hat{J}_\xi + i\hat{J}_\eta)\,\Theta_{kJM_J},$$

then

$$(\hat{J}_\xi + i\hat{J}_\eta)\,\Theta_{kJM_J} = a_k\,\Theta_{k-1JM_J}. \tag{4}$$

Inserting into (4) the explicit form of the operator $(\hat{J}_\xi + i\hat{J}_\eta)$ from (1), we arrive at the equation

$$\frac{d\Theta_{kJM_J}}{d\theta} + \frac{k\cos\theta - M_J}{\sin\theta}\,\Theta_{kJM_J} = i a_k\,\Theta_{k-1,JM_J},$$

which is transformed by a change of variable $x = \cos\theta$ into

$$\sqrt{1-x^2}\,\frac{dP_{kJM_J}(x)}{dx} + \frac{M_J - kx}{\sqrt{1-x^2}}\,P_{kJM_J}(x) = -i a_k P_{k-1JM_J}(x),$$

where

$$P_{kJM_J}(x) = \Theta_{kJM_J}\,(\text{arc}\cos x).$$

Setting

$$P_{kJM_J} = (1-x)^{-\frac{(k-M_J)}{2}}(1+x)^{-\frac{(k+M_J)}{2}}\,v_{kJM_J}, \tag{5}$$

we find a simple relation for v_{kJM_J}:

$$\frac{dv_{kJM_J}}{dx} = -i a_k v_{k-1\,JM_J}. \tag{6}$$

The functions P_{JJM_J} found earlier (see (3)) can be written in the form (5)

$$P_{JJM_J}(x) = (1-x)^{-\frac{(J-M_J)}{2}}(1+x)^{-\frac{(J+M_J)}{2}}\,v_{JJM_J},$$

where we denote by v_{JJM_J} the expression

$$v_{JJM_J}(x) = c(1-x)^{J-M_J}(1+x)^{J+M_J}. \tag{7}$$

From (7) and from the recursion relation (6) it follows that

$$v_{kJM_J} = c\,\frac{d^{J-k}}{dx^{J-k}}\{(1-x)^{J-M_J}(1+x)^{J+M_J}\}.$$

Consequently,

$$\Phi_{kJM_J}(\theta, \psi, \varphi) = ce^{ik\varphi}e^{iM_J\psi}(1-\cos\theta)^{-\frac{k-M_J}{2}} \times$$

$$\times(1+\cos\theta)^{-\frac{k+M_J}{2}}\left(\frac{d}{d\cos\theta}\right)^{J-k}\{(1-\cos\theta)^{J-M_J}\cdot(1+\cos\theta)^{J+M_J}\},$$

where for $M_J = 0$ this generalized spherical function passes over, as should be expected, into an ordinary spherical function and represents the wave function of a rotator:

$$\Phi_{kJ0}(\theta, \varphi) = ce^{ik\varphi}\frac{1}{\sin^k\theta}\frac{d^{J-k}}{(d\cos\theta)^{J-k}}(\sin^{2J}\theta).$$

20.

$$\hat{H}_{kk} = \frac{\hbar^2}{4}\left(\frac{1}{A}+\frac{1}{B}\right)\{J(J+1)-k^2\}+\frac{\hbar^2k^2}{2c},$$

$$\hat{H}_{k,k+2} = \hat{H}_{k+2,k} = \frac{\hbar^2}{8}\left(\frac{1}{A}-\frac{1}{B}\right)\sqrt{(J-k)(J-k-1)(J+k+1)(J+k+2)}.$$

21. In the case of an asymmetrical top there is still a degeneracy with respect to the directions of the angular momentum in the rest system. There is no degeneracy with respect to the quantum number k, since to a given J there correspond $(2J+1)$ different levels. In the case $J = 1$ the energy levels are given by the solution of the secular equation

$$\begin{vmatrix} H_{11}-E & H_{10} & H_{1,-1} \\ H_{10} & H_{00}-E & H_{0-1} \\ H_{-1,1} & H_{-10} & H_{-1,-1}-E \end{vmatrix} = \begin{vmatrix} H_{11}-E & 0 & H_{1,-1} \\ 0 & H_{00}-E & 0 \\ H_{-1,1} & 0 & H_{-1,-1}-E \end{vmatrix} = 0.$$

Since $H_{-1,-1} = H_{11}$, we have

$$(H_{00}-E)(H_{11}^2+E^2-2H_{11}E-H_{1,-1}^2) = 0,$$

rom which it follows that

$$E_1 = \frac{\hbar^2}{2}\left(\frac{1}{A}+\frac{1}{B}\right),$$

$$E_2 = \frac{\hbar^2}{2}\left(\frac{1}{C}+\frac{1}{A}\right),$$

$$E_3 = \frac{\hbar^2}{2}\left(\frac{1}{B}+\frac{1}{C}\right).$$

22.

E		m	
E_1	$\Phi_{k=0,\,m}$	1	$\dfrac{\sqrt{3}}{4\pi}\,e^{i\psi}\sin\theta$
		0	$\dfrac{1}{\pi}\sqrt{\dfrac{3}{8}}\,\cos\theta$
		-1	$\dfrac{\sqrt{3}}{4\pi}\,e^{-i\psi}\sin\theta$
E_2	$c\,(\Phi_{k=1,\,m}+$ $+\Phi_{k=-1,\,m})$	1	$\dfrac{1}{\pi}\sqrt{\dfrac{3}{16}}\,e^{i\psi}(\cos\varphi+i\cos\theta\sin\varphi)$
		0	$\dfrac{1}{\pi}\sqrt{\dfrac{3}{8}}\,\cos\varphi\sin\theta$
		-1	$\dfrac{1}{\pi}\sqrt{\dfrac{3}{16}}\,e^{-i\psi}(\cos\varphi-i\cos\theta\sin\varphi)$
E_3	$c\,(\Phi_{k=1,\,m}-$ $-\Phi_{k=-1,\,m})$	1	$\dfrac{1}{\pi}\sqrt{\dfrac{3}{16}}\,e^{i\psi}(\cos\theta\cos\varphi+i\sin\varphi)$
		0	$\dfrac{1}{\pi}\sqrt{\dfrac{3}{8}}\,\sin\theta\sin\varphi$
		-1	$\dfrac{1}{\pi}\sqrt{\dfrac{3}{16}}\,e^{-i\psi}(-\cos\theta\cos\varphi+i\sin\varphi)$

24. The splitting of the terms is caused by the spin-spin interaction. In order to find the splitting we must take the mean of the operator of the spin-spin interaction $a(\hat{\mathbf{S}}\mathbf{n})^2$ over the rotational state. For a given K the quantum number J takes the values

$$J = K+1,\ K,\ K-1.$$

The non-vanishing matrix elements $(\mathbf{n}\hat{\mathbf{S}})$ are

$$(\mathbf{n}\hat{\mathbf{S}})^{K-1\,J=K}_{K\ \ J=K} = (\mathbf{n}\hat{\mathbf{S}})^{K\ \ J=K}_{K-1\,J=K} = \sqrt{\frac{K+1}{2K+1}},$$

$$(\mathbf{n}\hat{\mathbf{S}})^{K+1\,J=K}_{K\ \ J=K} = (\mathbf{n}\hat{\mathbf{S}})^{K\ \ J=K}_{K+1\,J=K} = \sqrt{\frac{K}{2K+1}},$$

$$(\mathbf{n}\hat{\mathbf{S}})^{K\ \ J=K+1}_{K+1\,J=K+1} = (\mathbf{n}\hat{\mathbf{S}})^{K+1\,J=K+1}_{K\ \ J=K+1} = \sqrt{\frac{K+2}{2K+3}},$$

$$(\mathbf{n}\hat{\mathbf{S}})^{K\ \ J=K-1}_{K-1\,J=K-1} = (\mathbf{n}\hat{\mathbf{S}})^{K-1\,J=K-1}_{K\ \ J=K-1} = \sqrt{\frac{K-1}{2K-1}}.$$

Making use of the above relations, we find the splitting of the triplet terms

$$\Delta E_{J=K+1}=\frac{K+2}{2K+3}\alpha, \quad \Delta E_{J=K}=\alpha, \quad \Delta E_{J=K-1}=\frac{K-1}{2K-1}\alpha.$$

25. First of all let us find the non-diagonal matrix elements of the operator w

$$\{w\}^{n\Lambda\Omega vJ}_{n\Lambda\Omega' vJ'}.$$

It is readily seen that only those matrix elements which correspond to the transitions $\Omega'=\Omega\pm 1$, $J=J'$ have values other than zero. Since

$$\{\hat{M}_\xi\}^{n\Lambda\Omega}_{n\Lambda(\Omega\pm 1)}=\mp i\{\hat{M}_\eta\}^{n\Lambda\Omega}_{n\Lambda(\Omega\pm 1)},$$

therefore

$$\{w\}^{n\Lambda\Omega vJ}_{n\Lambda(\Omega\pm 1)vJ}=\frac{\hbar^2}{M}\left\{\pm\frac{\partial}{\partial\theta}+\frac{i}{\sin\theta}\frac{\partial}{\partial\varphi}+\right.$$
$$\left.+(\Omega\pm 1)\cot\theta\right\}^{\Lambda\Omega J}_{\Lambda(\Omega\pm 1)J}\left\{\frac{1}{\varrho^2}\hat{M}_\eta\right\}^{n\Lambda\Omega v}_{n\Lambda(\Omega\pm 1)v}$$

In order to calculate the matrix element

$$\left\{\pm\frac{\partial}{\partial\theta}+\frac{i}{\sin\theta}\frac{\partial}{\partial\varphi}+(\Omega\pm 1)\cot\theta\right\}^{\Lambda\Omega J}_{\Lambda(\Omega\pm 1)J},$$

we note that if in the operator $(\hat{J}_\xi+i\hat{J}_\eta)$ (see Prob. 13, §8) we set $\varphi=0$ and $-i\dfrac{\partial}{\partial\varphi}\to\hat{M}_\zeta$, then

$$(\hat{J}_\xi\pm i\hat{J}_\eta)_{\substack{\varphi\to 0\\ -i\frac{\partial}{\partial\varphi}\to\hat{M}_\zeta}}=\mp i\left\{\pm\frac{\partial}{\partial\theta}+\frac{i}{\sin\theta}\frac{\partial}{\partial\varphi}+(\Omega\pm 1)\cot\theta\right\}.$$

We thus have

$$\left\{\pm\frac{\partial}{\partial\theta}+\frac{i}{\sin\theta}\frac{\partial}{\partial\varphi}+(\Omega\pm 1)\cot\theta\right\}^{\Lambda\Omega J}_{\Lambda(\Omega\pm 1)J}=\pm i\sqrt{(J\mp\Omega)(J\pm\Omega+1)}.$$

In the case of small vibrations about the position of equilibrium the matrix element

$$\left\{\frac{1}{\varrho^2}\hat{M}_\eta\right\}^{n\Lambda\Omega v}_{n\Lambda(\Omega\pm 1)v}$$

is approximately equal to

$$\frac{1}{\varrho_0^2}\{\hat{M}_\eta\}_{n\Lambda(\Omega\pm 1)}^{n\Lambda\Omega},$$

where ϱ_0 is the distance between the nuclei in the equilibrium position. Since

$$\{\hat{L}_\eta + \hat{S}_\eta\}_{n\Lambda\Omega\pm 1}^{n\Lambda\Omega} = \{\hat{M}_\eta\}_{n\Lambda\Omega\pm 1}^{n\Lambda\Omega} = \pm\frac{i}{2}\sqrt{S(S+1)-\Sigma(\Sigma\pm 1)},$$

we finally have

$$\{w\}_{n\Lambda(\Omega\pm 1)vJ}^{n\Lambda\Omega vJ} = B_0\sqrt{S(S+1)-\Sigma(\Sigma\pm 1)}\sqrt{J(J+1)-(\Lambda+\Sigma)(\Lambda+\Sigma\pm 1)}.$$

Here, $B_0 = \hbar^2/2M\varrho_0^2$ represents the value of the rotational constant in the state of equilibrium corresponding to $\varrho = \varrho_0$.

In the general case, the doublet splitting may be of the same order of magnitude as the calculated matrix elements. Therefore, in order to calculate the displacements of the levels of the doublet term we employ perturbation theory in a somewhat modified form. Instead of the functions

$$\psi_{n\Lambda\left(\Lambda+\frac{1}{2}\right)vJ}, \quad \psi_{n\Lambda\left(\Lambda-\frac{1}{2}\right)vJ}$$

we take their linear combination $\psi = c_1\psi_{n\Lambda\left(\Lambda+\frac{1}{2}\right)vJ} + c_2\psi_{n\Lambda\left(\Lambda-\frac{1}{2}\right)vJ}$ as the zero-order approximation. Substituting this expression in the perturbation equation and proceeding in the usual manner, we find the secular equation

$$\begin{vmatrix} E^{(0)}_{n\Lambda\left(\Lambda+\frac{1}{2}\right)vJ} - E & w^{n\Lambda\left(\Lambda+\frac{1}{2}\right)vJ}_{n\Lambda\left(\Lambda-\frac{1}{2}\right)vJ} \\ w^{n\Lambda\left(\Lambda-\frac{1}{2}\right)vJ}_{n\Lambda\left(\Lambda+\frac{1}{2}\right)vJ} & E^{(0)}_{n\Lambda\left(\Lambda-\frac{1}{2}\right)vJ} - E \end{vmatrix} = 0.$$

From the solution of this equation it follows that

$$E = \frac{1}{2}E^{(0)} \pm \frac{1}{2}\sqrt{\Delta E^{(0)} + 4B_0^2\left\{\left(J+\frac{1}{2}\right)^2 - \Lambda^2\right\}}, \tag{1}$$

where

$$E^{(0)} = E^{(0)}_{n\Lambda\left(\Lambda+\frac{1}{2}\right)vJ} + E^{(0)}_{n\Lambda\left(\Lambda-\frac{1}{2}\right)vJ},$$

$$\Delta E^{(0)} = E^{(0)}_{n\Lambda\left(\Lambda+\frac{1}{2}\right)vJ} - E^{(0)}_{n\Lambda\left(\Lambda-\frac{1}{2}\right)vJ}.$$

In case *a*, where the multiplet splitting is large in comparison to the distance between two neighbouring rotational levels, it follows, from Eq. (1) that, approximately,

$$E_1 = E^{(0)}_{n\Lambda\left(\Lambda+\frac{1}{2}\right)vJ} + \frac{B_0^2\left\{\left(J+\frac{1}{2}\right)^2 - \Lambda^2\right\}}{\Delta E^{(0)}},$$

$$E_2 = E^{(0)}_{n\Lambda\left(\Lambda-\frac{1}{2}\right)vJ} - \frac{B_0^2\left\{\left(J+\frac{1}{2}\right)^2 - \Lambda^2\right\}}{\Delta E^{(0)}}.$$

In case *b* we obtain from (1)

$$E_{1,2} = \frac{1}{2}E^{(0)} \pm B_0\left\{\left(J+\frac{1}{2}\right)^2 - \Lambda^2\right\} \pm \frac{\Delta E^{(0)}}{8B_0\left\{\left(J+\frac{1}{2}\right)^2 - \Lambda^2\right\}}.$$

26. $K = 0, 2, 4, \ldots$ if total spin is $S = 2$ or $S = 0$.
$K = 1, 3, 5, \ldots$ if total spin is $S = 1$.

27. The magnetic moment of the molecule is $\frac{e\hbar}{2mc}(\Lambda + 2\sum)\mathbf{n}$, where $\mathbf{n}$ is a unit vector directed along the axis of the molecule.

To determine the energy between the split levels it is necessary to find the average of the quantity

$$-\frac{e\hbar}{2mc}(\Lambda + 2\sum)\mathbf{n}\mathcal{H}$$

over the rotational state, i.e., we must determine the matrix elements

$$(\mathbf{n})^{JM_J}_{JM_J}.$$

Since $\hat{\mathbf{J}}$ is the only vector constant of motion, then the matrix elements of the vector $\mathbf{n}$ will obviously be proportional to the matrix elements of the vector $\hat{\mathbf{J}}$, i.e.,

$$(\mathbf{n})^{JM'_J}_{JM_J} \sim (\hat{\mathbf{J}})^{JM'_J}_{JM_J}.$$

Regarding $\mathbf{n}$ as operator, we have

$$\hat{\mathbf{n}} = \text{const}\,\hat{\mathbf{J}}.$$

To determine the constant we multiply this expression on the left and right by $\hat{\mathbf{J}}$. Since the eigenvalues of $\hat{\mathbf{J}}$ are $J(J+1)$ and the eigenvalues of $\hat{\mathbf{J}}\hat{\mathbf{n}}$ are Ω, then we have

$$\hat{\mathbf{n}} = \frac{\Omega}{J(J+1)}\hat{\mathbf{J}}.$$

Therefore the energy operator of the perturbation is

$$-\frac{e\hbar}{2mc}(\Lambda+2\Sigma)\frac{\Omega}{J(J+1)}\mathcal{H}\hat{\mathbf{J}}.$$

Calculating the diagonal matrix elements, we obtain the energy difference between the split levels

$$\Delta E_{M_J} = -\frac{e\hbar\mathcal{H}}{2mc}(\Lambda+2\Sigma)\frac{\Omega}{J(J+1)}M_J.$$

28. The perturbation operator, as we may readily show, is

$$-\frac{e\hbar}{2mc}\mathcal{H}\left\{\frac{\Lambda^2}{K(K+1)}\hat{\mathbf{K}}+2\hat{\mathbf{S}}\right\}.$$

From this it follows that the energy difference between the Zeeman levels is

$$\Delta E_{M_J} = -M_J\frac{e\hbar\mathcal{H}}{2mc}\times$$

$$\times\left\{\Lambda^2\frac{J(J+1)-S(S+1)+K(K+1)}{2K(K+1)J(J+1)}+\frac{J(J+1)+S(S+1)-K(K+1)}{J(J+1)}\right\}.$$

29. The energy difference between the Zeeman levels is

$$\Delta E_{M_K M_S} = -\frac{e\hbar}{2mc}\mathcal{H}\left[\frac{\Lambda^2}{K(K+1)}M_K+2M_S\right].$$

30. Since the interaction energy of the magnetic moment with the external magnetic field and the energy of the spin-axis coupling are of the same order, we must take them into account simultaneously when we use perturbation theory. The perturbation operator has the form

$$\hat{V} = A\hat{\mathbf{n}}\hat{\mathbf{S}} - \mu_0\Lambda\mathbf{n}\mathcal{H} - 2\mu_0\hat{\mathbf{S}}\mathcal{H}.$$

For the wave function in zero-order approximation we take the wave functions of states in which the angular momentum K and the components of K and S along the direction of the magnetic field have definite values. We assume that the direction of the z axis coincides with the direction of the magnetic field. The component of the total angular momentum along the direction of the magnetic field is a constant of motion, and therefore in the case of the doublet term we should make use of perturbation theory. In doing so we must take into account the double degeneracy. Calculation of the matrix elements of the perturbation operator gives

$$V^{M_K, -1/2}_{M_K, -1/2} = -M_K \frac{\Lambda}{2K(K+1)} A - M_K \frac{\Lambda^2}{K(K+1)} \mu_0 \mathcal{H} + \mu_0 \mathcal{H},$$

$$V^{M_K-1, 1/2}_{M_K-1, 1/2} = (M_K - 1) \frac{\Lambda}{2K(K+1)} A - $$

$$- (M_K - 1) \frac{\Lambda^2}{K(K+1)} \mu_0 \mathcal{H} - \mu_0 \mathcal{H},$$

$$V^{M_K, -1/2}_{M_K-1, 1/2} = \frac{1}{2} A \frac{\Lambda}{K(K+1)} \sqrt{(K-M+1)(K+M)}$$

$$= \frac{1}{2} A (n_x + i n_y)^{M_K}_{M_K-1}$$

$$V^{M_K-1, 1/2}_{M_K, -1/2} = \frac{1}{2} A \{n_x - i n_y\}^{M_K-1}_{M_K}$$

$$= \frac{1}{2} A \frac{\Lambda}{K(K+1)} \sqrt{(K-M+1)(K+M)}.$$

Solving the secular equation, we obtain

$$E^{(1)}_{1,2} = \frac{\Lambda^2}{K(K+1)} \left(M_K - \frac{1}{2} \right) \mu_0 \mathcal{H} - \frac{A\Lambda}{4K(K+1)} \pm \frac{1}{2K(K+1)} \times$$

$$\times \sqrt{ \left\{ A\Lambda \left(M - \frac{1}{2} \right) - \Lambda^2 \mu_0 \mathcal{H} + 2 \mu_0 \mathcal{H} (K+1) K \right\}^2 + A^2 \Lambda^2 (K + M_K)(K - M_K + 1)}.$$

Let us consider the limiting cases. If $\mu_0 \mathcal{H} \gg A$, then for $E^{(1)}_{1,2}$ we have

$$E^{(1)}_{1,2} = -\frac{\Lambda^2}{K(K+1)} \left\{ M_K - \frac{1}{2} \mp \frac{1}{2} \right\} \mu_0 \mathcal{H} \mp \mu_0 \mathcal{H},$$

in agreement with the formula obtained in Prob. 29, §8.
If $A \gg \mu_0 \mathcal{H}$, we obtain

$$E^{(1)}_1 = \frac{A\Lambda}{2(K+1)} - \left(M_K - \frac{1}{2} \right) \left\{ \frac{\Lambda^2}{(K+1)\left(K+\frac{1}{2}\right)} + \frac{1}{\left(K+\frac{1}{2}\right)} \right\} \mu_0 \mathcal{H},$$

$$E^{(1)}_2 = -\frac{A\Lambda}{2K} - \left(M_K - \frac{1}{2} \right) \left\{ \frac{\Lambda^2}{K\left(K+\frac{1}{2}\right)} - \frac{1}{K+\frac{1}{2}} \right\} \mu_0 \mathcal{H}.$$

The second terms of the last formulae, i.e., the terms depending linearly on $\mathcal{H}$, coincide with the corresponding expressions which result from substitution into the formula in Prob. 28, §8,

$$J = K \pm \frac{1}{2}, \quad S = \frac{1}{2} \quad \text{and} \quad M_j = M_K - \frac{1}{2}.$$

31. Because of the axial symmetry, the dipole magnetic moment is oriented along the straight line connecting the nuclei, i.e.,

$$\mathbf{p} = p\mathbf{n}.$$

Proceeding as in the solution of Prob. 27, §8, we find

$$\Delta E_{M_J} = -\mathcal{E} p \frac{\Omega}{J(J+1)} M_J.$$

32.

$$\Delta E_{M_J} = -\mathcal{E} p M_J \Lambda \frac{J(J+1) - S(S+1) + K(K+1)}{2K(K+1)\, J(J+1)}.$$

33. In the first-order approximation the correction to the energy vanishes. As we know, in the case of degeneracy the correction to the energy in the second-order approximation is found from the condition that the system of homogeneous linear equations

$$E_n^{(2)} c_{n\beta}^{(0)} = \sum_j c_{nj}^{(0)} \sum_{m\alpha} \frac{V_{m\alpha}^{n\beta} V_{nj}^{m\alpha}}{E_n^{(0)} - E_m^{(0)}}$$

has a non-zero solution.

In our case we therefore obtain

$$E_{lm}^{(2)} = \frac{p^2 \mathcal{E}^2 A}{\hbar^2} \left\{ \frac{l^2 - m^2}{(2l+1)(2l-1)l} - \frac{(l+1) - m^2}{(2l+3)(2l+1)(l+1)} \right\}.$$

Thus the energy of a rigid dipole is equal to

$$E_{lm} = \frac{\hbar^2}{2A} l(l+1) + E_{lm}^{(2)},$$

$$E_{00} = -\frac{1}{3} \frac{Ap^2 \mathcal{E}^2}{\hbar^2} \qquad (l = 0).$$

This result is, to a certain extent, paradoxical. Indeed, according to the last relation, the energy of a rigid dipole is proportional not to $\mathcal{E}$, but to $\mathcal{E}^2$. Therefore, formally, one can say that the rigid dipole has a definite "polarizability."

Let us consider the case $l = 1$. The level corresponding to the value $m = 0$ has a greater energy in an electric field than it would have had if there had been no electric field. The molecule therefore behaves as though its polarizability were negative. It behaves like a diamagnetic body in a magnetic field. A molecule in the state $m = \pm 1$ behaves "normally." The electric field removes the degeneracy only in part, since the energy depends solely on the absolute value of the magnetic quantum number.

We note that

$$\sum_{m=-l}^{m=+l} E_{lm}^{(2)} = 0.$$

34. For large values of R the exchange effects may be neglected, that is, it may be assumed that one electron is associated with nucleus a and the other with nucleus b (See Fig. 31).

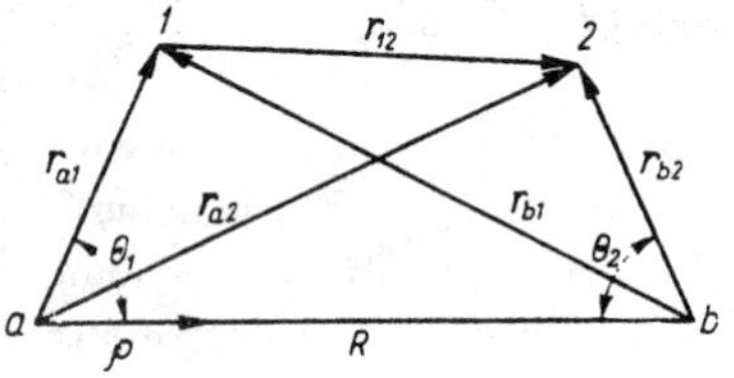

Fig. 31

The interaction between the two atoms has the form

$$V = \frac{1}{R} - \frac{1}{r_{a2}} - \frac{1}{r_{b1}} + \frac{1}{r_{12}} \tag{1}$$

and will be regarded as a small perturbation.

In the first-order approximation the interaction energy of the two atoms is equal to the diagonal matrix elements of V, i.e.,

$$\int \psi_0(r_{1a}) \psi_0(r_{2b}) V \psi_0(r_{1a}) \psi_0(r_{2b})\, d\tau_1\, d\tau_2,$$

where

$$\psi_0(r_{1a}) = 2e^{-r_{1a}}; \qquad \psi_0(r_{2b}) = 2e^{-r_{2b}}.$$

In the S state the diagonal matrix elements, i.e., the mean values of the dipole, quadrupole, and higher moments, are zero. Therefore, to calculate the interaction energy we must consider the second-order approximation of perturbation theory.

In the perturbation operator (1) we restrict ourselves to the dipole-dipole interaction, since this interaction decreases most slowly with distance. To obtain the dipole-dipole interaction operator we expand the potential V in decreasing powers of R. The expansion in spherical coordinates gives

$$\frac{1}{r_{b1}}=\frac{1}{|R\boldsymbol{\rho}-\mathbf{r}_{1a}|}=\sum_{\lambda=0}\frac{r_{a1}^{\lambda}}{R^{\lambda+1}}P_{\lambda}(\cos\theta)=\frac{1}{R}+\frac{\mathbf{r}_{a1}\boldsymbol{\rho}}{R^2}+\frac{3(\mathbf{r}_{a1}\boldsymbol{\rho})^2-\mathbf{r}_{a1}^2}{2R^3}+\ldots,$$

$$\frac{1}{r_{12}}=\frac{1}{|R\boldsymbol{\rho}+\mathbf{r}_{b2}-\mathbf{r}_{a1}|}=\frac{1}{R}+\frac{(\mathbf{r}_{a1}-\mathbf{r}_{b2}\ \boldsymbol{\rho})}{R^2}+$$

$$+\frac{3(\mathbf{r}_{a1}-\mathbf{r}_{b2},\boldsymbol{\rho})^2-(\mathbf{r}_{a1}-\mathbf{r}_{b2})^2}{2R^3}+\ldots,$$

$$\frac{1}{r_{a2}}=\frac{1}{R}+\frac{(\mathbf{r}_{b2}\boldsymbol{\rho})}{R^2}+\frac{3(\mathbf{r}_{b2}\boldsymbol{\rho})^2-\mathbf{r}_{b2}^2}{2R^3}+\ldots$$

Inserting the expansion into V, we obtain the expression for the dipole-dipole interaction

$$V=-\frac{2z_1z_2-x_1x_2-y_1y_2}{R^3},\tag{2}$$

where the z axis is directed along the line joining the nuclei. As already noted, the value of (2) averaged over the unperturbed eigenfunction $\psi=\psi_0(r_{a1})\psi_0(r_{b2})$ is zero. The nondiagonal elements of (2) corresponding to the transitions from the ground state to excited states can be represented in the form

$$V_{00}^{mn}=-\frac{2z_{0m}z_{0n}-x_{0m}x_{0n}-y_{0m}y_{0n}}{R^3}.$$

It follows from the selection rule that the matrix elements z_{0n}, x_{0n}, y_{n0} are different from zero for transitions from the ground state to the states

$$\psi_n(r)\cos\theta,\qquad \psi_n(r)\sin\theta\cos\varphi,$$

all three matrix elements being equal. The interaction energy in the second-order approximation is

$$E^{(2)}=\sum_{mn}\frac{(V_{00}^{mn})^2}{2E_0-E_m-E_n}=\frac{1}{R^6}\sum_{mn}\frac{4z_{0m}^2z_{0n}^2+x_{0m}^2x_{0n}^2+y_{0m}^2y_{0n}^2}{2E_0-E_m-E_n},$$

or

$$E^{(2)}=\frac{6}{R^6}\sum_{mn}\frac{z_{0m}^2z_{0n}^2}{2E_0-E_m-E_n}.\tag{3}$$

Since $E_0<E_m$ and $E_0<E_n$, then $E^{(2)}$ is negative, and consequently

the two atoms are in an unexcited state at a large distance from one another and attracted by a force that is inversely proportional to the seventh power of the distance. To calculate approximately the sum in (3) we note that the difference in energy between the different excited levels is small in comparison to the difference in energy between the excited levels and the ground state. Therefore (3) can be written approximately in the form

$$E^{(2)} = \frac{3}{R^6} \sum_m z_{0m}^2 \sum_n \frac{z_{0n}^2}{E_0 - E_n}.$$

From the theory of the quadratic Stark effect it follows that $\sum \frac{z_{0n}^2}{E_0 - E_n} = -\frac{1}{2}\alpha$, where α is the polarizability of the atom. For the ground state of the hydrogen atom $\alpha = 4 \cdot 5$ atomic units. In calculating the sum $\sum_{m \neq 0} z_{0m}^2$ we make use of the matrix multiplication rule

$$(AB)_{nk} = \sum_m A_{nm} B_{mk}$$

or

$$\int \psi_n^* (AB) \psi_k \, d\tau = \sum_m \left\{ \int \psi_n^* A \psi_m \, d\tau \int \psi_m^* B \psi_k \, d\tau \right\}.$$

Setting $A = B = z$, $n = k = 0$ in the last relation, we obtain

$$(z^2)_{00} = \sum_m z_{0m} z_{m0} = \sum_m z_{0m}^2$$

or

$$\sum_{m \neq 0} z_{0m}^2 = \sum_m z_{0m}^2 - z_{00}^2 = (z^2)_{00} - z_{00}^2.$$

Since in the S state we have, owing to symmetry, $z_{00} = 0$ and $(z^2)_{00} = \frac{1}{3}(r^2)_{00}$, then

$$\sum_{m \neq 0} z_{0m}^2 = \frac{1}{3}(r^2)_{00} = 1.$$

Thus we finally obtain for $V(r)$ the expression

$$V(r) = -\frac{6 \cdot 75}{R^6}.$$

To explain the source of the interaction force between two neutral spherically symmetric atoms of hydrogen we shall consider the wave function of the system. For the wave function of the system in the first-order approximation we obtain the expression

$$\psi = \psi_0(r_{a1})\psi_0(r_{b2})\left[1 + \frac{1}{2E_0R^3}(x_1x_2 + y_1y_2 - 2z_1z_2)\right].$$

If we neglect terms containing the factor $1/R^6$, the probability density $w\,(1,\ 2)$ will have the form

$$w(r_{a1}, r_{b2}) = w_0(r_{a1})w_0(r_{b2})\left[1 + \frac{1}{E_0R^3}(x_1x_2 + y_1y_2 - 2z_1z_2)\right].$$

If there is no interaction between the atoms the probability density is simply equal to the product of $w\,(1)$ and $w\,(2)$, i.e., in this case there is no correlation whatsoever between the electrons. In the case of an interaction, the position of the first electron is not independent of the position of the second. Statistically speaking, the electrons more often occupy positions in which their joint potential energy is the lowest.

Therefore the interaction forces in the first-order approximation can be explained not by the deformation of the electron shells, but by the correlation between the positions of the electrons.

35. We shall show that a system consisting of three atoms has additive properties. From the calculations it will be seen that we can apply these same methods to any number of atoms. The interaction energy can be written as

$$V = V(1, 2) + V(2, 3) + V(3, 1),$$

where we denote by 1, 2, 3 the coordinates of the first, second, and third atoms. Let us consider the interaction af atoms at a large distance from each other; in this case, the exchange forces can be neglected. For the wave function of the three atoms we take in this case, in the zero-order approximation, the function

$$\psi = \psi_{ai}(1)\psi_{bk}(2)\psi_{cl}(3),$$

where i, k, l characterize the quantum states of the atoms a, b, c. The functions $\psi_{ai}(1)$ for different i are orthogonal to one another. The same holds for the functions $\psi_{bk}(2)$ and $\psi_{cl}(3)$.

In the second-order approximation the perturbation energy has the form

$$\varepsilon = V_{000}^{000} + \sum{}' \frac{|V_{ikl}^{000}|^2}{E_{a0} + E_{b0} + E_{c0} - E_{ai} - E_{bk} - E_{cl}}. \qquad (1)$$

The prime after the summation sign indicates that i, k, l should not be zero at the same time. The first term represents the classical interaction of a multipole. In our case, this term vanishes. All of the terms in Eq. (1) for which, simultaneously, $i \neq 0$, $k \neq 0$, $l \neq 0$ vanish owing to the orthogonality of the functions.

The three partial sums for $i = k = 0$, $l \neq 0$; $i = l = 0$, $k \neq 0$; $k = l = 0$, $i \neq 0$ denote the polarization interaction for the lth, kth, and ith atoms respectively in the resultant field of the two other atoms. When the distribution of the charges in the atoms is spherically symmetric, these sums also vanish. It should be noted that these sums cannot be obtained by simply adding up the interaction energies of each pair of atoms. We must still consider those terms in which both indices differ from zero. Under the assumptions concerning the distribution of charge in the atoms, the interaction energy can be represented in the form of three partial sums

$$\varepsilon = \sum_{i\neq 0,\, k\neq 0}' \frac{|V_{ik0}^{000}|^2}{E_{a0}+E_{b0}-E_{ai}-E_{bk}} + \sum_{k\neq 0,\, l\neq 0}' \frac{|V_{0kl}^{000}|^2}{E_{b0}+E_{c0}-E_{bk}-E_{cl}} + \\ + \sum_{i\neq 0,\, l\neq 0}' \frac{|V_{i0l}^{000}|^2}{E_{a0}+E_{c0}-E_{ai}-E_{cl}}. \qquad (2)$$

From the conditions of the orthogonality and normalization of the eigenfunctions of the atom, the matrix element is

$$V_{ik0}^{000} = \int \psi_{a0}^*(1)\psi_{b0}^*(2)\psi_{c0}^*(3)\{V(1,2)+ \\ +V(2,3)+V(3,1)\}\psi_{ai}(1)\psi_{bk}(2)\psi_{c0}(3)\, d\tau_1\, d\tau_2 \\ = \int \psi_{a0}^*(1)\psi_{b0}^*(2)\, V(1,2)\psi_{ai}(1)\psi_{bk}(2)\, d\tau_1\, d\tau_2 = \{V(1,2)\}_{ik}^{00}.$$

Therefore the expression (2) has three terms, each of which represents the dispersive interaction of pairs of atoms. It is readily seen that we can generalize this calculation to any number of atoms.

When the distance between the atoms is small, we must take into account the motion of electrons from one atom to another, i.e., we must take into account the exchange forces.

§ 9. Scattering

1. The potential energy of the particle is

$$U(r) = -U_0 \quad (r < a),$$
$$U(r) = 0 \quad (r > a).$$

It is necessary to find the scattering phase shifts, i.e., the asymptotic form of the radial functions satisfying the equations

$$\text{for } r>a: \quad \chi_l''+\left[k^2-\frac{l(l+1)}{r^2}\right]\chi_l=0, \quad k^2=\frac{2\mu E}{\hbar^2};$$

$$\text{for } r<a: \quad \chi_l''+\left[k'^2-\frac{l(l+1)}{r^2}\right]\chi_l=0, \quad k'^2=\frac{2\mu(E+U_0)}{\hbar^2},$$

with the boundary condition $\chi_l(0)=0$.

In the case of a de Broglie wavelength considerably larger than the dimension of the well, the main contribution to the scattering comes from the S wave. The solution χ_0 satisfying the boundary condition has the form

$$\chi_0 = A \sin k'r \qquad (r<a),$$

$$\chi_0 = \sin(kr+\delta_0) \qquad (r>a).$$

The phase shift δ_0, like the coefficient A, is determined from the continuity conditions for the wave function and its first derivative at $r=a$. These conditions give

$$\delta_0 = \arctan\left(\frac{k}{k'}\tan k'a\right) - ka.$$

Thus the partial cross section for $l=0$ is

$$\sigma_0 = \frac{4\pi}{k^2}\sin^2\delta_0 = \frac{4\pi}{k^2}\sin^2\left[\arctan\left(\frac{k}{k'}\tan k'a\right)-ka\right]. \tag{1}$$

For incident particles of small velocity ($k\to 0$), the phase shift δ_0 is proportional to k:

$$\delta_0 \approx ka\left(\frac{\tan k_0 a}{k_0 a}-1\right), \quad k_0^2=\frac{2\mu U_0}{\hbar^2}, \tag{2}$$

but, owing to the factor $1/k^2$, the cross section σ_0 turns out to be non-vanishing

$$\sigma_0 = 4\pi a^2\left(\frac{\tan k_0 a}{k_0 a}-1\right)^2 \quad \text{(small } k\text{)}. \tag{3}$$

Let us consider the cross section σ_0 as a function of the well depth, which we characterize by k_0. If the well is not deep ($k_0 a \ll 1$), then

$$\sigma_0 = 4\pi a^2\frac{k_0^4 a^4}{9} = \frac{16\pi}{9}\frac{a^6 U_0^2 \mu^2}{\hbar^4}.$$

We note that, according to perturbation theory,

$$f(\vartheta) = -\frac{1}{4\pi}\frac{2\mu}{\hbar^2}\int U(r)\,d\tau = \frac{2\mu}{\hbar^2}U_0\frac{a^3}{3},$$

and consequently

$$\sigma = 4\pi|f(\vartheta)|^2 = \frac{16\pi}{9}\frac{a^6U_0^2\mu^2}{\hbar^4}.$$

As U_0 increases, the cross section increases and becomes infinitely large for $k_0a = \pi/2$. The condition $k_0a = \pi/2$ coincides with the condition for the appearance of the first level in the well. As the well becomes deeper, the cross section begins to decrease and vanishes for $\tan k_0a = k_0a$. With a further increase in U_0 the cross section oscillates between 0 and ∞ where the infinite values correspond to the appearance of new levels in the well. The rapid oscillation of the cross section for the scattering of slow particles explains qualitatively the fact that in the scattering of slow electrons by an atom the value of cross section can be considerably different from the geometrical one.

We note that if the value of k_0a is close to a multiple of $\pi/2$ then Eqs. (2) and (3) must be modified, since in this case $\tan k'a$ turns out to be a large number, and we cannot make the expansion in (1) that led to Eq. (2). In this case we can, as before, neglect $ka \ll 1$ in the brackets of Eq. (1). Then

$$\delta_0 = \arctan\left[\frac{k}{k'}\tan k'a\right],$$

We thus obtain for the cross section σ_0:

$$\sigma_0 = \frac{4\pi[1+O(\varkappa a)]}{\varkappa^2+k^2},$$

where

$$\varkappa = \frac{k'}{\tan k'a} \ll \frac{1}{a}.$$

This formula for "resonance" scattering gives the relation between the cross section and k for small k if a small change in the depth (or width) of the potential well leads to the appearance or disappearance of a discrete level.

2.

$$\sigma = 4\pi a^2\left(\frac{\tanh \varkappa a}{\varkappa a} - 1\right)^2, \quad \text{where} \quad \varkappa = \frac{\sqrt{2\mu U_0}}{\hbar}.$$

As $U_0 \to \infty$, we have $\sigma = 4\pi a^2$, i.e., σ is four times as large as the elastic scattering cross section for a rigid sphere in classical mechanics.

3.

$$\frac{d\sigma}{d\Omega} = \frac{1}{k^2}\sum_{l=0}^{\infty}(2l+1)\sin^2\delta_l +$$

$$+\frac{6\cos\vartheta}{k^2}\sum_{l=0}^{\infty}(l+1)\sin\delta_l\sin\delta_{l+1}\cos(\delta_{l+1}-\delta_l)+$$

$$+\frac{5}{k^2}\,\frac{3\cos^2\vartheta-1}{2}\sum_{l=0}^{\infty}\left\{\frac{l(l+1)(2l+1)}{(2l-1)(2l+3)}\sin^2\delta_l+\right.$$

$$\left.+\frac{3(l+1)(l+2)}{2l+3}\sin\delta_l\sin\delta_{l+2}\cos(\delta_{l+2}-\delta_l)\right\}+\ldots,$$

or

$$\int_0^\pi d\sigma = \frac{4\pi}{k^2}\sum_{l=0}^{\infty}(2l+1)\sin^2\delta_l,$$

$$\int_0^\pi \cos\vartheta\, d\sigma = \frac{8\pi}{k^2}\sum_{l=0}^{\infty}(l+1)\sin\delta_l\sin\delta_{l+1}\cos(\delta_{l+1}-\delta_l),$$

$$\int_0^\pi \frac{3\cos^2\vartheta-1}{2}\,d\sigma = \frac{4\pi}{k^2}\sum_{l=0}^{\infty}\frac{l(l+1)(2l+1)}{(2l+1)(2l+3)}\sin^2\delta_l+$$

$$+\frac{12\pi}{k^2}\sum_{l=0}^{\infty}\frac{(l+1)(l+2)}{2l+3}\sin\delta_l\sin\delta_{l+2}\cos(\delta_{l+2}-\delta_l).$$

4. In the Born approximation, the scattering amplitude has the form

$$f_{\text{Born}}(\vartheta) = -\frac{\mu}{2\pi\hbar^2}\int e^{i\mathbf{qr}}U(r)\,d\tau = -\frac{\pi\mu A}{\hbar^2 q},$$

where

$$\mathbf{q} = \mathbf{k}'-\mathbf{k}, \qquad q = 2k\sin\frac{\vartheta}{2},$$

Consequently

$$d\sigma_{\text{Born}} = |f(\vartheta)|^2\,d\Omega = \frac{\pi^3\mu A^2}{2\hbar^2 E}\cot\frac{\vartheta}{2}\,d\vartheta.$$

In classical mechanics the relation between the scattering angle and the impact parameter ϱ takes the form

$$\int_{r_0}^{\infty} \frac{\mu v \varrho \, dr}{r^2 \sqrt{2\mu(E-U) - \left(\frac{\mu v \varrho}{r}\right)^2}} = \frac{\pi - \vartheta}{2},$$

where r_0 is the root of the expression under the radical sign. Integrating, we find

$$\varrho^2 = \frac{A}{E} \frac{1}{\vartheta} \frac{(\pi-\vartheta)^2}{2\pi - \vartheta},$$

from which

$$d\sigma = -2\pi\varrho \frac{d\varrho}{d\vartheta} d\vartheta = \frac{2\pi^3 A}{E} \frac{\pi - \vartheta}{\vartheta^2 (2\pi - \vartheta)^2} d\vartheta.$$

If the condition

$$\frac{8\mu A}{\hbar^2} \ll 1$$

is satisfied, then the Born approximation is applicable for all angles.

In the opposite limiting case, if $8\mu A/\hbar^2 \gg 1$, then the classical formula can be employed for angles satisfying the equation

$$\vartheta \gtrsim \frac{\hbar^2}{8\mu A},$$

while for smaller angles

$$\vartheta \lesssim \frac{\hbar^2}{8\mu A},$$

the results obtained on the basis of the Born approximation are valid.

5. For the radial function we have the equation

$$\chi'' + \frac{2\mu}{\hbar^2}\left(E + U_0 e^{-\frac{r}{a}}\right)\chi = 0.$$

We introduce the notation $k^2 = 2\mu E/\hbar^2$, $\varkappa^2 = 2\mu U_0/\hbar^2$ and take $\xi = \exp(-r/2a)$ as the independent variable.

We then obtain the equation

$$\chi'' + \frac{1}{\xi}\chi' + 4a^2\left(\frac{k^2}{\xi^2} + \varkappa^2\right)\chi = 0$$

whose solutions are Bessel functions of imaginary order $\chi = J_{\pm 2a\kappa i}(2a\varkappa\xi)$. From the condition that χ vanishes at $r=0$, i.e., for $\xi = 1$, we obtain the unnormalized function:

$$\chi = J_{-2aki}(2a\varkappa)J_{2aki}(2a\varkappa\xi) - J_{2aki}(2a\varkappa)J_{-2aki}(2a\varkappa\xi). \qquad (1)$$

For $r \to \infty (\xi \to 0)$, the function χ has the following asymptotic form:

$$\chi = J_{-2aki}(2a\varkappa)\frac{e^{2aki \ln a\varkappa}}{\Gamma(2aki+1)}e^{-ikr} - J_{2aki}(2a\varkappa)\frac{e^{-2aki \ln a\varkappa}}{\Gamma(-2aki+1)}e^{ikr}.$$

The coefficients of exp $(-ikr)$ and exp (ikr) can be considered to be functions of complex k. If we denote them by $a(k)$ and $b(k)$ then, as we can readily see, the following relations are satisfied:

$$a(-k) = -b(k),$$
$$a^*(k) = -b(k)$$

(when we take the complex conjugate k does not become k^*).

The asymptotic formula for χ can be written

$$\chi = A(e^{-ikr-i\delta_0} - e^{+ikr+\delta_0}) = -2iA\sin(kr+\delta_0).$$

We find the phase shift δ_0 from the formula

$$e^{2i\delta_0} = \frac{J_{2aki}(2a\varkappa)}{J_{-2aki}(2a\varkappa)}\frac{\Gamma(2aki+1)}{\Gamma(-2aki+1)}e^{-4aki \ln a\varkappa}.$$

A pure imaginary number $k = ik_n$ corresponds to a bound state (in this case the energy is negative).

When $k_n > 0$ the coefficient $e^{-ikr} = e^{k_n r}$ in the first term of the asymptotic expression for χ should vanish. Therefore, either $J_{2ak_n}(2a\varkappa) = 0$, or $\frac{1}{\Gamma(2ak_n+1)} = 0$. From the second condition it follows that

$$2ak_n + 1 = -n \qquad (n = 0,\ 1,\ 2,\ \ldots),$$

i.e.,

$$E = -\frac{\hbar^2 k_n^2}{2\mu} = \frac{-\hbar^2(n+1)^2}{8\mu a^2};$$

however, the orders of the Bessel functions become integers, and since $J_n(x) = (-1)^n J_{-n}(x)$, the two solutions are linearly dependent; the wave function (1) then vanishes identically. Therefore, the energy levels obtained are not real ones. The first condition gives the true discrete spectrum:

$$J_{2ak_n}(2a\varkappa) = 0, \quad E_n = -\frac{\hbar^2 k_n^2}{2\mu}. \tag{2}$$

Hence, the zeros of the expression exp $[2i\delta_0(k)]$ lie on the imaginary axis and, apart from the values ik_n corresponding to the discrete levels (2) will assume still other values.

7. The radial wave function χ_l/r for the unperturbed potential satisfies the equation

$$\frac{d^2\chi_l}{dr^2} + \left[k^2 - \frac{l(l+1)}{r^2} - \frac{2\mu V(r)}{\hbar^2}\right]\chi_l = 0 \tag{1}$$

where $\chi_l^{(0)} = 0$ and $\chi_l \rightarrow \sin\left(kr - \frac{\pi l}{2} + \delta_l\right)$ as $r \rightarrow \infty$. We denote by χ_l' the corresponding function for the potential $V'(r)$ which differs very little from $V(r)$:

$$\frac{d^2\chi_l'}{dr^2}\left[k^2 - \frac{l(l+1)}{r^2} - \frac{2\mu V'(r)}{\hbar^2}\right]\chi_l' = 0. \tag{2}$$

We now multiply (1) by χ_l' and (2) by χ_l, subtract one from the other, and integrate over r in the limits from 0 to ∞:

$$\frac{d\chi_l}{dr}\chi_l' - \chi_l\frac{d\chi_l'}{dr} = \frac{-2\mu}{\hbar^2}\int_0^r (V' - V)\chi_l\chi_l'\, dr.$$

We now make use of the asymptotic expression for χ_l and χ_l' in the left-hand part of the equation and pass over on the right to the limit $r \rightarrow \infty$. We then obtain

$$k\sin(\delta_l' - \delta_l) = \frac{-2\mu}{\hbar^2}\int_0^\infty (V' - V)\chi_l\chi_l'\, dr.$$

This relation is exact; if now the change in the potential $\Delta V = V' - V$ is sufficiently small so that $\chi_l' \simeq \chi_l$ (here, of course, $\delta_l' - \delta_l \ll 1$), then

$$\delta_l' = \delta_l - \frac{2\mu}{\hbar^2 k}\int_0^\infty \Delta V \chi_l^2\, dr. \tag{3}$$

In particular, if the scattering potential itself is small (or more precisely, if perturbation theory is applicable), then, as the unperturbed problem, we may take the free motion

$$\chi_l = \sqrt{\frac{\pi kr}{2}} J_{l+\frac{1}{2}}(kr),$$

from which it follows that

$$\delta_l = -\frac{\pi\mu}{\hbar^2}\int_0^\infty V(r)\, J^2_{l+\frac{1}{2}}(kr)\, r dr. \tag{4}$$

8. We calculate the phase shifts from formula (4) of the preceding problem:

$$\delta_l = -\frac{\pi\mu a}{\hbar^2}\int_0^\infty J^2_{l+\frac{1}{2}}(kr)\frac{dr}{r^2} = -\frac{\mu a k}{\hbar^2 l(l+1)}.$$

Thus, for $l \geqslant 1$, the phase shifts are small and for $k \to 0$ are proportional to k (cf. discussion of this question in §108 of *Quantum Mechanics* by L. Landau and E. Lifshits). The condition for the applicability of this formula is that the phase δ_l be small in comparison with unity. The case for $l = 0$ must be considered separately. For $r \ll 1/k$, in the equation for χ_0

$$\chi_0'' + \left[k^2 - \frac{2\mu a}{\hbar^2 r^3}\right]\chi_0 = 0$$

we can discard the term k^2 and reduce the equation to the equation for Bessel functions of imaginary argument. The solution which vanishes for $r = 0$ then has the following form:

$$\chi_0 = c\sqrt{r}\, H_1^{(1)}\left(2i\sqrt{\frac{2\mu a}{\hbar^2 r}}\right).$$

If now $r \gg \frac{\mu a}{\hbar^2}$, but, as before, $r \ll 1/k$, then for $H_1^{(1)}$ we can take the expansion that is convenient in the region of small values of the argument. Thus

$$\chi_0 = c'\left(r - \frac{2\mu a}{\hbar^2}\ln\frac{\mu a}{\hbar^2 r}\right). \tag{1}$$

This solution must be matched to the solution in the outer region

$$\chi_0 = \sin(kr + \delta_0) \tag{2}$$

which, for $kr \ll 1$, reduces to (owing to δ_0 being small)

$$\chi_0 = kr + \delta_0.$$

To obtain δ_0 to a logarithmic accuracy it is sufficient to set $r \sim 1/k$ in the logarithmic expression in (1), and we then find

$$\delta_0 = -\frac{2\mu a k}{\hbar^2}\ln\frac{\beta\hbar^2}{\mu a k},$$

where β is a constant approximately equal to unity. This formula shows that $\delta_0 \ll 1$. For spherically symmetric scattering

$$\sigma_0 = \frac{16\pi\mu^2\alpha^2}{\hbar^4}\left(\ln\frac{\hbar^2}{\mu a k}\right)^2,$$

while for $l \gg 1$ we find

$$\sigma_l = \frac{4\pi\mu^2\alpha^2(2l+1)}{\hbar^4 l^2(l+1)^2}.$$

The fact that σ_l for $l \neq 0$ is comparatively large for slow particles is connected with the slow drop in the potential energy. The total scattering cross section (to logarithmic accuracy) is equal to σ_0, since the series $\sum_1^\infty \sigma_l$ converges.

9. The radial functions satisfying the boundary conditions $\chi_l(a) = 0$ are given in terms of Bessel functions by

$$\chi_l = \sqrt{r}\{J_{-l-1/2}(ka)J_{l+1/2}(kr) - J_{l+1/2}(ka)J_{-l-1/2}(kr)\}.$$

From the formulae for the asymptotic behaviour of Bessel functions we obtain the phase shifts

$$\cot\delta_l = (-1)^{l+1}\frac{J_{-l-\frac{1}{2}}(ka)}{J_{l+\frac{1}{2}}(ka)}.$$

The total cross section for elastic scattering is then

$$\sigma = \frac{4\pi}{k^2}\sum_{=0}^{\infty}(2l+1)\frac{J^2_{l+\frac{1}{2}}(ka)}{J^2_{l+\frac{1}{2}}(ka) + J^2_{-l-\frac{1}{2}}(ka)} \approx 2\pi a^2.$$

10.

(a) $$d\sigma = \left(\frac{g^2}{2E}\right)^2 \frac{d\Omega}{\left(\sin^2\frac{\vartheta}{2} + \frac{\hbar^2\alpha^2}{2\mu E}\right)^2}; \qquad \sigma = \frac{\pi\mu^2 g^4}{\hbar^2\alpha^2(2\mu E + \hbar^2\alpha^2)};$$

(b) $$d\sigma = \frac{\pi\mu^2 U_0}{4\hbar^4\alpha^6} e^{-\frac{4\mu E}{\hbar^2\alpha^2}\sin^2\frac{\vartheta}{2}} d\Omega; \qquad \sigma = \frac{\pi^2 U_0^2\mu}{4\hbar^2\alpha^4 E}\left(1 - e^{-\frac{4\mu E}{\hbar^2\alpha^2}}\right);$$

(c) $$d\sigma = \frac{16\mu^2 U_0^2}{\hbar^4}\frac{\alpha^2}{(\alpha^2+q^2)^4} d\Omega, \qquad \sigma = \frac{64\pi}{3}\frac{\mu^2 U_0}{\hbar^4}\frac{16k^4 + 12k^2\alpha^2 + 3\alpha^4}{\alpha^4(\alpha^2+4k^2)^3}.$$

11. (a) The atomic form factor for hydrogen is given by

$$F(q)=\int e^{i\mathbf{qr}}n(\mathbf{r})\,d\tau=\frac{1}{\pi a^3}\int e^{i\mathbf{qr}-\frac{2r}{a}}\,d\tau=\frac{1}{\left(1+\frac{q^2a^2}{4}\right)^2}$$

$\left(a=\frac{\hbar^2}{\mu e^2}\text{ is the Bohr radius}\right).$

We thus obtain the differential cross section

$$d\sigma=\frac{4a^2(8+q^2a^2)^2}{(4+q^2a^2)^4}\,d\Omega\quad\left(q=2k\sin\frac{\theta}{2}\right)\tag{1}$$

and the total cross section

$$\sigma=\frac{\pi a^2}{3}\,\frac{7k^4a^4+18k^2a^2+12}{(k^2a^2+1)^3}.$$

The condition of applicability of the Born approximation in this case is

$$ka\gg 1,$$

and therefore the expression for the total cross section reduces to

$$\sigma=\frac{7\pi}{3k^2}.\tag{2}$$

(b) For the helium atom the electron density distribution obtained with the aid of a variational calculation takes the form

$$n(r)=\frac{2}{\pi b^3}e^{-\frac{2r}{b}},\quad b=\frac{16}{27}a.$$

The differential and total cross sections for elastic scattering by a helium atom in this approximation have the same form as in the case of scattering by a hydrogen atom; in formulae (1) and (2), one must replace a by b and introduce a multiplicative factor $Z^2=4$. We therefore have

$$\sigma=\frac{28\pi}{3k^2}.$$

12. For $ka \ll 1$ it is necessary to consider only the S waves. Setting $\psi=\frac{\chi}{r}$, we obtain for χ the one-dimensional Schrödinger equation

$$\frac{d^2\chi}{dr^2}+\frac{2\mu}{\hbar^2}\left(E+V_0\cosh^{-2}\frac{r}{a}\right)\chi=0.\tag{1}$$

The function χ should satisfy the boundary conditions

$$\chi(0)=0, \qquad \chi(r)\underset{r\to\infty}{\simeq}\sin(kr+\delta_0), \tag{2}$$

where δ_0 is the phase of the S wave. The scattering amplitude is $f = \frac{1}{2ik}(l^{2i\delta_0}-1)$ and the total cross section σ is $4\pi|f|^2$. The solution of Eq. (1) was considered in Prob. 12, §1. It was shown that Eq. (1) has even and odd solutions with respect to r. The first of the boundary conditions (2) is satisfied by the odd solution, which has the form:

$$\chi(r)=\frac{\sinh\frac{r}{a}}{\cosh^3\frac{\hbar}{a}}F\left(\frac{-s+ika+1}{2}, \frac{-s-ika+1}{2}, \frac{3}{2}; -\sinh^2\frac{r}{a}\right), \tag{3}$$

where

$$s=\frac{1}{2}\sqrt{\frac{8\mu V_0a^3}{\hbar^2}+1}-\frac{1}{2}, \quad k=\sqrt{\frac{2\mu E}{\hbar}},$$

and F is the hypergeometric function.

For $r\to\infty$ we obtain from (3)

$$\chi(r)=Ae^{ikr}+Be^{-ikr}, \tag{4}$$

where (apart from an unessential real multiplicative factor)

$$A=2^{-ika}\Gamma(ika)|\Gamma\frac{(1-s+ika)}{2}\Gamma\left(1+\frac{s+ika}{2}\right) \tag{5}$$

and for real values of k we have $B=A^*$.

Comparing formulae (2) and (4), we find that

$$\delta_0=\frac{1}{2i}\ln\left(-\frac{A}{A^*}\right), \quad f=\frac{i}{2k}\left(\frac{A}{A^*}+1\right). \tag{6}$$

We note that for complex values of k the function $\chi(r)$ turns out to be, generally speaking, unbounded. For $\mathrm{Im}\cdot k>0$ the unbounded solution of Eq. (1) exists only under the condition $B=0$. According to (5), for this it is necessary that $\Gamma\left(\frac{1-s-ika}{2}\right)=\infty$. Setting $ak=i\varepsilon$ ($\varepsilon>0$), we thus find that

$$\varepsilon=s-2n-1 \quad \text{or} \quad E=\frac{-\hbar^2}{2\mu a^2}(s-2n-1)^2,$$

i.e., there are discrete levels of negative energy.

Very important for what follows is the fact that for s equal to an odd integer there is an energy level in the well whose energy is zero. For $s-1<0$, i.e., for $\frac{\mu V_0 a^2}{\hbar^2}<1$, there is not a single negative energy level. Let us consider this case in more detail. Since $ka \ll 1$, we have, approximately,

$$\left.\begin{aligned} &\Gamma(ika) \simeq \frac{1}{ika}\Gamma(1+ika) \simeq \frac{1+ika\,\psi(1)}{ika}\Gamma(1), \\ &\Gamma\left(\frac{1-s}{2}+\frac{ika}{2}\right) \simeq \Gamma\left(\frac{1-s}{2}\right)\left[1+\frac{ika}{2}\psi\left(\frac{1-s}{2}\right)\right], \\ &\Gamma\left(1+\frac{s}{2}+\frac{ika}{2}\right) \simeq \Gamma\left(1+\frac{s}{2}\right)\left[1+\frac{ika}{2}\psi\left(1+\frac{s}{2}\right)\right], \end{aligned}\right\} \tag{7}$$

where $\psi(x)=\frac{d}{dx}\ln\Gamma(x)$ is the logarithmic derivative of the Γ function.

Substituting expressions (7) into (5) and (6), we obtain

$$\delta_0 = ka\left[-\ln 2+\psi(1)-\frac{1}{2}\psi\left(\frac{1-s}{2}\right)-\frac{1}{2}\psi\left(1+\frac{s}{2}\right)\right]^2. \tag{8}$$

When the condition $s<1$ is satisfied, the expression in the brackets is bounded, so that $\delta_0 \ll 1$, and for the total cross section we obtain the expression

$$\delta = 4\pi a^2\left[-\ln 2+\psi(1)-\frac{1}{2}\psi\left(\frac{1-s}{2}\right)-\frac{1}{2}\psi\left(1+\frac{s}{2}\right)\right]^2. \tag{9}$$

The Born approximation for $ka \ll 1$ gives

$$\sigma = \frac{16\pi\mu^2}{\hbar^4}\left|\int_0^\infty V(r)r^2\,dr\right|^2 = \frac{16\pi\mu^2 V_0^2 a^6}{\hbar^4}\left|\int_0^\infty \frac{x^2\,dx}{\cosh^2 x}\right|^2.$$

It may be shown that $\int_0^\infty x^2\cosh^{-2}x\,dx = \pi^2/12$, so that the expression for the total cross section readily reduces to the form

$$\sigma = 4\pi a^2 s^2(s+1)^2\frac{\pi^4}{144}. \tag{10}$$

In order for the Born approximation to be applicable it is necessary, as we know, that the inequality $2\mu a^2 V_0/\hbar^2 \ll 1$ be satisfied, which in our notation means $s \ll 1$.

Expanding the function ψ into a series, we obtain

$$\delta \simeq \frac{kas}{4}\left\{\psi'\left(\frac{1}{2}\right) - \psi'(1)\right\} = \frac{kas\,\pi^2}{12}.$$

Hence expressions (9) and (10) in this limiting case coincide, as should have been expected. We note, however, that already for $s = \frac{1}{2}$ the exact expression (9) gives for the cross section a value approximately seven times that of the approximate formula (10).

Now let $s = 1 + \varepsilon$, $|\varepsilon| \ll 1$. For $\varepsilon > 0$ there is a level in the well whose energy is $E = \dfrac{-\hbar^2\varepsilon^2}{2\mu a^2}$. For $\varepsilon < 0$ there is no real level; however, we say that there is a virtual level, since a slight change in the external field may prove to be sufficient for a level to appear.

In this case we set

$$\Gamma\left(\frac{1-s+ika}{2}\right) \simeq \Gamma\left(\frac{-s+ika}{2}\right) \simeq \frac{2}{-s+ika},$$

$$\Gamma\left(1+\frac{s}{2}+\frac{ika}{2}\right) \simeq \Gamma\left(\frac{3}{2}\right), \quad \Gamma(ika) \simeq \frac{1}{ika},$$

so that

$$A \simeq \frac{-s+ika}{ika}.$$

Inserting this expression into (6), we obtain

$$f = -\frac{a}{s+ika},$$

so that for the cross section we obtain the well-known Wigner formula

$$\sigma = \frac{2\pi\hbar^2}{\mu}\cdot\frac{1}{E+E_0}, \tag{11}$$

where $E_0 = \dfrac{\hbar^2\varepsilon^2}{2\mu a^2}$, $E = \dfrac{\hbar^2 k^2}{2\mu}$. For $E \gg E_0$, $\sigma \simeq \dfrac{4\pi}{k^2} \gg 4\pi a^2$. For $E \ll E_0$ the cross section ceases to depend on the energy and tends toward the value $\sigma_0 = \dfrac{2\pi\hbar^2}{\mu E_0}$.

Now let s be greater than unity and not close to any odd number. Then, as before, the expansions (7) and formulae (8) and (9), which result from them, will be valid.

If $s = 1 + 2n + \varepsilon$, $|\varepsilon| \ll 1$, then, using the recursion relations for the Γ functions, we may readily show that in this case the cross section is given by formula (11).

13. The problem is solved similarly to the previous problem. The main difference between this case and that of the preceding problem is that in the case of a repulsive field the negative energy levels are absent.

The asymptotic form of the wave function is

$$\chi(r) = Ae^{ikr} + Be^{-ikr},$$

where A is given by expression (5) and (see the preceding problem)

$$B = 2^{ika}\,\Gamma(-ika)\left[\Gamma\left(\frac{1-s-ika}{2}\right)\Gamma\left(1+\frac{s-ika}{2}\right)\right]^{-1},$$

where

$$s = \frac{1}{2}\sqrt{1-\frac{8\mu V_0 a^2}{\hbar^2}} - \frac{1}{2}.$$

For any values $V_0 > 0$ and a, parameter s remains either smaller than unity or becomes complex; in both cases the conditions of applicability of the approximation formulae (7) are fulfilled. Therefore formulae (8) and (9) remain valid.

14. The wave function for a system of two identical particles has the form of the product of the orbital and spin functions. Regardless of whether the spin of the particles is a whole or half integer, an even total spin corresponds to a symmetric orbital wave function and an odd total spin to an antisymmetric one.

We introduce the coordinates of the centre-of-mass system and separate the variables. We then obtain the orbital wave function in the form

$$\Psi(\mathbf{r}_1, \mathbf{r}_2) = \varphi(\mathbf{R})\psi(\boldsymbol{\rho}),$$

where

$$\mathbf{R} = \frac{\mathbf{r}_1 + \mathbf{r}_2}{2}, \qquad \boldsymbol{\rho} = \mathbf{r}_1 - \mathbf{r}_2.$$

If we interchange $\mathbf{r}_1$ and $\mathbf{r}_2$ the function $\varphi(\mathbf{R})$ describing the motion of the centre of mass obviously does not change. Thus the wave function of the relative motion of the two particles should be even

$$\psi(\boldsymbol{\rho}) = \psi(-\boldsymbol{\rho})$$

if the total spin S is even, and should be odd

$$\psi(\boldsymbol{\rho}) = -\psi(-\boldsymbol{\rho}),$$

if the spin S is odd.

The unperturbed wave function can be written in the form

$$\psi(\boldsymbol{\rho}) = e^{i\mathbf{k}_0\boldsymbol{\rho}} + e^{-i\mathbf{k}_0\boldsymbol{\rho}}, \tag{1}$$

if S is even and

$$\psi(\boldsymbol{\rho}) = e^{i\mathbf{k}_0\boldsymbol{\rho}} - e^{-i\mathbf{k}_0\boldsymbol{\rho}}, \tag{2}$$

if S is odd. As a result of the interaction of the particles we have a scattered wave $\frac{F(\vartheta)}{\varrho} e^{ik\varrho}$, where ϑ is the angle between $\mathbf{k}_0$ and the direction in which the particles move in the centre-of-mass system. The scattering amplitude is the same as that for a particle of mass equal to the reduced mass of both particles scattered in the field $U(r)$. In fact, ψ $(\boldsymbol{\rho})$ satisfies the equation.

$$\left\{-\frac{\hbar^2}{2\mu}(\Delta_\varrho + k^2) - U(\varrho)\right\}\psi(\boldsymbol{\rho}) = 0.$$

If the scattered wave $\frac{f(\vartheta)}{r} e^{ik\varrho}$ corresponds to the incident wave $e^{i\mathbf{k}_0\boldsymbol{\rho}}$, then for the incident wave (1)

$$F_0(\vartheta) = f(\vartheta) + f(\pi - \vartheta) \quad \text{(even spin)},$$

and for (2)

$$F_1(\vartheta) = f(\vartheta) - f(\pi - \vartheta) \quad \text{(odd spin)}.$$

The ratio of the probability that one of the particles of the incident stream is scattered into the solid angle $d\Omega$ (the other particle moving in the opposite direction) to the density of the incident stream is

$$d\sigma_0 = |f(\vartheta) + f(\pi - \vartheta)|^2 \, d\Omega,$$

$$d\sigma_1 = |f(\vartheta) - f(\pi - \vartheta)|^2 \, d\Omega.$$

The amplitude $f(\vartheta)$ is obtained from the phase shift δ_l by means of the relation

$$f(\vartheta) = \frac{1}{2ik} \sum_{l=0}^{\infty} (2l+1) P_l(\cos\vartheta) [e^{2i\delta_l} - 1].$$

Since

$$P_l[\cos(\pi - \vartheta)] = P_l(-\cos\vartheta) = (-1)^l P_l(\cos\vartheta),$$

we obtain

$$F_0(\vartheta) = \frac{1}{ik} \sum_{\text{even}_l} (2l+1) P_l(\cos\vartheta) [e^{2i\delta_l} - 1],$$

$$F_1(\vartheta) = \frac{1}{ik} \sum_{\text{odd}_l} (2l+1) P_l(\cos\vartheta) [e^{2i\delta_l} - 1].$$

For slow particles the main contribution to the scattering comes from small l. In the event of an even total spin the cross section (as in the case of different particles) is spherically symmetric and does not vanish for $k \to 0$.

If the total spin is odd, then the cross section is determined from the term with $l=1$. Since $\delta_l \sim k^{2l+1}$, for small wave numbers k, the cross section decreases to zero like E^2 (as $E \to 0$) and varies with the angle ϑ approximately as $\cos^2\vartheta$.

15. In the case of a Coulomb field the scattering amplitude* is

$$f(\vartheta) = -\frac{1}{2k^2 \sin^2 \frac{\vartheta}{2}} e^{-\frac{2i}{k} \ln \sin \frac{\vartheta}{2}} \frac{\Gamma\left(1+\frac{i}{k}\right)}{\Gamma\left(1-\frac{i}{k}\right)}.$$

Making use of the results of the preceding problem, we find the differential scattering cross section in the case of even spin:

$$d\sigma_0 = |f(\vartheta)+f(\pi-\vartheta)|^2 \, d\Omega$$

$$= \frac{1}{4k^4}\left\{\frac{1}{\sin^4\frac{\vartheta}{2}} + \frac{1}{\cos^4\frac{\vartheta}{2}} + \frac{2\cos\left(\frac{2}{k}\ln\tan\frac{\vartheta}{2}\right)}{\sin^2\frac{\vartheta}{2}\cos^2\frac{\vartheta}{2}}\right\} d\Omega.$$

This formula gives the scattering cross section of α-particles, which have a spin equal to zero.

In the case of two electrons it is possible to have a state with total spin equal to 1. Then the differential cross section is

$$d\sigma_1 = |f(\vartheta)-f(\pi-\vartheta)|^2 \, d\Omega$$

$$= \frac{1}{4k^4}\left\{\frac{1}{\sin^4\frac{\vartheta}{2}} + \frac{1}{\cos^4\frac{\vartheta}{2}} - \frac{2\cos\left(\frac{2}{k}\ln\tan\frac{\vartheta}{2}\right)}{\sin^2\frac{\vartheta}{2}\cos^2\frac{\vartheta}{2}}\right\} d\Omega.$$

If the scattered electrons are not polarized, then the component of the total spin along a given direction z can have three values: $S_z = 0$, $S_z = 1$, $S_z = -1$, and the total spin can take on the values $S = 0$ and $S = 1$. The probability of the component having each of these values is: $W_{-1} = 1/4$, $W_0 = 1/2$, $W_{+1} = 1/4$. The values $S_z = +1$ or $S_z = -1$

* In Coulomb units.

obviously correspond to the spin $S = 1$. Since the various values of the components for $S = 1$ are equally probable, then the probability that $S_z = 0$ for a total spin of 1, as in the case of $S_z = \pm 1$, is equal to $\frac{1}{4}$. Therefore, the probability that the total spin is zero is $W_0 - \frac{1}{4} = \frac{1}{4}$, whereas the probability that $S = 1$ is $\frac{3}{4}$. Hence for an unpolarized beam of electrons

$$d\sigma = \frac{1}{4} d\sigma_0 + \frac{3}{4} d\sigma_1$$

$$= \frac{1}{4k^4} \left\{ \frac{1}{\sin^4 \frac{\vartheta}{2}} + \frac{1}{\cos^4 \frac{\vartheta}{2}} - \frac{\cos\left(\frac{2}{k} \ln \tan \frac{\vartheta}{2}\right)}{\sin^2 \frac{\vartheta}{2} \cos^2 \frac{\vartheta}{2}} \right\} d\Omega.$$

The last interference term in the brackets is characteristic of the scattering of identical particles. As $\hbar \to 0$ the formula for $d\sigma$ should go over into the classical Rutherford formula, which, in the centre-of-mass system, has the form

$$d\sigma = \frac{1}{4k^4} \left\{ \frac{1}{\sin^4 \frac{\vartheta}{2}} + \frac{1}{\cos^4 \frac{\vartheta}{2}} \right\} d\Omega.$$

The transition to this formula, however, is not obvious. If the condition $e^2/\hbar v \gg 1$, is satisfied, in other words, when we can employ classical theory, the interference term, which in conventional units has the form

$$\frac{\cos\left(\frac{2e^2}{\hbar v} \ln \tan \frac{\vartheta}{2}\right)}{\sin^2 \frac{\vartheta}{2} \cos^2 \frac{\vartheta}{2}},$$

oscillates rapidly. Therefore the quantum differential cross section for a strictly defined value of ϑ differs significantly from the classical cross section even for large values of $e^2/\hbar v$. However, when the average is taken over a small range of angles $\Delta\vartheta \sim \hbar v/e^2$, the interference term vanishes and the quantum formula passes over into the classical formula.

17. The radial function χ/r of a slow neutron inside the nucleus has the form

$$\chi = A \sin \varkappa r, \quad \varkappa = \sqrt{2MV_0}/\hbar$$

(since the neutrons are slow $E \ll V_0$).

Outside the nucleus

$$\chi = \sin (kr + \delta_0), \quad k = \sqrt{2ME}/\hbar.$$

Matching these solutions at the boundary, we have

$$\frac{\delta_0}{k} \simeq \frac{\tan \varkappa R}{\varkappa R} - 1,$$

from which we obtain the scattering length

$$a = -\left(\frac{\delta_0}{k}\right)_{k\to 0} = \frac{\varkappa R - \tan \varkappa R}{\varkappa R}.$$

Thus the sign of the scattering length will be negative if $\varkappa R > \tan \varkappa R$. If $\varkappa R \gg 1$ (this inequality holds for nuclei that are not very light), then the regions in which $a < 0$ constitute a small part of the entire region of variation of $\varkappa R$, and the probability of negative a is

$$W \simeq \frac{\hbar}{\pi R\sqrt{2MV_0}} \sim \frac{1}{4A^{1/3}}.$$

18.

$$\hat{a}^2 = \frac{3}{4}a_1^2 + \frac{1}{4}a_0^2 + \frac{1}{4}(a_1^2 - a_0^2)(\hat{\sigma}_n\hat{\sigma}_p).$$

The mean value of the operator $(\hat{\sigma}_n\hat{\sigma}_p)$ in the state described by the spin function

$$\begin{pmatrix} e^{-i\alpha}\cos\beta \\ e^{i\alpha}\sin\beta \end{pmatrix}_n \begin{pmatrix} 1 \\ 0 \end{pmatrix}_p$$

is equal to

$$\overline{(\hat{\sigma}_n\hat{\sigma}_p)} = \cos^2\beta - \sin^2\beta.$$

Consequently, for the scattering cross section we obtain the expression

$$\sigma = \pi\{3a_1^2 + a_0^2 - (a_1^2 - a_0^2)\cos 2\beta\},$$

or

$$\sigma = \left\{\frac{3}{4}\sigma^{\text{tripl}} + \frac{1}{4}\sigma^{\text{singl}} - \frac{\cos 2\beta}{4}(\sigma^{\text{tripl}} - \sigma^{\text{singl}})\right\}.$$

In the case of an unpolarized beam of neutrons the cross section is

$$\sigma = \frac{3}{4}\sigma^{\text{tripl}} + \frac{1}{4}\sigma^{\text{singl}},$$

since $\overline{\cos 2\beta} = 0$.

19. The spin state of the neutron and proton prior to the interaction is described by the function

$$\begin{pmatrix}1\\0\end{pmatrix}_n\begin{pmatrix}0\\1\end{pmatrix}_p.$$

Let us expand this in terms of the spin functions of the singlet and triplet states:

$$\begin{pmatrix}1\\0\end{pmatrix}_n\begin{pmatrix}0\\1\end{pmatrix}_p=\frac{1}{\sqrt{2}}\left[\frac{1}{\sqrt{2}}\left\{\begin{pmatrix}1\\0\end{pmatrix}_n\begin{pmatrix}0\\1\end{pmatrix}_p+\begin{pmatrix}0\\1\end{pmatrix}_n\begin{pmatrix}1\\0\end{pmatrix}_p\right\}+\right.$$
$$\left.+\frac{1}{\sqrt{2}}\left\{\begin{pmatrix}1\\0\end{pmatrix}_n\begin{pmatrix}0\\1\end{pmatrix}_p-\begin{pmatrix}0\\1\end{pmatrix}\begin{pmatrix}1\\0\end{pmatrix}_p\right\}\right].$$

The scattered wave has the form

$$\frac{e^{ikr}}{r}\frac{1}{\sqrt{2}}\left[a_1\frac{1}{\sqrt{2}}\left\{\begin{pmatrix}1\\0\end{pmatrix}_n\begin{pmatrix}0\\1\end{pmatrix}_p+\begin{pmatrix}0\\1\end{pmatrix}_n\begin{pmatrix}1\\0\end{pmatrix}_p\right\}+\right.$$
$$\left.+a_0\frac{1}{\sqrt{2}}\left\{\begin{pmatrix}1\\0\end{pmatrix}_n\begin{pmatrix}0\\1\end{pmatrix}_p-\begin{pmatrix}0\\1\end{pmatrix}_n\begin{pmatrix}1\\0\end{pmatrix}_p\right\}\right],$$

or

$$\frac{e^{ikr}}{r}\left\{\frac{a_1+a_0}{2}\begin{pmatrix}1\\0\end{pmatrix}_n\begin{pmatrix}0\\1\end{pmatrix}_p+\frac{a_1-a_0}{2}\begin{pmatrix}0\\1\end{pmatrix}_n\begin{pmatrix}1\\0\end{pmatrix}_p\right\},$$

from which it follows that the probability of a change in the direction of the spin is

$$\frac{1}{2}\frac{(a_1-a_0)^2}{a_1^2+a_0^2}.$$

20. We introduce the operator of the total spin of two protons

$$\frac{1}{2}(\hat{\sigma}_{p_1}+\hat{\sigma}_{p_2})=\hat{\mathbf{S}}.$$

As is readily shown

$$(\hat{\sigma}_n\hat{\mathbf{S}})^2=\hat{\mathbf{S}}^2-(\hat{\sigma}_n\hat{\mathbf{S}}).$$

With the help of the last relation we obtain

$$\hat{a}^2=\tfrac{1}{4}\{(a_0+3a_1)^2+(5a_1^2-2a_1a_0-3a_0^2)(\hat{\sigma}_n\mathbf{S})+(a_1-a_0)^2\mathbf{S}^2\}.$$

In the case of scattering on parahydrogen the cross section is

$$\sigma^{\text{para}}=\pi(a_0+3a_1)^2.$$

It is to be expected that the cross section does not depend on the polarization of the neutrons, since there is no physically preferred direction in space.

The scattering cross section on orthohydrogen is

$$\sigma^{\text{ortho}} = \pi\{(a_0+3a_1)^2+(5a_1^2-2a_1a_0-3a_0^2)\cos 2\beta+2(a_1-a_0)^2\},$$

where 2β is the angle between the direction of the total spin of the two protons and the spin of the neutron.

If the neutron stream is not polarized, then the average value of $\cos 2\beta$ for a mixed ensemble is zero; σ^{ortho} takes the form

$$\sigma^{\text{ortho}} = \pi\{(a_0+3a_1)^2+2(a_1-a_0)^2\},$$

and the ratio of the cross sections is

$$\frac{\sigma^{\text{ortho}}}{\sigma^{\text{para}}} = 1+2\left(\frac{a_1-a_0}{a_0+3a_1}\right)^2.$$

21. The scattering length depends on the total angular momentum, and therefore the scattering of the neutrons is characterized by two scattering lengths $a_{I-1/2}$ and $a_{I+1/2}$.

Here the probability of a total angular momentum $I-\frac{1}{2}$ and $I+\frac{1}{2}$ are $\frac{I}{2I+1}$ and $\frac{I+1}{2I+1}$ respectively

We thus find the effective cross section

$$\frac{d\sigma}{d\Omega} = \frac{I}{2I+1}\left|a_{I-\frac{1}{2}}\right|^2+\frac{I+1}{2I+1}\left|a_{I+\frac{1}{2}}\right|^2.$$

22. At a large distance from the target (the target consists of the scalar particles) the wave function of the incident particles is given by

$$e^{ikz}\begin{pmatrix}1\\0\end{pmatrix} \approx \frac{1}{2kr}\sum_{l=0}^{\infty} i^{l+1}(2l+1)\begin{pmatrix}1\\0\end{pmatrix}P_l(\cos\vartheta)\times$$

$$\times\left\{e^{-i\left(kr-\frac{\pi l}{2}\right)}-e^{i\left(kr-\frac{\pi l}{2}\right)}\right\}. \quad (1)$$

We expand the function $\begin{pmatrix}1\\0\end{pmatrix}P_l(\cos\vartheta)$ in terms of eigenfunctions of the operator $\hat{J}^2$ and obtain

$$\begin{pmatrix}1\\0\end{pmatrix}P_l\cos(\theta) = \frac{2\sqrt{\pi}}{\sqrt{2l+1}}\begin{pmatrix}1\\0\end{pmatrix}Y_{l0}(\vartheta)$$

$$= \frac{2\sqrt{\pi}}{2l+1}\{\sqrt{l+1}\,\Psi_l^+ + \sqrt{l}\,\Psi_l^-\}. \quad (2)$$

Here Ψ_l^+ and Ψ_l^- denote the Pauli functions

$$\Psi_l^+ = \frac{1}{\sqrt{2l+1}}\begin{pmatrix}\sqrt{l+1}\,Y_{l0}\\ \sqrt{l}\,Y_{l1}\end{pmatrix}\quad \left(j=l+\frac{1}{2},\ l,\ j_z=\frac{1}{2}\right),$$

$$\Psi_l^- = \frac{1}{\sqrt{2l+1}}\begin{pmatrix}\sqrt{l}\,Y_{l0}\\ -\sqrt{l+1}\,Y_{l1}\end{pmatrix}\quad \left(j=l-\frac{1}{2},\ l,\ j_z=\frac{1}{2}\right)$$

(see Prob. 21, §4).

Inserting (2) into (1) we obtain

$$e^{ikz}\begin{pmatrix}1\\0\end{pmatrix} \approx \frac{\sqrt{\pi}}{kr}\sum_{l=0}^{\infty} i^{l+1}\{\sqrt{l+1}\,\Psi_l^+ + \sqrt{l}\,\Psi_l^-\}\times$$

$$\times\left\{e^{-i\left(kr-\frac{\pi l}{2}\right)} - e^{i\left(kr-\frac{\pi l}{2}\right)}\right\}. \qquad (3)$$

As a result of the interaction only the outgoing wave e^{ikr}/r undergoes change.

Since for any type of interaction of these particles, $\hat{j}^2$, $\hat{l}^2$, and $\hat{j}_z$ will be constants of motion (see Prob. 48, §4), the change in the outgoing wave will, in the general case, be different for states with different quantum numbers j, l. For the outgoing wave we obtain the following expression:

$$\Psi_s \approx \frac{e^{ikr}}{r}\frac{\sqrt{\pi}}{ik}\sum_{l=0}^{\infty}\{\sqrt{l+1}\,(\eta_l^+-1)\,\Psi_l^+ + \sqrt{l}\,(\eta_l^- -1)\,\Psi_l^-\},$$

$$\eta_l^+ = \eta\left(j=l+\frac{1}{2},\ l\right),$$

$$\eta_l^- = \eta\left(j=l-\frac{1}{2},\ l\right),$$

or

$$\Psi_s \approx \frac{e^{ikr}}{r}\frac{\sqrt{\pi}}{ik}\sum_{l=0}^{\infty}\frac{1}{2l+1}\times$$

$$\times\left\{\begin{pmatrix}1\\0\end{pmatrix}Y_{l0}[(l+1)(\eta_l^+-1)+l(\eta_l^- -1)] + \begin{pmatrix}0\\1\end{pmatrix}Y_{l1}(\eta_l^+-\eta_l^-)\right\}.$$

It thus follows that the reorientation of the spin of the particle can take place if $\eta_l^+ \neq \eta_l^-$. We shall now express the scattering cross section in terms of η_l^+ and η_l^-. The differential cross section $d\sigma_1$ for scattering with a change of polarization is

$$d\sigma_1 = \frac{\pi}{k^2}\left|\sum_{l=0}^{\infty}\sqrt{\frac{l(l+1)}{2l+1}}(\eta_l^+ - \eta_l^-)Y_{l1}\right|^2 d\Omega,$$

and the differential cross section $d\sigma_2$ for scattering with no change of polarization is

$$d\sigma_2 = \frac{\pi}{k^2}\left|\sum_{l=0}^{\infty}\frac{1}{\sqrt{2l+1}}Y_{l0}\{(l+1)(\eta_l^+ - 1) + l(\eta_l^- - 1)\}\right|^2 d\Omega.$$

When the relative velocity of the particles is not large, only the S and P waves need be considered:

$$(|\eta_l^+ - 1| \ll 1, \quad |\eta_l^- - 1| \ll 1 \quad \text{if } l > 1).$$

In this case

$$d\sigma_1 \approx \frac{1}{4k^2}(\eta_1^+ - \eta_1^-)\sin^2\vartheta\, d\Omega,$$

$$d\sigma_2 \approx \frac{1}{4k^2}|\cos\vartheta(2\eta_1^+ + \eta_1^- - 1) + \eta_0^+ - 1|^2\, d\Omega.$$

From the expression for $d\sigma_1$ it is seen that the particles whose spin orientation is changed are scattered mainly in a direction perpendicular to the z axis.

24. Since $R \gg \lambda$ we can employ quasi-classical theory. The nucleus absorbs all the particles with $l < R/\lambda\!\!\!^-$, hence

$$\eta_l = 0 \quad \text{for} \quad l < \frac{R}{\lambda\!\!\!^-},$$

$$\eta_l = 1 \quad \text{for} \quad l > \frac{R}{\lambda\!\!\!^-}.$$

Inserting the value of η_l into the formula for the total cross sections

$$\sigma_r = \pi\lambda\!\!\!^{-2}\sum(2l+1)(1 - |\eta_l|^2),$$

$$\sigma_s = \pi\lambda\!\!\!^{-2}\sum(2l+1)|1 - \eta_l|^2,$$

we have

$$\sigma_r = \sigma_s = \pi\lambda\!\!\!{}^-\sum_{l=0}^{R/\lambda\!\!\!{}^-}(2l+1) \approx \pi R^2.$$

Therefore the total cross section $\sigma = \sigma_r + \sigma_s$ is equal to twice the geometric cross section of the nucleus.

25. In all three cases the distribution is isotropic.

26. The eigenfunctions of the operator I_z for the meson-nucleon system are written in the form of all possible products of the functions φ and ψ. Here φ refers to the various charge states of the meson ($\varphi_+, \varphi_0, \varphi_-$), and ψ to those of the nucleon (ψ_p, ψ_n). All told, there are six such functions:

$(p^+) = \varphi_+\psi_p$ $I_z = \frac{3}{2}$	$(p^0) = \varphi_0\psi_p$ $I_z = \frac{1}{2}$	$(p^-) = \varphi_-\psi_p$ $I_z = -\frac{1}{2}$
$(n^+) = \varphi_+\psi_n$ $I_z = \frac{1}{2}$	$(n^0) = \varphi_0\psi_n$ $I_z = -\frac{1}{2}$	$(n^-) = \varphi_-\psi_n$ $I_z = -\frac{3}{2}$

Here (p^+) denotes the function of a system consisting of a π^+-meson and proton; (n^+) denotes the function of a system consisting of a π^+-meson and neutron, etc. Generally speaking, these functions are not eigenfunctions of the operator of the square of the system's total isotopic spin $\mathbf{I}^2$.

The eigenfunctions of $\mathbf{I}^2$ belonging simultaneously to a given eigenvalue of I_z are linear combinations of the functions above with Clebsch-Gordan coefficients. The Clebsch-Gordan coefficients for $m' = \pm\frac{1}{2}$ have the form (see Prob. 21, §4)

	$m' = \frac{1}{2}$	$m' = -\frac{1}{2}$
$I = j + \frac{1}{2}$	$\sqrt{\dfrac{j+M+\frac{1}{2}}{2j+1}}$	$\sqrt{\dfrac{j-M+\frac{1}{2}}{2j+1}}$
$I = j - \frac{1}{2}$	$-\sqrt{\dfrac{j-M+\frac{1}{2}}{2j+1}}$	$\sqrt{\dfrac{j+M+\frac{1}{2}}{2j+1}}$

In our case $M = I_z$, $j = 1$, $m' = \tau_z$.

Making use of this table, we obtain the eigenfunctions $\Phi^I_{I_z}$ of the operators $\mathbf{I}^2$ and $\mathbf{I}_z$:

$$\Phi^{3/2}_{3/2} = (p^+),$$

$$\Phi^{3/2}_{1/2} = \sqrt{\frac{2}{3}}(p^0) + \sqrt{\frac{1}{3}}(n^+),$$

$$\Phi^{3/2}_{-1/2} = \sqrt{\frac{1}{3}}(p^-) + \sqrt{\frac{2}{3}}(n^0),$$

$$\Phi^{3/2}_{-3/2} = (n^-),$$

$$\Phi^{1/2}_{1/2} = -\sqrt{\frac{1}{3}}(p^0) + \sqrt{\frac{2}{3}}(n^+),$$

$$\Phi^{1/2}_{-1/2} = -\sqrt{\frac{2}{3}}(p^-) + \sqrt{\frac{1}{3}}(n^0).$$

From this we can easily express the eigenfunctions of the meson-nucleon system by means of eigenfunctions of $\mathbf{I}^2$ and $\mathbf{I}_z$. These functions are equal to

$$(p^+) = \Phi^{3/2}_{3/2}, \quad (n^+) = \sqrt{\frac{1}{3}}\Phi^{3/2}_{1/2} + \sqrt{\frac{2}{3}}\Phi^{1/2}_{1/2},$$

$$(p^0) = \sqrt{\frac{2}{3}}\Phi^{3/2}_{1/2} - \sqrt{\frac{1}{3}}\Phi^{1/2}_{1/2}, \quad (n^0) = \sqrt{\frac{2}{3}}\Phi^{3/2}_{-1/2} + \sqrt{\frac{1}{3}}\Phi^{1/2}_{-1/2},$$

$$(p^-) = \sqrt{\frac{1}{3}}\Phi^{3/2}_{-1/2} - \sqrt{\frac{2}{3}}\Phi^{1/2}_{-1/2}, \quad (n^-) = \Phi^{3/2}_{-3/2}.$$

27. The expansion of the incident wave is

$$\psi = e^{ikz}\begin{pmatrix}1\\0\end{pmatrix}\delta(\pi-\pi_i)\,\delta(n-\tau_z)$$

$$= \sum_{l=0}^{\infty} i^l(2l+1)P_l(\cos\theta)\begin{pmatrix}1\\0\end{pmatrix}\delta(\pi-\pi_i)\,\delta(n-\tau_z)\frac{\sin\left(kr-\frac{\pi l}{2}\right)}{kr}$$

$$= 2\sqrt{\pi}\sum_{l=0}^{\infty}\sum_{I} C^{I\tau_z}_{I_z}\Phi^I_{I_z} i^l\sqrt{2l+1}\,Y_{l0}\begin{pmatrix}1\\0\end{pmatrix}\frac{\sin\left(kr-\frac{\pi l}{2}\right)}{kr}, \tag{1}$$

where $C_{I_z}^{I\tau_z}$ are the Clebsch-Gordan coefficients given in the table below:

	$\tau_z = \frac{1}{2}$	$\tau_z = -\frac{1}{2}$
$C_{I_z}^{3/2}$	$\sqrt{\frac{1}{2}+\frac{I_z}{3}}$	$\sqrt{\frac{1}{2}-\frac{I_z}{3}}$
$C_{I_z}^{1/2}$	$-\sqrt{\frac{1}{2}-\frac{I_z}{3}}$	$\sqrt{\frac{1}{2}+\frac{I_z}{3}}$

Here we have made use of the relation (see Prob. 26)

$$\delta(\pi-\pi_i)\,\delta(n-\tau_z) = \sum_I C_{I_z}^{I\tau_z}\Phi_{I_z}^I.$$

We introduce the Pauli functions for $m_j = \frac{1}{2}$ (see Prob. 22, §9)

$$Y_l^+ = \begin{pmatrix} \sqrt{\dfrac{l+1}{2l+1}}\,Y_{l0} \\ -\sqrt{\dfrac{l}{2l+1}}\,Y_{l1} \end{pmatrix}; \quad Y_l^- = \begin{pmatrix} \sqrt{\dfrac{l}{2l+1}}\,Y_{l0} \\ -\sqrt{\dfrac{l+1}{2l+1}}\,Y_{l1} \end{pmatrix}$$

and we expand the function $Y_{l0}\begin{pmatrix}1\\0\end{pmatrix}$ in terms of them:

$$Y_{l0}\begin{pmatrix}1\\0\end{pmatrix} = \frac{1}{\sqrt{2l+1}}(\sqrt{l+1}\,Y_l^+ + \sqrt{l}\,Y_l^-). \tag{2}$$

Inserting (2) into (1), we obtain

$$\begin{aligned}\psi &= \frac{2\sqrt{\pi}}{kr}\sum_{l=0}^{\infty}\sum_I i^l(\sqrt{l+1}\,Y_l^+ + \sqrt{l}\,Y_l^-)\sin\left(kr-\frac{\pi l}{2}\right)C_{I_z}^{I\tau_z}\Phi_{I_z}^I \\ &= \frac{\sqrt{\pi}}{ikr}\sum_{l=0}^{\infty}\sum_I(\sqrt{l+1}\,Y_l^+ + \sqrt{l}\,Y_l^-)C_{I_z}^{I\tau_z}\Phi_{I_z}^I[e^{ikr}-(-1)^l e^{-ikr}] \\ &= \frac{\sqrt{\pi}}{ik}\sum_{l=0}^{\infty}\sum_I C_{I_z}^{I\tau_z}\Phi_{I_z}^I(\sqrt{l+1}\,Y_l^+ + \sqrt{l}\,Y_l^-)\frac{e^{ikr}}{r} - \\ &\quad - \frac{\sqrt{\pi}}{ik}\sum_{l=0}^{\infty}\sum_I C_{I_z}^{I\tau_z}\Phi_{I_z}^I(\sqrt{l+1}\,Y_l^+ + \sqrt{l}\,Y_l^-)(-1)^l\frac{e^{-ikr}}{r}.\end{aligned} \tag{3}$$

28. We make use of the expansion of the incident wave in terms of the eigenfunctions of the conservative operators (see Eq. (3) of the preceding problem).

Each term of this sum corresponding to given values of l, J, I will be scattered independently of the others.

Consequently, the number of particles with given values of l, J, I will not change in the process of elastic scattering, and therefore the influence of the scattering centre reduces to the multiplication by some phase factor $e^{2i\delta^I_{l\pm}}$. Here $\delta^I_{l\pm} = \delta^I_{l, J=l\pm\frac{1}{2}}$ is a function of l, J, I and not of I_z, owing to the assumption of isotopic spin invariance.

It should be noted that the scattering centre does not affect the incoming waves, but only the outgoing waves. Then the wave function of the system, after taking into account the scattering, can be written

$$\psi = \frac{\sqrt{\pi}}{ik}\sum_I \sum_{l=0}^{\infty} C^{I\tau_z}_{I_z} \cdot \Phi^I_{I_z} (\sqrt{l+1}\, Y^+_l e^{2i\delta^I_{l+}} + \sqrt{l}\, Y^-_l e^{2i\delta^I_{l-}}) \frac{e^{ikr}}{r} -$$

$$- \frac{\sqrt{\pi}}{ik}\sum_I \sum_{l=0}^{\infty} C^{I\tau_z}_{I_z} \Phi^I_{I_z} (\sqrt{l+1}\, Y^+_l + \sqrt{l}\, Y^-_l) \frac{e^{-ikr}}{r}. \qquad (1)$$

At large distances from the scattering centre this relation can be written in the form

$$\psi = \psi_{\text{inc}} + f\frac{e^{ikr}}{r},$$

Here f is the scattering amplitude. Subtracting from (1) the expression for ψ_{inc} (see Eq. 3 of Prob. 27, §9), we obtain

$$f = \frac{\sqrt{\pi}}{ik}\sum_I \sum_{l=0}^{\infty} C^{I\tau_z}_{I_z} \Phi^I_{I_z} \{\sqrt{l+1}\, Y^+_l (e^{2i\delta^I_{l+}} - 1) + \sqrt{l}\, Y^-_l (e^{2i\delta^I_{l-}} - 1)\}. \qquad (2)$$

We expand the functions $\Phi^I_{I_z}$ in terms of eigenfunctions of the operator I_z:

$$\Phi^I_{I_z} = C^{I\tau_z}_{I_z}\, \delta(\pi - \pi_i)\, \delta(n - \tau_z) + C^{I,-\tau_z}_{I_z}\, \delta(\pi - \pi_k)\, \delta(n + \tau_z), \qquad (3)$$

where the Clebsch-Gordon coefficients $C^{I\tau_z}_{I_z}$ are given in the table of Problem 27, and where $\pi_k = \pi_i + 2\tau_z$.

Inserting (3) into (2), we obtain

$$f=\frac{\sqrt{\pi}}{ik}\sum_{I}\sum_{l=0}^{\infty}\{G^{I\tau_z}_{I\tau_z}\delta(\pi-\pi_i)\,\delta(n-\tau_z)+G^{I\tau_z}_{I_{z'}-\tau_z}\,\delta(\pi-\pi_k)\,\delta(n+\tau_z)\}\times$$
$$\times\{\sqrt{l+1}\,Y_l^+(e^{2i\delta^I_{l+}}-1)+\sqrt{l}\,Y_l^-(e^{2i\delta^I_{l-}}-1)\},$$

where

$$G^{I\tau_z}_{I_z\tau_z'}=C^{I\tau_z}_{I_z}\,C^{I\tau_z'}_{I_z}$$

and τ_z' denotes the final state of the nucleon.

Then setting

$$f=f^{\tau_z}_{\tau_z}\delta(\pi-\pi_i)\,\delta(n-\tau_z)+f^{\tau_z}_{\tau_z}\delta(\pi-\pi_k)\,\delta(n+\tau_z),$$

we obtain

$$f^{\tau_z}_{\tau_z'}=\frac{\sqrt{\pi}}{ik}\sum_{I}\sum_{l=0}^{\infty}G^{I\tau_z}_{I_z\tau_z'}\{\sqrt{l+1}\,Y_l^+(e^{2i\delta^I_{l+}}-1)+\sqrt{l}\,Y_l^-(e^{2i\delta^I_{l-}}-1)\},\qquad(4)$$

where for the reaction under consideration $G^{I\tau_z}_{I_z\tau_z'}$ are given in the table below:

Reaction	$p^+\to p^+$	$p^-\to p^-$	$p^-\to n^0$
$G^{3/2\tau_z}_{I_z\tau_z'}$	1	$\frac{1}{3}$	$\frac{\sqrt{2}}{3}$
$G^{1/2\tau_z}_{I_z\tau_z'}$	0	$\frac{2}{3}$	$-\frac{\sqrt{2}}{3}$

Inserting these values of $G^{I\tau_z}_{I_z\tau_z'}$ into (4), we finally obtain

$$\left.\begin{aligned}
f(p^+,p^+)&=\frac{\sqrt{\pi}}{ik}\sum_{l=0}^{\infty}\{\sqrt{l+1}\,Y_l^+(e^{2i\delta^{3/2}_{l+}}-1)+\sqrt{l}\,Y_l^-(e^{2i\delta^{3/2}_{l-}}-1)\},\\
f(p^-,p^-)&=\frac{\sqrt{\pi}}{3ik}\sum_{l=0}^{\infty}\{\sqrt{l+1}\,Y_l^+(e^{2i\delta^{3/2}_{l+}}+2e^{2i\delta^{1/2}_{l+}}-3)+\\
&\qquad+\sqrt{l}\,Y_l^-(e^{2i\delta^{3/2}_{l-}}+2e^{2i\delta^{1/2}_{l-}}-3)\},\\
f(p^-,n^0)&=\frac{\sqrt{2\pi}}{3ik}\sum_{l=0}^{\infty}\{\sqrt{l+1}\,Y_l^+(e^{2i\delta^{3/2}_{l+}}-e^{2i\delta^{1/2}_{l+}})+\\
&\qquad+\sqrt{l}\,Y_l^-(e^{2i\delta^{3/2}_{l-}}-e^{2i\delta^{1/2}_{l-}})\}.
\end{aligned}\right\}\quad(5)$$

29. The table of coefficients of $G^{I\tau_z}_{I_z\tau'_z}$ for all possible reactions of mesons with nucleons is given below:

No.	Reaction	I_z	$G^{3/2\,\tau_z}_{I_z\tau'_z}$	$G^{1/2\,\tau_z}_{I_z\tau'_z}$
1	$p^+ \to p^+$	$\frac{3}{2}$	1	0
2	$p^0 \to p^0$	$\frac{1}{2}$	$\frac{2}{3}$	$\frac{1}{3}$
3	$p^0 \to n^+$	$\frac{1}{2}$	$\frac{\sqrt{2}}{3}$	$-\frac{\sqrt{2}}{3}$
4	$p^- \to n^0$	$-\frac{1}{2}$	$\frac{\sqrt{2}}{3}$	$-\frac{\sqrt{2}}{3}$
5	$p^- \to p^-$	$-\frac{1}{2}$	$\frac{1}{3}$	$\frac{2}{3}$
6	$n^+ \to n^+$	$\frac{1}{2}$	$\frac{1}{3}$	$\frac{2}{3}$
7	$n^0 \to p^-$	$-\frac{1}{2}$	$\frac{\sqrt{2}}{3}$	$-\frac{\sqrt{2}}{3}$
8	$n^+ \to p^0$	$\frac{1}{2}$	$\frac{\sqrt{2}}{3}$	$-\frac{\sqrt{2}}{3}$
9	$n^0 \to n^0$	$-\frac{1}{2}$	$\frac{2}{3}$	$\frac{1}{3}$
10	$n^- \to n^-$	$-\frac{3}{2}$	1	0

Because of the assumption of isotopic spin invariance, the phases do not depend on I_z, and it then follows directly from the table and from Eq. (4) of Prob. 28, §9 that

(1) $f(p^+, p^+) = f(n^-, n^-)$,

(2) $f(p^-, p^-) = f(n^+, n^+)$,

(3) $f(p^0, n^+) = f(n^+, p^0) = f(p^-, n^0) = f(n^0, p^-)$,

(4) $f(p^0, p^0) = f(n^0, n^0)$.

The expressions for the first three amplitudes are given in Prob. 28. From the table it is readily seen that

$$f(p^0, p^0) = \frac{1}{2}[f(p^+, p^+) + f(p^-, p^-)].$$

We then obtain with the help of the table:

(1) $f(p^+, p^+) = f(n^-, n^-) = f^{3/2}$,

(2) $f(p^-, p^-) = f(n^+, n^+) = \frac{1}{3}[f^{3/2} + 2f^{1/2}]$,

(3) $f(p^0, n^+) = f(n^+, p^0) = f(p^-, n^0) = f(n^0, p^-) = \frac{\sqrt{2}}{3}[f^{3/2} - f^{1/2}]$,

(4) $f(p^0, p^0) = f(n^0, n^0) = \frac{1}{3}[2f^{3/2} + f^{1/2}]$.

30. The differential scattering cross section is

$$\frac{d\sigma}{d\Omega} = |f|^2, \quad \text{where} \quad d\Omega = \sin\theta \, d\theta \, d\varphi.$$

The total scattering cross section is

$$\sigma = \int_0^{2\pi}\int_0^{\pi} |f|^2 \sin\theta \, d\theta \, d\varphi.$$

Inserting here the scattering amplitudes found in Prob. 28 and taking into account the fact that the Pauli functions are orthonormal

$$\iint (Y_l^+)^+ (Y_{l'}^+) \, d\Omega = \iint (Y_l^-)^+ (Y_{l'}^-) \, d\Omega = \delta_{ll'},$$

$$\iint (Y_l^+)^+ (Y_{l'}^-) \, d\Omega = \iint (Y_l^-)^+ (Y_{l'}^+) \, d\Omega = 0,$$

we obtain

$$\sigma(p^+, p^+) = \frac{4\pi}{k^2} \sum_{l=0}^{\infty} \{(l+1) \sin^2 \delta_{l+}^{3/2} + l \sin^2 \delta_{l-}^{3/2}\},$$

$$\sigma(p^-, p^-) = \frac{4\pi}{3k^2} \sum_{l=0}^{\infty} \{(l+1) \times$$

$$\times \left[\sin^2 \delta_{l+}^{3/2} + 2 \sin^2 \delta_{l+}^{1/2} - \frac{2}{3} \sin^2 (\delta_{l+}^{3/2} - \delta_{l+}^{1/2})\right] +$$

$$+ l \left[\sin^2 \delta_{l-}^{3/2} + 2 \sin^2 \delta_{l-}^{1/2} - \frac{2}{3} \sin^2 (\delta_{l-}^{3/2} - \delta_{l-}^{1/2})\right],$$

$$\sigma(p^-, n^0) = \frac{8\pi}{9k^2} \sum_{l=0}^{\infty} \{(l+1) \sin^2 (\delta_{l+}^{3/2} - \delta_{l+}^{1/2}) + l \sin^2 (\delta_{l-}^{3/2} - \delta_{l-}^{1/2})\}.$$

31. We give a detailed solution for the reactions (p^+, p^+). The scattering amplitude for the S and P waves (see Prob. 28, §9) has the form:

$$f(p^+, p^+) = \frac{\sqrt{\pi}}{ik}\{\alpha_0 Y_0^+ + \sqrt{2}\alpha_1 Y_1^+ + \beta_1 Y_1^-\},$$

where

$$\alpha_0 = e^{2i\delta_0^{3/2}} - 1; \quad \alpha_1 = e^{2i\delta_{1+}^{3/2}} - 1; \quad \beta_1 = e^{2i\delta_{1-}^{3/2}} - 1.$$

Then the differential scattering cross section for the S and P waves is (here we take into account the explicit expression for the Pauli function; see Prob. 27, §9)

$$\frac{k^2}{\pi}\frac{d\sigma}{d\Omega} = |Y_{00}|^2|\alpha_0^2|^2 +$$

$$+ \frac{1}{\sqrt{3}}|Y_{00}|\,|Y_{10}|\{2(\alpha_0\alpha_1^* + \alpha_0^*\alpha_1) + (\alpha_0\beta_1^* + \alpha_0^*\beta_1)\} +$$

$$+ \frac{1}{3}|Y_{10}|^2\{4|\alpha_1|^2 + |\beta_1|^2 + 2(\alpha_1\beta_1^* + \alpha_1^*\beta_1)\} +$$

$$+ \frac{2}{3}|Y_{11}|^2\{|\alpha_1|^2 + |\beta_1|^2 - (\alpha_1\beta_1^* + \alpha_1^*\beta_1)\}.$$

Since the spherical functions Y_{lm} are

$$Y_{00} = \frac{1}{\sqrt{4\pi}}; \qquad Y_{10} = \sqrt{\frac{3}{4\pi}}\cos\theta; \qquad Y_{11} = \sqrt{\frac{3}{8\pi}}\sin\theta\, e^{i\varphi},$$

the differential cross section can be written in the form

$$k^2\frac{d\sigma}{d\Omega} = A + B\cos\theta + C\cos^2\theta, \tag{1}$$

where the coefficients A, B and C are given by the equations

$$A = \frac{1}{4}\{|\alpha_0|^2 + |\alpha_1|^2 + |\beta_1|^2 - (\alpha_1\beta_1^* + \alpha_1^*\beta_1)\},$$

$$B = \frac{1}{4}\{2(\alpha_0\alpha_1^* + \alpha_0^*\alpha_1) + (\alpha_0\beta_1^* + \alpha_0^*\beta_1)\},$$

$$C = \frac{3}{4}\{|\alpha_1|^2 + (\alpha_1\beta_1^* + \alpha_1^*\beta_1)\}.$$

To express coefficients A, B and C in terms of the phase shifts we make use of the identities

$$|e^{2ix} - 1)|^2 = 4\sin^2 x,$$

$$(e^{2ix} - 1)(e^{-2iy} - 1) + (e^{-2ix} - 1)(e^{2iy} - 1) = 4[\sin^2 x + \sin^2 y - \sin^2(x - y)].$$

Then

$$\left.\begin{aligned} A(p^+,\, p^+) &= \sin^2 \delta_0^{3/2} + \sin(\delta_{1+}^{3/2} - \delta_{1-}^{3/2}), \\ B(p^+,\, p^+) &= 3\sin^2 \delta_0^{3/2} + 2\sin^2 \delta_{1+}^{3/2} + \sin^2 \delta_{1-}^{3/2} - \\ &\quad - 2\sin^2(\delta_0^{3/2} - \delta_{1+}^{3/2}) - \sin^2(\delta_0^{3/2} - \delta_{1-}^{3/2}), \\ C(p^+,\, p^+) &= 3\{2\sin^2 \delta_{1+}^{3/2} + \sin^2 \delta_{1-}^{3/2} - \sin^2(\delta_{1+}^{3/2} - \delta_{1-}^{3/2})\}. \end{aligned}\right\} \quad (2)$$

In the manner indicated we can readily calculate the coefficients A, B and C for the reactions (p^-, p^-) and (p^-, n^0). We give here only the final results:

$$\begin{aligned} A(p^-,\, p^-) = {} & \frac{1}{3}\sin^2 \delta_0^{3/2} + \frac{2}{3}\sin^2 \delta_0^{1/2} - \frac{2}{9}\sin^2(\delta_0^{3/2} - \delta_0^{1/2}) + \\ & + \frac{1}{9}\sin^2(\delta_{1+}^{3/2} - \delta_{1-}^{3/2}) - \frac{2}{9}\sin^2(\delta_{1-}^{3/2} - \delta_{1-}^{1/2}) + \\ & + \frac{2}{9}\sin^2(\delta_{1-}^{3/2} - \delta_{1+}^{1/2}) + \frac{2}{9}\sin^2(\delta_{1+}^{3/2} - \delta_{1-}^{1/2}) + \\ & + \frac{4}{9}\sin^2(\delta_{1-}^{1/2} - \delta_{1+}^{1/2}) - \frac{2}{9}\sin^2(\delta_{1+}^{3/2} - \delta_{1+}^{1/2}), \end{aligned}$$

$$\begin{aligned} B(p^-,\, p^-) = {} & \sin^2 \delta_0^{3/2} + 2\sin^2 \delta_0^{1/2} + \frac{2}{3}\sin^2 \delta_{1+}^{3/2} + \frac{4}{3}\sin^2 \delta_{1+}^{1/2} + \frac{1}{3}\sin^2 \delta_{1-}^{3/2} + \\ & + \frac{2}{3}\sin^2 \delta_{1-}^{1/2} - \frac{2}{9}\sin^2(\delta_0^{3/2} - \delta_{1+}^{3/2}) - \frac{4}{9}\sin^2(\delta_0^{3/2} - \delta_{1+}^{1/2}) - \\ & - \frac{1}{9}\sin^2(\delta_0^{3/2} - \delta_{1-}^{3/2}) - \frac{2}{9}\sin^2(\delta_0^{3/2} - \delta_{1-}^{1/2}) - \\ & - \frac{4}{9}\sin^2(\delta_0^{1/2} - \delta_{1+}^{1/2}) - \frac{8}{9}\sin^2(\delta_0^{1/2} - \delta_{1+}^{1/2}) - \\ & - \frac{2}{9}\sin^2(\delta_0^{1/2} - \delta_{1-}^{3/2}) - \frac{4}{9}\sin^2(\delta_0^{1/2} - \delta_{1-}^{1/2}), \end{aligned}$$

$$\begin{aligned} C(p^-,\, p^-) = {} & 2\sin^2 \delta_{1+}^{3/2} + 4\sin^2 \delta_{1+}^{1/2} + \sin^2 \delta_{1-}^{3/2} + 2\sin^2 \delta_{1-}^{1/2} - \\ & - \frac{2}{3}\sin^2(\delta_{1+}^{3/2} - \delta_{1+}^{1/2}) - \frac{1}{3}\sin^2(\delta_{1+}^{3/2} - \delta_{1-}^{3/2}) - \\ & - \frac{2}{3}\sin^2(\delta_{1+}^{3/2} - \delta_{1-}^{1/2}) - \frac{2}{3}\sin^2(\delta_{1+}^{1/2} - \delta_{1-}^{3/2}) - \frac{2}{3}\sin^2(\delta_{1+}^{1/2} - \delta_{1-}^{1/2}), \end{aligned} \quad (3)$$

$$A(p^-,\ n^0)=\frac{2}{9}\{\sin^2(\delta_0^{3/2}-\delta_0^{1/2})+\sin^2(\delta_{1+}^{3/2}-\delta_{1+}^{1/2})+\sin^2(\delta_{1+}^{3/2}-\delta_{1-}^{3/2})-$$
$$-\sin^2(\delta_{1+}^{3/2}-\delta_{1-}^{1/2})-\sin^2(\delta_{1+}^{1/2}-\delta_{1-}^{3/2})+\sin^2(\delta_{1+}^{1/2}-\delta_{1-}^{1/2})+$$
$$+\sin^2(\delta_{1-}^{3/2}-\delta_{1-}^{1/2})\},$$

$$B(p^-,\ n^0)=\frac{2}{9}\{-2\sin^2(\delta_0^{3/2}-\delta_{1+}^{3/2})+2\sin^2(\delta_0^{3/2}-\delta_{1+}^{1/2})-$$
$$-\sin^2(\delta_0^{3/2}-\delta_{1-}^{3/2})+\sin^2(\delta_0^{3/2}-\delta_{1-}^{1/2})+2\sin^2(\delta_0^{1/2}-\delta_{1+}^{3/2})-$$
$$-2\sin^2(\delta_0^{1/2}-\delta_{1+}^{1/2})+\sin^2(\delta_0^{1/2}-\delta_{1-}^{3/2})-\sin^2(\delta_0^{1/2}-\delta_{1-}^{1/2})\},$$

$$C(p^-,\ n^0)=\frac{2}{3}\{\sin^2(\delta_{1+}^{3/2}-\delta_{1+}^{1/2})+\sin^2(\delta_{1+}^{3/2}-\delta_{1-}^{1/2})-\sin^2(\delta_{1+}^{3/2}-\delta_{1-}^{3/2})+$$
$$+\sin^2(\delta_{1+}^{1/2}-\delta_{1-}^{3/2})-\sin^2(\delta_{1+}^{1/2}-\delta_{1-}^{1/2})\}. \tag{4}$$

The results obtained are of great importance in the study of the angular distribution of pions scattered on protons. It is possible to verify formula (1) experimentally and find the coefficients A, B, C for the reactions (p^+, p^+), (p^-, p^-), (p^-, n^0). We then find the six unknown phase shifts $\delta_0^{3/2}$, $\delta_0^{1/2}$, $\delta_{1+}^{3/2}$, $\delta_{1-}^{3/2}$, $\delta_{1+}^{1/2}$, $\delta_{1-}^{1/2}$ from the nine equations (2), (3) and (4), which prove to be compatible.

However, the phase shifts thus calculated are not single-valued for two reasons: firstly, owing to the fact that equations (2), (3), (4) contain squares of the sines of the phase shifts and the differences of the phase shifts, the signs of the latter are undetermined; secondly, there are several different sets of phase shifts satisfying the experimental data. Of these, in best agreement with experiment is the Fermi solution in which the greatest contribution to the scattering is made by the phase shift $\delta_{1+}^{3/2}$ which passes through 90° at a meson energy of $E \approx 195$ MeV in the laboratory system, while the phase shifts $\delta_{1-}^{3/2}$, $\delta_{1+}^{1/2}$, $\delta_{1-}^{1/2}$ are small.

A number of supplementary criteria allow the elimination of these ambiguities. The signs of the phase shifts can be determined from considerations based on the principle of casualty and also from experiment, especially if one takes into account the Coulomb interaction. Experiments with polarized recoil nucleons could be of assistance in the selection of the correct solution, but such experiments have not yet been performed. The expected magnitudes of the polarization for the reactions (p^+, p^+), (p^-, p^-), (p^-, p^0) will be found in the next problem.

32. We shall consider the reaction (p^+, p^+) in detail. We assume that the spin functions of the protons have the form

$$\alpha = \begin{pmatrix} 1 \\ 0 \end{pmatrix} \quad \text{for } s_z = \frac{1}{2},$$

$$\beta = \begin{pmatrix} 0 \\ 1 \end{pmatrix} \quad \text{for } s_z = -\frac{1}{2}.$$

If the proton, initially, had $s_z = \frac{1}{2}$, the scattering amplitude can be written in the form

$$f_{1/2} = f_{\alpha\alpha}\alpha + f_{\alpha\beta}\beta, \tag{1}$$

On the other hand, if, initially, $s_z = -\frac{1}{2}$, the scattering amplitude is

$$f_{-1/2} = f_{\beta\alpha}\alpha + f_{\beta\beta}\beta, \tag{2}$$

Here $f_{\alpha\alpha}$, $f_{\beta\beta}$ are the scattering amplitudes without change in the spin direction, while $f_{\alpha\beta}$, $f_{\beta\alpha}$ are the amplitudes with change in spin direction. The amplitudes $f_{\alpha\alpha}$ and $f_{\alpha\beta}$ are given by Eq. (5) of Prob. 28, §9 as coefficients of the matrices $\begin{pmatrix} 1 \\ 0 \end{pmatrix}$ and $\begin{pmatrix} 0 \\ 1 \end{pmatrix}$, only the S and P waves being taken:

$$f_{\alpha\alpha} = \frac{\sqrt{\pi}}{ik}\left\{\alpha_0 Y_{00} + \frac{1}{\sqrt{3}}[2\alpha_1 + \beta_1]\, Y_{10}\right\}, \tag{3}$$

$$f_{\alpha\beta} = \frac{\sqrt{\pi}}{ik}\sqrt{\frac{2}{3}}(\beta_1 - \alpha_1)\, Y_{11}, \tag{4}$$

where α_0, α_1, β_1 are given in Prob. 31, §9. Making use of the formulae for the Pauli functions for $m_j = \frac{1}{2}$, we readily obtain

$$f_{\beta\alpha} = \frac{\sqrt{\pi}}{ik}\sqrt{\frac{2}{3}}(\alpha_1 - \beta_1)\, Y_{1,-1},$$

$$f_{\beta\beta} = f_{\alpha\alpha}.$$

If we assume that the pions undergo scattering in the xz plane, then the polar angle $\varphi = 0$, and consequently

$$f_{\alpha\beta} = -f_{\beta\alpha}.$$

Then the amplitude for scattering by a proton with $s_z = -\frac{1}{2}$ [see (2)] is

$$f_{-1/2} = -f_{\beta\alpha}\alpha + f_{\alpha\alpha}\beta. \tag{5}$$

Since, the protons were not initially polarized, then from formulae (1) and (5) it follows that after polarization they will remain unpolarized with respect to the z axis.

We can show that after scattering there is no polarization of the protons in the xz plane. Indeed, the proton spin functions γ_θ and δ_θ corresponding to the spin components $\frac{1}{2}$ and $-\frac{1}{2}$ along the z' axis lying in the xz plane at an angle of θ to the z axis have the form

$$\begin{pmatrix}\gamma_\theta\\ \delta_\theta\end{pmatrix} = e^{-i\hat{s}_y\theta}\begin{pmatrix}\alpha\\ \beta\end{pmatrix} = \begin{pmatrix}\cos\frac{\theta}{2} & -\sin\frac{\theta}{2}\\ \sin\frac{\theta}{2} & \cos\frac{\theta}{2}\end{pmatrix}\begin{pmatrix}\alpha\\ \beta\end{pmatrix}$$

(see Prob. 20, §4).

Then

$$f_{1/2} = f_{\alpha\alpha}\alpha + f_{\alpha\beta}\beta = \left(f_{\alpha\alpha}\cos\frac{\theta}{2} - f_{\alpha\beta}\sin\frac{\theta}{2}\right)\gamma_\theta + \left(f_{\alpha\alpha}\sin\frac{\theta}{2} + f_{\alpha\beta}\cos\frac{\theta}{2}\right)\delta_\theta,$$

$$_{-1/2} = -f_{\alpha\beta}\alpha + f_{\alpha\alpha}\beta = -\left(f_{\alpha\alpha}\sin\frac{\theta}{2} + f_{\alpha\beta}\cos\frac{\theta}{2}\right)\gamma_\theta + \left(f_{\alpha\alpha}\cos\frac{\theta}{2} - f_{\alpha\beta}\sin\frac{\theta}{2}\right)\delta_\theta,$$

i.e., there is no polarization in any direction lying in the xz plane.

The protons will, however, be polarized along the y axis perpendicular to the plane of scattering. In order to find the magnitude of the polarization we express $f'_{1/2}$ and $f_{-1/2}$ in terms of the spin eigenfunctions

$$\gamma = \frac{\alpha + i\beta}{\sqrt{2}}, \quad \delta = \frac{\alpha - i\beta}{\sqrt{2}}$$

corresponding to the direction of the spin parallel and antiparallel to the y axis. Then

$$f_{1/2} = \frac{1}{\sqrt{2}}(f_{\alpha\alpha} - if_{\alpha\beta})\gamma + \frac{1}{\sqrt{2}}(f_{\alpha\alpha} + if_{\alpha\beta})\delta,$$

$$f_{-1/2} = \frac{-i}{\sqrt{2}}(f_{\alpha\alpha} - if_{\alpha\beta})\gamma + \frac{i}{\sqrt{2}}(f_{\alpha\alpha} + if_{\alpha\beta})\delta.$$

From this we obtain

$$W_+ \sim |f_{\alpha\alpha} - if_{\alpha\beta}|^2, \qquad W_- \sim |f_{\alpha\alpha} + if_{\alpha\beta}|^2, \tag{6}$$

where W_+ and W_- are the probabilities that the spin, after scattering, will be oriented parallel or antiparallel to the y axis. We note that (6) is valid regardless of the initial values of s_z of the proton. Inserting into (6) the expressions (3) and (4) for $f_{\alpha\alpha}$ and $f_{\alpha\beta}$, we obtain

$$W_\pm \sim |\alpha_0 + (2\alpha_1 + \beta_1)\cos\theta \mp i(\beta_1 - \alpha_1)\sin\theta|^2,$$

or

$$W_{\pm} \sim \left| \left(e^{2i\delta_0^{3/2}} - 1\right) + \left(2e^{2i\delta_{1+}^{3/2}} - 3 + e^{2i\delta_{1-}^{3/2}}\right)\cos\theta \pm i\left(e^{2i\delta_{1+}^{3/2}} - e^{2i\delta_{1-}^{3/2}}\right)\sin\theta \right|^2.$$

Similarly, we find for the reactions (p^-, p^-) and $(p^- \ n^0)$

$$W_{\pm}(p^-, p^-) \sim \left| \left(e^{2i\delta_0^{3/2}} - 3 + 2e^{2i\delta_0^{1/2}}\right) + \right.$$

$$+ \left(2e^{2i\delta_{1+}^{3/2}} - 9 + 4e^{2i\delta_{1+}^{1/2}} + e^{2i\delta_{1-}^{3/2}} + 2e^{2i\delta_{1-}^{1/2}}\right)\cos\theta \pm$$

$$\left. \pm i\left(e^{2i\delta_{1+}^{3/2}} + 2e^{2i\delta_{1+}^{1/2}} - e^{2i\delta_{1-}^{3/2}} - 2e^{2i\delta_{1-}^{1/2}}\right)\cos\vartheta \right|^2;$$

$$W_{\pm}(p^-, n^0) \sim \left| \left(e^{2i\delta_0^{3/2}} - e^{2i\delta_0^{1/2}}\right) + \left(2e^{2i\delta_{1+}^{3/2}} - 2e^{2i\delta_{1+}^{1/2}} + e^{2i\delta_{1-}^{3/2}} - e^{2i\delta_{1-}^{1/2}}\right)\cos\theta \pm \right.$$

$$\left. \pm i\left(e^{2i\delta_{1+}^{3/2}} - e^{2i\delta_{1+}^{1/2}} - e^{2i\delta_{1-}^{3/2}} + e^{2i\delta_{1-}^{1/2}}\right)\sin\theta \right|^2.$$

APPENDIX I

The quasi-classical approximation is used in the solution of a large number of problems in quantum mechanics. But the quasi-classical solution is valid only in a region sufficiently far from the turning point, which is defined by the condition

$$V(x) = E.$$

Since the quasi-classical approach gives solutions only to the right and left of the turning point, it is necessary to bring both ends together at this point in order to obtain a solution over all of space.

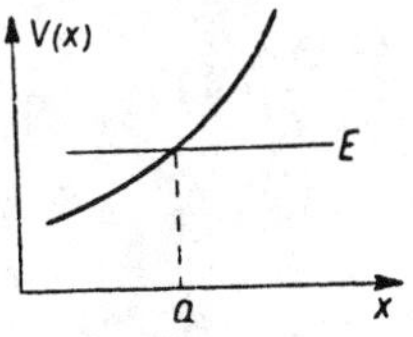

Fig. 32

If the potential function $V(x)$ close to the turning point behaves like the curve shown in Fig. 32, the overall solution has the form

$$\psi_1 = \begin{cases} \dfrac{1}{\sqrt{p}} \sin\left(\dfrac{1}{\hbar}\displaystyle\int_x^a p\,dx + \dfrac{\pi}{4}\right) & \text{for} \quad x < a, \qquad (1) \\[2ex] \dfrac{1}{2\sqrt{|p|}} e^{-\frac{1}{\hbar}\int_a^x |p|\,dx} & \text{for} \quad x > a. \qquad (2) \end{cases}$$

For the potential function $V(x)$ shown in Fig. 33 the solution will be

$$\psi_1 = \begin{cases} \dfrac{1}{\sqrt{p}} \sin\left(\dfrac{1}{\hbar}\displaystyle\int_a^x p\,dx + \dfrac{\pi}{4}\right) & \text{for} \quad x > a, \quad (1') \\[2ex] \dfrac{1}{2\sqrt{|p|}} e^{-\frac{1}{\hbar}\int_x^a |p|\,dx} & \text{for} \quad x < a. \quad (2') \end{cases}$$

(See Landau and Lifshitz, *Quantum Mechanics*, Pergamon Press, 1958.)

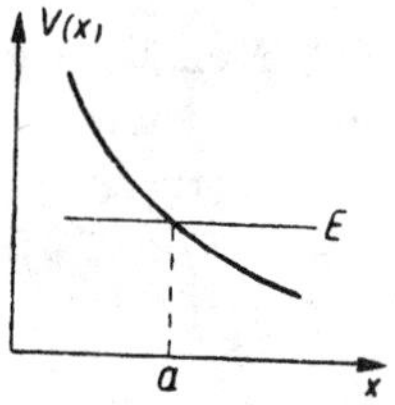

Fig. 33

We shall obtain in this way a solution of the one-dimensional Schrödinger equation

$$-\frac{\hbar^2}{2\mu}\frac{d^2\psi}{dx^2} + V(x)\psi = E\psi$$

which, to the left of the turning point (see Fig. 32), will take the quasi-classical form

$$\frac{1}{\sqrt{p}} e^{\pm\frac{i}{\hbar}\int_x^a p\,dx}$$

To find the solution it is necessary to find another solution which is linearly independent of the solutions (1), (2). We shall seek it in the form

$$\psi_2 = \begin{cases} \dfrac{1}{\sqrt{p}} \cos\left(\dfrac{1}{\hbar}\displaystyle\int_x^a p\,dx + \dfrac{\pi}{4}\right) & \text{for} \quad x < a, \quad (3) \\[2ex] \dfrac{c}{\sqrt{|p|}} e^{\frac{1}{\hbar}\int_a^x |p|\,dx} & \text{for} \quad x > a. \quad (4) \end{cases}$$

To determine c we make use of the fact that for the Schrödinger equation

$$W = \begin{vmatrix} \psi_1 & \psi_1' \\ \psi_2 & \psi_2' \end{vmatrix} = \text{const.}$$

Making use of solutions (1) and (3), we write the Wronskian determinant for our case:

$$W = \begin{vmatrix} \frac{1}{\sqrt{p}} \sin\left(\frac{1}{\hbar}\int_x^a p\,dx + \frac{\pi}{4}\right); & -\frac{\sqrt{p}}{\hbar}\cos\left(\frac{1}{\hbar}\int_x^a p\,dx + \frac{\pi}{4}\right) \\ \frac{1}{\sqrt{p}} \cos\left(\frac{1}{\hbar}\int_x^a p\,dx + \frac{\pi}{4}\right); & \frac{\sqrt{p}}{\hbar}\sin\left(\frac{1}{\hbar}\int_x^a p\,dx + \frac{\pi}{4}\right) \end{vmatrix} = \frac{1}{\hbar}.$$

(It is sufficient to differentiate with respect to the trigonometric functions, since $d(\hbar/p)/dx \ll 1$.)

Similarly, we find that in the region $x > a$, i.e., for solutions (2) and (4), we have for the Wronskian determinant

$$W = \frac{c}{\hbar}.$$

From the condition $W(x < a) = W(x > a)$ we find that $c = 1$. We take for the solution the linear combination of the functions ψ_1 and ψ_2:

$$\psi = (\psi_2 \pm i\psi_1)\, e^{\mp i\frac{\pi}{4}}.$$

Finally, we have

$$\psi = \begin{cases} \dfrac{1}{\sqrt{p}}\, e^{\pm\frac{i}{\hbar}\int_x^a p\,dx} & \text{for } x < a, \\ \dfrac{1}{\sqrt{|p|}}\, e^{\frac{1}{\hbar}\int_a^x |p|\,dx \mp i\frac{\pi}{4}} + \dfrac{1}{2\sqrt{|p|}}\, e^{-\frac{1}{\hbar}\int_a^x |p|\,dx \pm i\frac{\pi}{4}} & \text{for } x > a. \end{cases}$$

We shall now find a solution of the one-dimensional Schrödinger equation

$$-\frac{\hbar^2}{2\mu}\frac{d^2\psi}{dx^2} + V(x)\,\psi = E\psi$$

which to the left of the turning point (see Fig. 33) passes over into the quasi-classical solution of the form

$$\frac{1}{\sqrt{|p|}}e^{\pm\frac{i}{\hbar}\int_x^a |p|\,dx}.$$

To find this solution we make use of the above results. The solution we are seeking will be a linear combination of the solutions ψ_1' and ψ_2', where

$$\psi_1' = \begin{cases} \frac{1}{\sqrt{p}}e^{\frac{i}{\hbar}\int_a^x p\,dx} & \text{for } x > a, \\ \frac{1}{\sqrt{|p|}}e^{\frac{i}{\hbar}\int_x^a |p|\,dx - i\frac{\pi}{4}} + \frac{1}{2\sqrt{|p|}}e^{-\frac{1}{\hbar}\int_x^a |p|\,dx + i\frac{\pi}{4}} & \text{for } x < a, \end{cases}$$

$$\psi_2' = \begin{cases} \frac{1}{\sqrt{p}}e^{-\frac{i}{\hbar}\int_a^x p\,dx} & \text{for } x > a, \\ \frac{1}{\sqrt{|p|}}e^{\frac{1}{\hbar}\int_x^a |p|\,dx + i\frac{\pi}{4}} + \frac{1}{2\sqrt{|p|}}e^{-\frac{1}{\hbar}\int_x^a |p|\,dx - i\frac{\pi}{4}} & \text{for } x < a. \end{cases}$$

We thus obtain

$$\psi = \begin{cases} \frac{1}{\sqrt{|p|}}e^{\frac{1}{\hbar}\int_x^a |p|\,dx} & \text{for } x < a, \\ \frac{1}{2\sqrt{p}}e^{\frac{i}{\hbar}\int_a^x p\,dx + i\frac{\pi}{\hbar}} + \frac{1}{2\sqrt{p}}e^{-\frac{i}{\hbar}\int_a^x p\,dx - i\frac{\pi}{4}} & \text{for } x > a, \end{cases}$$

$$\psi = \begin{cases} \frac{1}{\sqrt{|p|}}e^{-\frac{1}{\hbar}\int_x^a |p|\,dx} & \text{for } x < a, \\ \frac{1}{\sqrt{p}}e^{\frac{i}{\hbar}\int_a^x p\,dx - i\frac{\pi}{4}} + \frac{1}{\sqrt{p}}e^{-\frac{i}{\hbar}\int_a^x p\,dx + i\frac{\pi}{4}} & \text{for } x > a. \end{cases}$$

APPENDIX II

A whole series of experimental data (e.g., the scattering of pions on protons and neutrons) indicate that the proton and neutron can be transformed into each other. We can therefore regard the proton and neutron as one particle, a nucleon, which can exist in two states: proton and neutron. These states differ as to the value of a charge variable: the charge of the proton in units of e is 1, while that of the neutron is 0. The nucleon can therefore be described by a two-component wave function in accordance with the two values of the charge variable. These functions can be written in the form

$$\psi = \begin{pmatrix} \psi_p \\ \psi_n \end{pmatrix}.$$

Taking into account the normalization condition $|\psi_p|^2 + |\psi_n|^2 = 1$, we take

$$\psi_p = \begin{pmatrix} 1 \\ 0 \end{pmatrix}; \qquad \psi_n = \begin{pmatrix} 0 \\ 1 \end{pmatrix}.$$

We introduce the operators acting on these two-component functions

$$\tau_+ = \begin{pmatrix} 0 & 1 \\ 0 & 0 \end{pmatrix}; \qquad \tau_- = \begin{pmatrix} 0 & 0 \\ 1 & 0 \end{pmatrix}.$$

It is readily verified that

$$\tau_+ \psi_p = 0, \qquad \tau_- \psi_p = \psi_n,$$

$$\tau_+ \psi_n = \psi_p, \qquad \tau_- \psi_n = 0.$$

From these formulae it follows that τ_+ is the charge creation operator, i.e., it takes the nucleon from the neutron to the proton state; τ_- is the charge annihilation operator.

Next we introduce the operators

$$\tau_x = \frac{1}{2}\left(\tau_+ + \tau_-\right) = \frac{1}{2}\begin{pmatrix} 0 & 1 \\ 1 & 0 \end{pmatrix},$$

$$\tau_y = -\frac{i}{2}\left(\tau_+ + \tau_-\right) = \frac{1}{2}\begin{pmatrix} 0 & -i \\ i & 0 \end{pmatrix},$$

$$\tau_z = \frac{1}{2}\left(\tau_+ \tau_- - \tau_- \tau_+\right) = \frac{1}{2}\begin{pmatrix} 1 & -0 \\ 0 & 1 \end{pmatrix}.$$

These operators are identical to the Pauli matrices known from the theory of spin and therefore have the same formal properties. By analogy with spin theory we assume that τ_x, τ_y, τ_z are operator components of a vector $\boldsymbol{\tau}$ in some three-dimensional space.

We call this space the isotopic spin space and the vector $\boldsymbol{\tau}$ the isotopic spin vector of the nucleon. It should be noted that the notion of isotopic spin space is introduced for convenience and has no direct physical meaning. The absolute value of the vector $\boldsymbol{\tau}$ is $\frac{1}{2}$ and the two charge states of the nucleon can be regarded as states of different values of isotopic spin components along the z axis in isotopic spin space. When $\tau_z = \frac{1}{2}$ we have a proton, when $\tau_z = -\frac{1}{2}$ we have a neutron. We note that since the isotopic spin space and the isotopic spin bear a formal character, then it is not the operator τ_z, but the charge exchange operator

$$q = \tau_z + \frac{1}{2} I = \begin{pmatrix} 1 & 0 \\ 0 & 0 \end{pmatrix}.$$

that has physical meaning. The π^+, π^0, π^- mesons can also be regarded as one particle which can appear in three charge states corresponding to the values of the charge variable 1, 0, −1. The wave function of the pions will, of course, be a three-component function in accordance with three possible charge states:

$$\varphi = \begin{pmatrix} \varphi_+ \\ \varphi_0 \\ \varphi_- \end{pmatrix}.$$

Taking into account the normalization condition, we set

$$\varphi_+ = \begin{pmatrix} 1 \\ 0 \\ 0 \end{pmatrix}, \quad \varphi_0 = \begin{pmatrix} 0 \\ 1 \\ 0 \end{pmatrix}, \quad \varphi_- = \begin{pmatrix} 0 \\ 0 \\ 1 \end{pmatrix}.$$

We can also introduce the operators of creation and annihilation of the pion charge

$$T_+ = \begin{pmatrix} 0 & 1 & 0 \\ 0 & 0 & 1 \\ 0 & 0 & 0 \end{pmatrix}, \quad T_- = \begin{pmatrix} 0 & 0 & 0 \\ 1 & 0 & 0 \\ 0 & 1 & 0 \end{pmatrix}.$$

These operators satisfy the relations

$$T_+\varphi_+ = 0; \quad T_+\varphi_0 = \varphi_+; \quad T_+\varphi_- = \varphi_0;$$
$$T_-\varphi_+ = \varphi_0; \quad T_-\varphi_0 = \varphi_-; \quad T_-\varphi_- = 0.$$

Let us go over to the operators

$$T_x = \frac{1}{\sqrt{2}}(T_+ + T_-) = \frac{1}{\sqrt{2}}\begin{pmatrix} 0 & 1 & 0 \\ 1 & 0 & 1 \\ 0 & 1 & 0 \end{pmatrix},$$

$$T_y = \frac{i}{\sqrt{2}}(T_+ - T_-) = \frac{i}{\sqrt{2}}\begin{pmatrix} 0 & 1 & 0 \\ -1 & 0 & 1 \\ 0 & -1 & 0 \end{pmatrix},$$

$$T_z = T_+T_- - T_-T_+ = \begin{pmatrix} 1 & 0 & 0 \\ 0 & 0 & 0 \\ 0 & 0 & -1 \end{pmatrix},$$

which are the operators of the isotopic spin vector components $\mathbf{T}$ in isotopic spin space. The absolute value of $\mathbf{T}$ is 1. The different charge states will be states of different values of the z component of the isotopic spin in isctopic spin space. Note that for the pion the charge operator Q is identical to the operator T_z

$$Q = T_z.$$

Let us now investigate the isotopic spin properties of the nucleon-meson system. We shall describe this system by the total isotopic spin angular momentum $\mathbf{I}$

$$\mathbf{I} = \boldsymbol{\tau} + \mathbf{T}$$

and by its z component I_z.

From exprimental data to date we can assume that the hypothesis of isotopic spin invariance (charge independence), i.e., the hypothesis stating that the properties of the system are independent of the total charge of the system if the Coulomb forces are negligible, is valid for a meson-nucleon system. Mathematically we can express this fact as the invariance of the interaction Hamiltonian with respect to rotations in three-dimensional isotopic space. From this it directly follows that the total isotopic spin angular momentum $\mathbf{I}$ and its z component I_z are constants of motion in the meson-nucleon system.

Mathematics

FUNCTIONAL ANALYSIS (Second Corrected Edition), George Bachman and Lawrence Narici. Excellent treatment of subject geared toward students with background in linear algebra, advanced calculus, physics and engineering. Text covers introduction to inner-product spaces, normed, metric spaces, and topological spaces; complete orthonormal sets, the Hahn-Banach Theorem and its consequences, and many other related subjects. 1966 ed. 544pp. 6⅛ x 9¼. 0-486-40251-7

ASYMPTOTIC EXPANSIONS OF INTEGRALS, Norman Bleistein & Richard A. Handelsman. Best introduction to important field with applications in a variety of scientific disciplines. New preface. Problems. Diagrams. Tables. Bibliography. Index. 448pp. 5⅜ x 8½. 0-486-65082-0

VECTOR AND TENSOR ANALYSIS WITH APPLICATIONS, A. I. Borisenko and I. E. Tarapov. Concise introduction. Worked-out problems, solutions, exercises. 257pp. 5⅝ x 8¼. 0-486-63833-2

AN INTRODUCTION TO ORDINARY DIFFERENTIAL EQUATIONS, Earl A. Coddington. A thorough and systematic first course in elementary differential equations for undergraduates in mathematics and science, with many exercises and problems (with answers). Index. 304pp. 5⅜ x 8½. 0-486-65942-9

FOURIER SERIES AND ORTHOGONAL FUNCTIONS, Harry F. Davis. An incisive text combining theory and practical example to introduce Fourier series, orthogonal functions and applications of the Fourier method to boundary-value problems. 570 exercises. Answers and notes. 416pp. 5⅜ x 8½. 0-486-65973-9

COMPUTABILITY AND UNSOLVABILITY, Martin Davis. Classic graduate-level introduction to theory of computability, usually referred to as theory of recurrent functions. New preface and appendix. 288pp. 5⅜ x 8½. 0-486-61471-9

ASYMPTOTIC METHODS IN ANALYSIS, N. G. de Bruijn. An inexpensive, comprehensive guide to asymptotic methods–the pioneering work that teaches by explaining worked examples in detail. Index. 224pp. 5⅜ x 8½ 0-486-64221-6

APPLIED COMPLEX VARIABLES, John W. Dettman. Step-by-step coverage of fundamentals of analytic function theory–plus lucid exposition of five important applications: Potential Theory; Ordinary Differential Equations; Fourier Transforms; Laplace Transforms; Asymptotic Expansions. 66 figures. Exercises at chapter ends. 512pp. 5⅜ x 8½. 0-486-64670-X

INTRODUCTION TO LINEAR ALGEBRA AND DIFFERENTIAL EQUATIONS, John W. Dettman. Excellent text covers complex numbers, determinants, orthonormal bases, Laplace transforms, much more. Exercises with solutions. Undergraduate level. 416pp. 5⅜ x 8½. 0-486-65191-6

RIEMANN'S ZETA FUNCTION, H. M. Edwards. Superb, high-level study of landmark 1859 publication entitled "On the Number of Primes Less Than a Given Magnitude" traces developments in mathematical theory that it inspired. xiv+315pp. 5⅜ x 8½. 0-486-41740-9

Physics

OPTICAL RESONANCE AND TWO-LEVEL ATOMS, L. Allen and J. H. Eberly. Clear, comprehensive introduction to basic principles behind all quantum optical resonance phenomena. 53 illustrations. Preface. Index. 256pp. 5⅜ x 8½. 0-486-65533-4

QUANTUM THEORY, David Bohm. This advanced undergraduate-level text presents the quantum theory in terms of qualitative and imaginative concepts, followed by specific applications worked out in mathematical detail. Preface. Index. 655pp. 5⅜ x 8½. 0-486-65969-0

ATOMIC PHYSICS (8th EDITION), Max Born. Nobel laureate's lucid treatment of kinetic theory of gases, elementary particles, nuclear atom, wave-corpuscles, atomic structure and spectral lines, much more. Over 40 appendices, bibliography. 495pp. 5⅜ x 8½. 0-486-65984-4

A SOPHISTICATE'S PRIMER OF RELATIVITY, P. W. Bridgman. Geared toward readers already acquainted with special relativity, this book transcends the view of theory as a working tool to answer natural questions: What is a frame of reference? What is a "law of nature"? What is the role of the "observer"? Extensive treatment, written in terms accessible to those without a scientific background. 1983 ed. xlviii+172pp. 5⅜ x 8½. 0-486-42549-5

AN INTRODUCTION TO HAMILTONIAN OPTICS, H. A. Buchdahl. Detailed account of the Hamiltonian treatment of aberration theory in geometrical optics. Many classes of optical systems defined in terms of the symmetries they possess. Problems with detailed solutions. 1970 edition. xv + 360pp. 5⅜ x 8½. 0-486-67597-1

PRIMER OF QUANTUM MECHANICS, Marvin Chester. Introductory text examines the classical quantum bead on a track: its state and representations; operator eigenvalues; harmonic oscillator and bound bead in a symmetric force field; and bead in a spherical shell. Other topics include spin, matrices, and the structure of quantum mechanics; the simplest atom; indistinguishable particles; and stationary-state perturbation theory. 1992 ed. xiv+314pp. 6⅛ x 9¼. 0-486-42878-8

LECTURES ON QUANTUM MECHANICS, Paul A. M. Dirac. Four concise, brilliant lectures on mathematical methods in quantum mechanics from Nobel Prize-winning quantum pioneer build on idea of visualizing quantum theory through the use of classical mechanics. 96pp. 5⅜ x 8½. 0-486-41713-1

THIRTY YEARS THAT SHOOK PHYSICS: THE STORY OF QUANTUM THEORY, George Gamow. Lucid, accessible introduction to influential theory of energy and matter. Careful explanations of Dirac's anti-particles, Bohr's model of the atom, much more. 12 plates. Numerous drawings. 240pp. 5⅜ x 8½. 0-486-24895-X

ELECTRONIC STRUCTURE AND THE PROPERTIES OF SOLIDS: THE PHYSICS OF THE CHEMICAL BOND, Walter A. Harrison. Innovative text offers basic understanding of the electronic structure of covalent and ionic solids, simple metals, transition metals and their compounds. Problems. 1980 edition. 582pp. 6⅛ x 9¼. 0-486-66021-4

CATALOG OF DOVER BOOKS

A TREATISE ON ELECTRICITY AND MAGNETISM, James Clerk Maxwell. Important foundation work of modern physics. Brings to final form Maxwell's theory of electromagnetism and rigorously derives his general equations of field theory. 1,084pp. 5⅜ x 8½. Two-vol. set. Vol. I: 0-486-60636-8 Vol. II: 0-486-60637-6

QUANTUM MECHANICS: PRINCIPLES AND FORMALISM, Roy McWeeny. Graduate student-oriented volume develops subject as fundamental discipline, opening with review of origins of Schrödinger's equations and vector spaces. Focusing on main principles of quantum mechanics and their immediate consequences, it concludes with final generalizations covering alternative "languages" or representations. 1972 ed. 15 figures. xi+155pp. 5⅜ x 8½. 0-486-42829-X

INTRODUCTION TO QUANTUM MECHANICS With Applications to Chemistry, Linus Pauling & E. Bright Wilson, Jr. Classic undergraduate text by Nobel Prize winner applies quantum mechanics to chemical and physical problems. Numerous tables and figures enhance the text. Chapter bibliographies. Appendices. Index. 468pp. 5⅜ x 8½. 0-486-64871-0

METHODS OF THERMODYNAMICS, Howard Reiss. Outstanding text focuses on physical technique of thermodynamics, typical problem areas of understanding, and significance and use of thermodynamic potential. 1965 edition. 238pp. 5⅜ x 8½. 0-486-69445-3

THE ELECTROMAGNETIC FIELD, Albert Shadowitz. Comprehensive undergraduate text covers basics of electric and magnetic fields, builds up to electromagnetic theory. Also related topics, including relativity. Over 900 problems. 768pp. 5⅝ x 8¼. 0-486-65660-8

GREAT EXPERIMENTS IN PHYSICS: FIRSTHAND ACCOUNTS FROM GALILEO TO EINSTEIN, Morris H. Shamos (ed.). 25 crucial discoveries: Newton's laws of motion, Chadwick's study of the neutron, Hertz on electromagnetic waves, more. Original accounts clearly annotated. 370pp. 5⅜ x 8½. 0-486-25346-5

EINSTEIN'S LEGACY, Julian Schwinger. A Nobel Laureate relates fascinating story of Einstein and development of relativity theory in well-illustrated, nontechnical volume. Subjects include meaning of time, paradoxes of space travel, gravity and its effect on light, non-Euclidean geometry and curving of space-time, impact of radio astronomy and space-age discoveries, and more. 189 b/w illustrations. xiv+250pp. 8⅜ x 9¼. 0-486-41974-6

STATISTICAL PHYSICS, Gregory H. Wannier. Classic text combines thermodynamics, statistical mechanics and kinetic theory in one unified presentation of thermal physics. Problems with solutions. Bibliography. 532pp. 5⅜ x 8½. 0-486-65401-X

Paperbound unless otherwise indicated. Available at your book dealer, online at **www.doverpublications.com**, or by writing to Dept. GI, Dover Publications, Inc., 31 East 2nd Street, Mineola, NY 11501. For current price information or for free catalogues (please indicate field of interest), write to Dover Publications or log on to **www.doverpublications.com** and see every Dover book in print. Dover publishes more than 500 books each year on science, elementary and advanced mathematics, biology, music, art, literary history, social sciences, and other areas.